H · O · L · T

LIFE SCIENCE

WILLIAM L. RAMSEY

LUCRETIA A. GABRIEL

JAMES F. McGUIRK

CLIFFORD R. PHILLIPS

FRANK M. WATENPAUGH

HOLT, RINEHART AND WINSTON, PUBLISHERS
New York • Toronto • Mexico City • London • Sydney • Tokyo

THE AUTHORS

William L. Ramsey
Former Head of the Science Department
Helix High School
La Mesa, California

Lucretia A. Gabriel
Science Consultant
Guilderland Central Schools
Guilderland, New York

James F. McGuirk
Head of the Science Department
South High Community School
Worcester, Massachusetts

Clifford R. Phillips
Former Head of the Science Department
Monte Vista High School
Spring Valley, California

Frank M. Watenpaugh
Former Chairman, Science Department
Helix High School
La Mesa, California

About the Cover:
These Gentoo penguins are coming ashore to a nesting colony on an island near Antarctica. Gentoo penguins, distinguished by the white patch over their eyes, live in Antarctica and on more temperate islands in the southern hemisphere. Like most penguins, they feed on krill, tiny shrimplike ocean animals. Krill are also a food source for many whales. As whales have decreased in numbers, the worldwide populations of penguins have increased.

Picture credits appear on page 563.
Cover photograph by Jen and Des Bartlett / Bruce Coleman.

Art Credits:
Andres Acosta, Monika Babackova, Robin Brickman, D. L. Cramer, Ph.D., Leslie Dunlap, Robert Frank, Jean Helmer, Network Graphics, Ray Srugis, Craig Zuckerman.

ISBN 0-03-001917-6

34 040 12

ACKNOWLEDGMENTS

Teacher Consultants

Donald Bell
Sykesville Middle School
Carroll County Schools
Sykesville, Maryland

Mary Bley
Science Department Chairman
Dulles Junior High School
Sugar Land, Texas

Barbara Casey
Briscoe Middle School
Beverly, Massachusetts

Cindy J. Harris
Hall-McCarter Junior High School
Blue Springs, Missouri

Marilyn Plowman
Eric Smith Middle School
Ramsey, New Jersey

William F. Roberts, Jr.
District Science Specialist
Portland Public Schools
Portland, Oregon

Content Critic

Welton L. Lee
Curator of Invertebrate Zoology
California Academy of Sciences
San Francisco, California

Safety Consultant

Franklin D. Kizer
Executive Secretary
Council of State Science Supervisors
Lancaster, Virginia

Readability Consultant

Jane Kita Cooke
Assistant Professor of Education
College of New Rochelle
New Rochelle, New York

Computer Consultant

Nicholas Paschenko
Computer Coordinator
Englewood Cliffs Schools
Englewood Cliffs, New Jersey

Computer Features Written by

James Congelli
Science Team Leader
Monroe–Woodbury Middle School
Central Valley, New York

TO THE STUDENT

Scientists have great curiosity. They want to know *why, what if, how,* and *when.* In short, they seek to understand the world in which we live. As you use HOLT LIFE SCIENCE, you will learn how scientists seek answers to their questions.

This text will describe how scientists study living things and their surroundings. Some of the questions that will be asked are: How can living things exist in the darkness of the deep ocean floor? What things does life need to exist anywhere? Could life exist on other planets? How are living things classified so that they can be better understood? How do the systems of the human body work? What happens to the body when disease strikes? How have human activities threatened the other living things in our environment?

Many of these questions involve solving problems. Throughout HOLT LIFE SCIENCE, we will stress the use of problem-solving skills such as those described below. These skills are important tools for scientists. More than that, they can be important tools for anyone.

Basic Science Skills

The most basic *problem-solving* skill is *observing.* When you observe, you use your five senses to notice details about the world around you. When exact observations are required, scientists use *measuring* skills. Observations and measurements are organized and saved when you *record data.* In order to make your observations more meaningful, you use the skills of *comparing and contrasting, classifying,* and *sequencing.* All of this information can then be used to *hypothesize, predict,* and *infer* about future results. To do this, scientists (and you) learn to *use laboratory equipment* to carry out experiments that will answer your questions.

You will read about how scientists use these skills in Chapter 1. There are also Skill-Building Activities throughout the text. Each one will give you practice in using a particular skill. These are skills you can use in all areas of your school work and even your nonschool life. Be sure that you learn to use these skills effectively.

How to Use This Book

HOLT LIFE SCIENCE is divided into seven main parts. The first part, Chapter 1, is called *What is Science?* This chapter will help you understand what life science is and how scientists work. Unit 1 deals with what living things are made of and how they are organized. Units 2, 3, and 4 examine the four major groups of living things. In Unit 5, you will study the human body and heredity. The last unit is concerned with how living things interact with their environments.

Each chapter begins with a list of chapter goals. These goals provide a framework for studying the chapter. The chapter is divided into sections. Each section begins with objectives that help you set your goal for that section. When you can answer the Questions that follow the Summary, you have reached your goals for that section. The Chapter Review provides an additional way of checking your progress. Here you will also find additional activities and thought-provoking questions to help you apply the ideas you learned in the chapter to other situations.

Science is a "hands-on" subject. Throughout the text are Investigations in which you will work like a scientist. In your study of life science, you will learn many fascinating things about the natural world. However, the study of science can also involve many potential dangers. Science classrooms and laboratories contain equipment and chemicals that can be dangerous if not handled properly. Following the directions and cautions given in the book and by your teacher is *your* responsibility when carrying out an activity or investigation. Safety always begins with you.

Finally, science is a constantly *changing* subject. Throughout HOLT LIFE SCIENCE, you will find special features. One kind of feature, called ¡Compute!, will show you how to use computers to learn about science. Another feature shows how the work of scientists affects you. This feature is called Technology. We hope this book will interest some of you in a career in science. The Careers in Science features will make you more familiar with some of the interesting job and career opportunities you might choose to pursue in the field of science.

CONTENTS

A life scientist's work may take him or her to the bottom of the sea. This deep-sea submersible allows the scientist to examine the coral animals that live deep underwater on the ocean's floor.

WHAT IS SCIENCE?

CHAPTER GOALS

1. List and describe the environmental factors that define the biosphere.
2. List and define the steps in scientific problem solving.
3. Explain how scientists set up a controlled experiment.
4. Describe several ways scientists record observations and analyze data.
5. Identify the metric units used to measure distance, mass, volume, and temperature.

1-1. What Is Life Science?

At the end of this section you will be able to:

- ☐ List and describe the conditions that are necessary for life on earth.
- ☐ Describe the limits of the biosphere.
- ☐ Describe the kinds of work that life scientists do.

Wanted: Men and women willing to work nine and one-half hours at a time inside a small, metal sphere on the ocean bottom. Must have a background in biological sciences. Duties include measuring water depth and temperature, taking water samples, and photographing and collecting living things from the ocean floor. Must be able to withstand cramped areas, cold, and long periods of crouching. Does such a job interest you? If so, you may be interested in a career as a *biologist.* A biologist is a scientist who studies living things.

SEARCHING FOR LIVING THINGS

Why would anyone want a job like the one above? To a scientist, it could be a rewarding opportunity. Dr. J. Frederick Grassle is one scientist who has been on trips to the ocean bottom. His goal is to study the living things found there.

In 1977, scientists were on board the research submarine *Alvin.* They were exploring the bottom of the Pacific Ocean near the Galapagos Islands. The scientists were studying the hot water springs that escaped from cracks in the ocean

Organism A complete living thing.

Energy The ability to produce motion or cause change.

bottom. These cracks are called *vents*. They were surprised to discover a large community of creatures living around the hot springs. Here they found blood-red tube worms up to three meters long. See Fig. 1–1. There were clams 30 centimeters wide. Several types of **organisms,** or living things, they saw had never been seen before.

Dr. Grassle is a marine biologist. He works at the Woods Hole Oceanographic Institute in Massachusetts. Marine biologists study life in the oceans of the world. To biologists like Dr. Grassle, such vent communities are "as strange as a lost valley of prehistoric dinosaurs." See Fig. 1–2. Biologists have long known that there is life on even the deepest parts of the ocean floor. The organisms living there use the dead material that settles from the surface as their source of food and **energy.** *Energy* is the ability to cause change or produce motion. Living things need energy to carry on all the activities that make them alive.

The sun had long been thought to be the source of energy for all living things on earth. The organisms around the hot springs, however, have a different source of energy. The water around the vents is rich in tiny organisms called *bacteria*. The bacteria are able to use chemicals in the hot water as a source of energy. The larger animals get energy by feeding on the bacteria. See Fig. 1–2. For the first time, a whole community

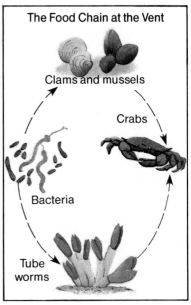

The Food Chain at the Vent

Clams and mussels

Crabs

Bacteria

Tube worms

Fig. 1–1 (left) *Giant tube worms were discovered in the hot water around ocean vents.*

Fig. 1–2 (right) *Chemicals provide a source of energy for a whole community.*

of living things had been discovered that didn't get its energy from the sun.

Scientists like Dr. Grassle are always searching for new types of living things. See Fig. 1-3. New types of plants, insects, and microorganisms are discovered every day. Scientists are seeking answers to questions such as "What conditions are necessary for life to exist?" and "Where can organisms find these conditions?"

CONDITIONS FOR LIFE

Living things have been found in some strange places. Some organisms can survive under very harsh conditions. Biologists have discovered bacteria floating high above the earth's surface. Living things are found in hot-water springs and geysers.

An organism needs certain things in order to exist. First is a source of energy outside of itself. Second is water. Also important are certain gases and minerals. The temperature around the organisms is important. And for plants, light is needed. These conditions are known as *environmental factors*. Different combinations of these environmental factors form the variety of **environments** in which life exists. Can you name some organisms that live in *environments* such as a desert or ocean? The factors in one place may favor the survival of some organisms and not others.

THE BIOSPHERE

The total of all the places on earth where life is found is called the **biosphere.** The *biosphere* reaches about eight to ten kilometers above sea level. In addition, the biosphere reaches ten kilometers below the surface of the sea. In places like jungles, life is also found to a depth of 700 meters in the soil. See Fig. 1-4 on page 4.

Energy from the sun controls many environmental factors. The sun warms the earth. The earth, in turn, warms the **atmosphere.** The *atmosphere* is a protective layer of air. Most of the atmosphere is part of the earth's biosphere. Without it the earth would be very hot where the sun is shining and very cold where it is not. It would be either too hot or too cold for most living things to survive for very long. The earth's temperature range is about −88°C to 58°C. Of course, these are the extreme temperatures. Most living things need a much

Fig. 1–3 Dr. J. Frederick Grassle is a marine biologist.

Environment Everything in the surroundings that affects the way an organism grows, lives, and behaves.

Biosphere That area at or near the earth's surface where life can exist.

Atmosphere A layer of gases surrounding a planet.

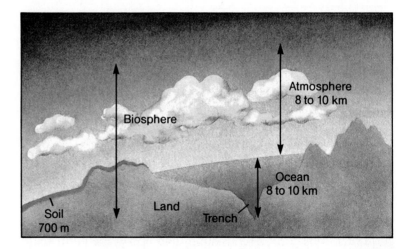

Fig. 1–4 The biosphere includes all the places where living things exist.

narrower temperature range. A range of about 0°C to 50°C is favorable for most living things. People seem most comfortable when the temperature is about 25°C.

Very few types of life can survive for long periods of time without a supply of water. All living things contain water in their bodies. Your own body is about two-thirds water. Water is needed to carry on an organism's activities. The earth's temperature range allows water to exist as a liquid.

The *biosphere* contains all the gases needed by living things. About 20 percent of our atmosphere is oxygen. Most of the rest is nitrogen. Only a very small amount is carbon dioxide. These three gases are very important to living things in the following ways. Organisms need oxygen to release energy from their food. Plants need carbon dioxide to make food. Nitrogen is a substance used to make living matter.

The atmosphere has another function. It protects the biosphere. Some of the sun's rays are harmful. The atmosphere acts like a screen. It filters out most of the harmful rays. The light that does reach the earth is used by green plants to make their food and to grow. Animals, in turn, depend on plants for food. Until the discovery of the hot-water vents on the ocean bottom, light energy from the sun was believed to be the only source of energy for life in the biosphere.

WHAT DO LIFE SCIENTISTS DO?

Life scientists, or biologists, are people who study all forms of life. *Botanists* have searched for plants and *zoologists* for animals from one end of the biosphere to the other. *Biochemists* study the chemicals that make up and control living things. Other life scientists concentrate on the growth, reproduction, or behavior of organisms. They may apply what they learn to a goal, such as developing new drugs.

All of the careers in health care and medicine are part of the life sciences. They range from understanding how the body works to building artificial hearts or knees. See Fig. 1–5. The life scientist may help produce more food for the hungry of this world. See Fig. 1–6. Or, he or she might search for life on other planets.

Fig. 1–5 (left) *Careers in surgery and nursing require a background in biology.*

Fig. 1–6 (right) *An agronomist studies how soil affects the growth of food crops.*

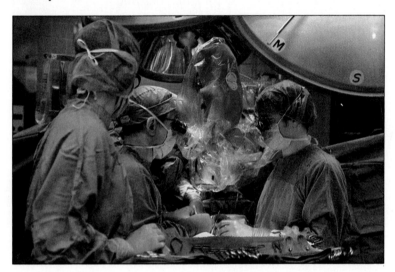

SUMMARY

Living things are found at or near the surface of the earth. They live in a wide range of environmental conditions. Each of these environments provides those factors that the organisms need to exist. People who work as life scientists study living things in their environments.

QUESTIONS

Use complete sentences to write your answers.

1. Describe the biosphere and its boundaries.
2. How does the atmosphere help life survive on earth?
3. Why is the earth's temperature so important to life?
4. What areas of research might a life scientist study?

Fig. 1–7 While other large North American animals are disappearing, coyotes seem to be thriving.

1-2. Thinking Scientifically

At the end of this section you will be able to:

- ☐ Identify the steps in scientific problem solving.
- ☐ Discuss why the state of scientific knowledge is constantly changing.

The mournful howl of a lone coyote, like the one in Fig. 1–7, broke the stillness of the night. At a distant sheep ranch, a rancher heard the howl. He reached for his rifle. "That pest won't kill any of my sheep tonight!" he said.

IDENTIFYING THE PROBLEM

For over a century, the coyote has been the focus of a dispute. Ranchers and farmers claim it kills domestic animals, such as sheep and poultry, for food. Environmental groups say that doesn't happen often. In fact, very little is known about coyotes. If more could be learned about them, perhaps the dispute could be settled.

Marc Bekoff and Michael Wells are scientists. See Fig. 1–8. They are trying to learn about the behavior of coyotes. They carried out a three-year study in Grand Teton National Park, Wyoming. One of their goals was to learn why some coyotes live alone and others form packs. Bekoff and Wells suspected that some coyotes changed from one type of living to the other.

Scientists tackle different problems in different ways. Some problems need a definite solution. For example, a new type of disease appears. Its cause must be found and a cure de-

Fig. 1–8 Dr. Marc Bekoff uses a spotting scope to observe coyotes in the wild.

veloped. At other times, the purpose of research is to find out more about a general subject. This was what Bekoff and Wells wanted to do in Wyoming.

In either case, the first step is to identify and state a problem clearly. Scientists state the problem in the form of a question. This is is an important step in the **scientific method,** or scientific problem solving. All scientists do not use the same steps in problem solving. Nor are the steps always in the same order. The steps of the *scientific method* are better thought of as skills that are common to all scientists. These skills are also used by anyone who solves problems in an orderly, thoughtful way.

Scientific method The set of skills used to solve problems in an orderly way.

GATHERING INFORMATION

First, the problem is stated as a clearly worded question. The next step is to find out what information is already known. Bekoff and Wells went to a library. They found some earlier studies done by scientists on coyotes. They also read about the behavior of animals such as jackals and hyenas. These animals are related to coyotes. Information from these earlier studies, based on the **observations** of other scientists, would help Bekoff and Wells answer their questions. They would also watch the coyotes, to make their own *observations.*

Observation Any information that we gather by using our senses.

FORMING HYPOTHESES

After collecting information, scientists form **hypotheses.** A *hypothesis* is a possible answer to the question. It is sometimes called an "educated guess." An "educated guess" is based on the information the scientist has already gathered. It seems to be the best answer to the problem. More than one hypothesis may be formed from the same information. Each possible answer must then be tested.

Bekoff and Wells made a hypothesis. They felt that the food sources available to coyotes determined whether they lived alone or in packs.

Hypothesis A statement that explains a group of related observations.

TESTING THE HYPOTHESIS

Stating a hypothesis is one step. Proving that a hypothesis is correct or incorrect is another. Scientists test their hypotheses by doing **experiments.**

Experiment An activity designed to test a hypothesis.

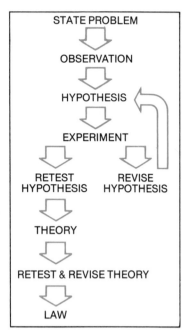

STATE PROBLEM

⬇

OBSERVATION

⬇

HYPOTHESIS

⬇

EXPERIMENT

⬇   ⬇

RETEST   REVISE
HYPOTHESIS   HYPOTHESIS

⬇

THEORY

⬇

RETEST & REVISE THEORY

⬇

LAW

Fig. 1–9 The steps of the scientific method.

When you think of *experimenting,* do you think of a scientist in a laboratory mixing chemicals? There are also other types of experiments. Imagine that you are an *astronomer* trying to find out how stars form. There isn't much that you can do in a laboratory. Your experiment would involve making many observations of stars. This was also the case with the coyote study. The two scientists would have to watch the coyotes over a long period of time.

To test their hypothesis, Bekoff and Wells went to Grand Teton National Park. From the top of a hill, they could observe the behavior of several lone coyotes. They could also see packs of the animals in the valley below.

Observations of the coyotes were made over several years. During the summer months, the coyotes' main foods were field mice, gophers, and ground squirrels. They hunted and killed these small animals. See Fig. 1–10. During the winters, however, they ate the remains of deer, elk, and moose. These larger animals had died from other causes. Throughout the study, coyotes were never seen attacking any live, large animal. This does not prove that they never do.

Fig. 1–10 In summer, a lone coyote hunts for small animals called meadow voles.

Data Information collected from observations.

DRAWING CONCLUSIONS

First, **data** is collected and studied. Then conclusions can be made. The results may lead the scientist to believe that the hypothesis is a good one. On the other hand, they may not. Then the scientist has to make a decision. Was there an error in the experimental setup? Or was the hypothesis not correct to begin with? In either case, experiments are run many times

Fig. 1–11 In winter, a pack of coyotes feeds on the carcass of a mule deer.

to check the results. If the hypothesis is not supported by the results, it must be changed. Sometimes it is thrown out. Then new hypotheses must be developed.

The *data* gathered by Bekoff and Wells showed several things. In the summer, the coyotes hunted and killed small animals for food. There were fewer coyotes in packs. More were hunting alone over wider areas. In the winter, the weather was very cold and snowy. There were more large, dead animals available. More coyotes could feed on the same food source. The coyotes came together in packs.

Living together in packs seemed to have advantages for the coyotes. The animals worked together. It was easier to find food. Food could be defended from stray coyotes more easily. The pack shared the work of caring for the young.

Where the winter supply of large, dead animals was plentiful, the pack size was larger. See Fig. 1–11. Where the food was scarce, the packs were smaller. These observations tended to support the scientists' hypothesis: Food supply determines if packs will form and how large they will be.

THEORY AND LAW

From their results Bekoff and Wells were able to make a general statement. They said that the social behavior of coyotes seems to be related to the food supply. They wrote an article in a scientific magazine. This sharing of results is an important part of science. See Fig. 1–12. In this way other

Fig. 1–12 A scientist presents the results of an experiment at a meeting.

scientists learn what has been done. Then they can decide what still needs to be studied. They can also repeat the experiments the scientists did to check their accuracy. They may see ways to apply other scientists' results to their own work.

The hypothesis made by Bekoff and Wells will have to be tested again and again. The experiment will be done with other coyotes in other places. It will also be tested with similar animals like wolves, jackals, and hyenas. If these animals show the same behavior, the hypothesis may be accepted by scientists as a **theory.** A *theory* is a hypothesis that has been supported by lots of experiments and data. Even theories are always being tested. Better equipment or experiments may produce new data. The theory may turn out to be wrong. The theory that the sun is the only source of energy for life on earth was proved wrong. It had to be changed when the communities around the ocean hot-water vents were discovered.

Theory A general statement based on hypotheses that have been tested many times.

Scientific law A theory that has been proven correct over a long period of time.

The most accepted and proven ideas are called **scientific laws.** Even these can sometimes be broken or proven wrong. A scientist once said that science moves ahead by correcting errors it made earlier. The body of information we call scientific knowledge is always changing.

SUMMARY

All scientists answer questions by using problem-solving skills. The skills include identifying the problem, gathering data, forming a hypothesis, experimenting, and drawing conclusions. The answers to questions may change as scientists design better experiments and make new observations.

QUESTIONS

Use complete sentences to write your answers.

1. List the steps in scientific problem solving.
2. What are observations? Why are accurate observations necessary in problem solving?
3. Your teacher has just told you that you are in danger of failing science this semester. Suggest three hypotheses to explain your low mark. How would you test each of these hypotheses?
4. Explain the statement, "The only constant thing in science is change."

WHAT ENVIRONMENTAL FACTORS DO EARTHWORMS NEED?

PURPOSE: To suggest hypotheses and ways to test them.

MATERIALS:

rectangular	soil
aluminum foil tray	light source
aluminum foil sheet	hot plate
(for dividing the	beaker with ice
tray into small	beaker with water
sections)	earthworms

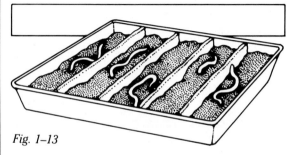

Fig. 1–13

PROCEDURE:

A. To obtain money for class activities, your class has decided to raise earthworms. They will be sold to people who fish. You already know that the worms should be kept in wooden boxes with rich soil. However, not everyone in the class agrees about some of the other environmental factors. Do earthworms like light or darkness? Do they like wet, moist, or dry soil? Do they prefer warm or cool temperatures? Choose *one* of these three problems to solve.

B. State your hypothesis by completing the sentence:

1. My hypothesis is that earthworms prefer _____.

C. Plan an experiment to test your hypothesis. Choose your equipment from the list. Your teacher will tell you if it is possible to carry out your experiment.

2. Describe how your equipment should be set up. Make a labeled sketch of the setup. Plan to place one worm in each of the five sections of the tray.

3. List the steps that should be carried out in the experiment.

CONCLUSION:

1. Why is it important to form a hypothesis before experimenting?

2. Why is it important to keep a record of your materials and how you used them?

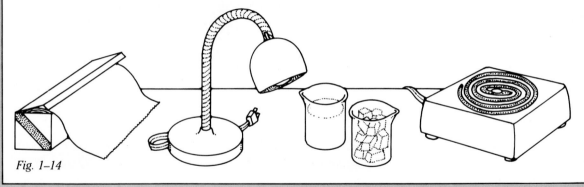

Fig. 1–14

1-3. Experimenting

At the end of this section you will be able to:

☐ Explain how to set up an experiment with a control.

☐ Define the term "variable."

☐ Explain how to put data in the form of a chart or graph.

Bekoff and Wells were interested in another aspect of coyote behavior. They wanted to find out how lone coyotes find and capture their prey. The two scientists believed that coyotes used all of their senses while hunting. But were some senses more important than others? How could the importance of sight be compared to the importance of hearing, or of smell?

The scientists would have liked to watch the animals from their hilltop spot. But the range of a lone coyote may be more than 30 square kilometers. The scientists would not be able to observe animals that traveled over such a large area to find food. They also realized that just watching the animals would not give them enough information to answer their questions. They decided to run experiments under more controllable conditions.

To do this, several coyotes were captured. See Fig. 1–15. They were brought to a university for study. Each coyote was placed in a large, fenced-in outdoor area. See Fig. 1–16. Some-

Fig. 1–15 The scientists captured several coyotes. Some, like this one, were fitted with radio tracking collars. Others were brought to a university for study.

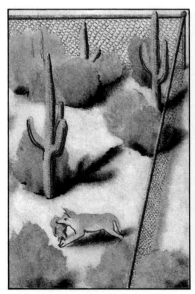

Fig. 1–16 Bekoff and Wells created a fenced-in piece of land where they could observe one coyote and rabbit at a time. At left, coyotes used all their senses. At right, they could not use their sight.

where in that area was a hidden rabbit. The coyotes were able to use all of their senses to find the rabbit. The amount of time it took each coyote was measured. This part of the experiment is called the **control.** The *control* conditions were as close to the natural setting as possible. It is the control against which all other parts of the experiment would be compared.

Control The part of an experiment in which all factors remain the same.

VARIABLES

The part of the setup that is different from the control is called the **variable.** Scientists change only one part of the experiment at a time. In this way they know that the different results probably are caused by the *variable.*

Variable Any factor in an experiment that could affect the results.

For example, the scientists wanted to check how important sight is to the coyote. To do this, the test was done on a dark night when there was no moon. Infrared photography was used to track the coyote's movement in the dark.

In this case, the variable was the sense of sight. To eliminate the variable of sounds, dead rabbits were hidden. There would be no noise to help the coyote locate the rabbit. To eliminate the variable of smells, the coyote's nostrils were flushed with a nasal spray. The spray temporarily stopped the coyote's sense of smell. One at a time, each of the three variables was tested.

ORGANIZING DATA

The time it took for each coyote to find the rabbit was measured. All this data had to be organized. Then it could be studied. Data, in the form of measurements, can be organized in data tables. Table 1–1 is a data table that could be used in this experiment.

Coyote Number	Sight Only		Hearing Only		Smell Only	
	Trials	Avg.	Trials	Avg.	Trials	Avg.
1						
2						
3						
etc.						

Table 1–1

Each part of the experiment was run many times. This was done to find an average time. Why is this necessary?

Once the data was collected and averaged, the results were analyzed. The results showed that sight was most important in locating food. Smell was next in importance, and then hearing.

Sometimes results are more easily understood in the form of graphs. Fig. 1–17 shows the results of the coyote experiment in the form of a bar graph. The average time in seconds is given on the left side of the graph. Bar A shows the average time it took the coyotes to find the rabbits while using all three senses. This was about 30 seconds. This is the control time. It is the control against which the other parts of the experiment are compared.

Bar B shows the average time using sight only. The variables of sound and smell had been removed. Using sight only, the

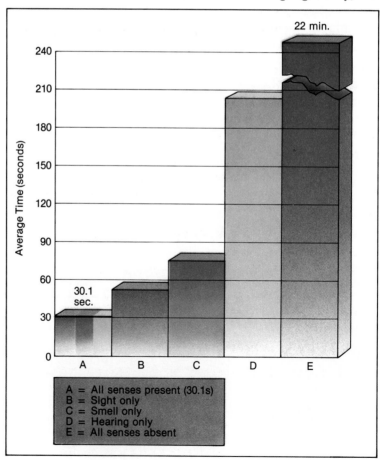

Fig. 1–17 What does the height of the bars signify on this graph?

time increased to about 50 seconds. This proved that coyotes use more senses than sight to find their food.

When the coyotes were able to use only smell, as shown by bar C, their average time was more than double the control time, about 73 seconds. How were the variables of sight and sound removed in this experimental test?

Bar D shows the results when using hearing only. The time increased to 209 seconds. This is almost seven times as long as the control situation.

Finally, all three senses were removed. Now it took the coyotes an average of 22 *minutes* to find the rabbit. The bar representing 22 minutes would be too long to fit on the graph. It had to be broken and labeled so that readers would understand that it should really be much longer.

You can see from the graph that the loss of smell and hearing only slightly increased the times. Therefore, vision is the most important sense when coyotes search for food. Hearing is the least important. This is probably because the wind can distort sounds.

As you use this textbook, you will be constructing and interpreting many data tables and graphs. Be sure to read them carefully so you understand what they show.

SUMMARY

Scientists test their hypotheses by doing experiments. The control setup is compared to all other parts of the experiment. The effect of only one variable is tested at a time. Measurements are often recorded in data tables. These measurements are then presented for analysis as charts or graphs.

QUESTIONS

Use complete sentences to write your answers.

1. A farmer wants to test the effect of a new fertilizer on the growth of corn plants. An experiment is being designed. How should the control be set up?
2. What is the variable that will be changed in the rest of the setups?
3. What types of observations would you make during this experiment? How would you organize your data?
4. How would you decide if the fertilizer was helpful?

SKILL-BUILDING ACTIVITY

RECORDING AND ORGANIZING DATA

PURPOSE: To practice recording observations in a table and constructing a graph.

MATERIALS:

soaked bean seeds   paper towels
plastic or glass con-   water
   tainers with lids   graph paper

PROCEDURE:

A. If you are a gardener, you know that seeds sprout well under some conditions but not under others. You will test three temperature conditions: (1) under 10°C, (2) about 20°C, and (3) over 32°C. State a hypothesis as to which of these three conditions is best for sprouting seeds. Complete the following statement:

   1. My hypothesis is that the bean seeds will sprout best under _____ temperature conditions.

B. Place a paper towel in each of the three containers. Moisten the paper with water. Sprinkle 50 seeds on the paper in each container. Place the lids on top of the containers, but do not seal them. See Fig. 1–18. Label the jars 1, 2, and 3.

C. Place each container in a different temperature environment: (1) in a refrigerator, (2) in a box in a room, and (3) in a box in a warm place. Your teacher will show you where to place the containers.

   2. What is the single variable being tested in this experiment?

D. Leave the containers in place for 24 hours. At the end of the experiment, examine the seeds. Count the number of seeds that sprouted in each container. A sprouted seed should have a part of the root showing.

Temperature Conditions	Number of Sprouted Seeds	Percent That Sprouted
Under 10°C		
About 20°C		
Over 32°C		

Table 1–2

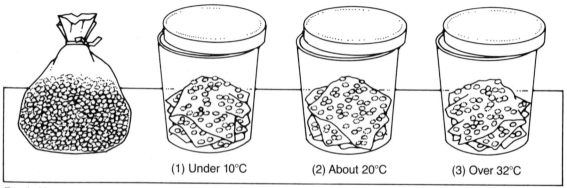

(1) Under 10°C   (2) About 20°C   (3) Over 32°C

Fig. 1–18

E. Copy data table 1–2. List the three conditions used to sprout the seeds. Next to each condition, write the number of seeds that sprouted. Now find the percentage that sprouted. Do this by dividing the number that sprouted by the total number of seeds (50). Write the percent of sprouted seeds in the correct boxes.

3. Which jar produced the greatest number of sprouted seeds?

4. Were your results the same as or different from your classmates?

F. Data tables are a useful way to record observations. The same information also can be used to make a graph. A graph is somewhat like a picture. Graphs present the data in a way that lets the reader compare all the parts of the experiment easily. Hold the graph paper so that it is in a vertical position. Draw the X axis, a horizontal line, about ten lines up from the bottom of the page. Draw the Y axis, a vertical line, about ten lines in from the left side of the page. See Fig. 1–19. Then label the X and Y axes as shown in the picture.

G. Draw a vertical bar for each of the three conditions. See Fig. 1–19. Make each bar as tall as the number of seeds that sprouted. Label each bar.

5. Why might it be helpful to color each bar differently?

6. Not all graphs are bar graphs. How many other different types of graphs can you think of?

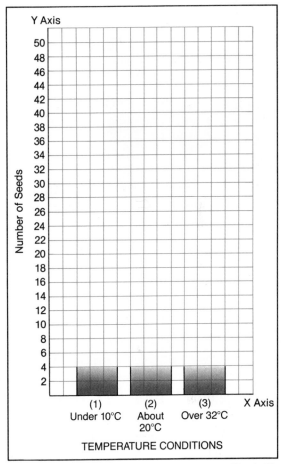

Fig. 1–19

CONCLUSIONS:

1. State your conclusion. What are the best temperature conditions for sprouting seeds?

2. Was your hypothesis supported by the results? Should it be changed?

3. During which part of an experiment is a data table helpful?

4. Why do scientists create graphs of their results?

1-4. Measuring

At the end of this section you will be able to:

- ☐ Identify the metric units used for measuring distance, mass, volume, and temperature.
- ☐ Identify and explain how to use the instruments for measuring distance, mass, volume, and temperature.
- ☐ Explain and use the prefixes of the metric system.

Fig. 1–20 Measurements are an essential part of all scientific observations. This researcher is measuring trees in Costa Rica.

Measuring is a vital part of all scientific work. See Fig. 1–20. On the ocean bottom, Dr. Grassle measured the sizes of tube worms, crabs, and clams. He measured the temperature of the water. In their coyote study, Drs. Bekoff and Wells measured the distance the coyote traveled. They also measured the weight of the captured animals. They measured how much time it took each coyote to find a rabbit.

It is not enough to say that the tube worms were "really huge." It is not helpful to know that a coyote can run "very fast." Measurements make observations more exact. Measurements are needed to show *how* big, *how* fast, or *how* many. The more exact the measurements are, the more exact our understanding of nature will be. It has often been said that mathematics is the language of science. Why is this so?

Scientists all over the world use the SI metric system of measurement. The letters SI come from its French name, *le Système International.* In the metric system, sizes and distances are measured in *meters.* Weights are given in *grams.* Volume is recorded in *liters* or *cubic centimeters.* Temperature is recorded in *degrees Celsius.* Which of these units have you already used in your science or math classes? Which ones have you noticed outside of school?

THE METRIC ADVANTAGE

The metric system has many advantages over the English system of inches, pounds, and so on. First of all, the metric system is used worldwide. Therefore, there are no problems in understanding measurements made in Russia, Germany, China, or any other country. The next time you watch an international sporting event, such as the Olympics, pay attention to the units used. See Fig. 1–21.

.00 M LIBRE   DAMAS
.00 M FREESTYLE   WOMEN
  1 STEINSIEFER C.   USA
  2 KERR J.   CAN
  3 BALD K.   CAN
  4 TREIBLE K.   USA
  5 SACHERO V.   ARG
  6 RIVERA T.   MEX

Fig. 1–21 The distance of an international swim race is measured in meters.

How many inches are there in 13 1/2 feet? If something weighs 743 ounces, how many pounds does it weigh? In order to answer each question, you must remember a different set of relationships. There are 12 inches in each foot. You would multiply 13 1/2 feet by 12. There are 16 ounces in a pound. You would divide 743 ounces by 16.

The metric system is much simpler. The metric system is based on multiples of ten. It is easy to change measurements to larger or smaller units. All you have to do is multiply or divide by some multiple of ten. All metric units are changed from one form to another in the same way.

The set of prefixes is the same for all metric units. Each of the prefixes given in Table 1–3 comes from the French and Latin languages.

PREFIXES AND METRIC UNITS				
Prefix	Definition	Metric Distances	Metric Mass	Metric Volume
kilo (k)	× 1,000	kilometer (km)	kilogram (kg)	kiloliter (kL)
hecto (h)	× 100	hectometer (hm)	hectogram (hg)	hectoliter (hL)
deca (dk)	× 10	decameter (dkm)	decagram (dkg)	decaliter (dkL)
	× 1	meter (m)	gram (g)	liter (L)
deci (d)	1/10	decimeter (dm)	decigram (dg)	deciliter (dL)
centi (c)	1/100	centimeter(cm)	centigram (cg)	centiliter (cL)
milli (m)	1/1,000	millimeter (mm)	milligram (mg)	milliliter (mL)

Table 1–3

MEASURING DISTANCES

Meter (m) The basic unit of length in the metric system.

Distance in the metric system is measured by the **meter.** A *meter* is about 39 inches long. Sometimes the meter is too long a unit for the distance you want to measure. The meter (m) is divided into ten equal, smaller units, which are called *decimeters* (dm). The prefix deci- means "one-tenth." Our word dime, one-tenth of a dollar, comes from the same prefix. If the dm is still too long a unit, it too can be divided into ten equal, smaller units. These are called *centimeters* (cm). There are 10 cm in one dm. There are 100 cm in a meter. The prefix centi- means one-hundredth, just as our "cent" is one hundredth of a dollar. Fig. 1–22 shows a picture of a centimeter ruler.

If the cm is too long, it can also be divided into ten equal, smaller units called *millimeters* (mm). Milli- means one-thousandth. There are 1,000 mm in a meter. We do have a unit of money called the mill. There is no coin this small, but we do use the unit. The nine-tenths fraction at the end of the price of a gallon of gasoline is nine-tenths of a cent, or nine mills.

There are also units for measurements longer than a meter. These are also based on multiples of ten. The one that is used most often is the *kilometer* (km). The prefix kilo- means one thousand. There are 1,000 meters in a kilometer. Kilometers are used to measure long distances such as the distance between cities or in long distance races.

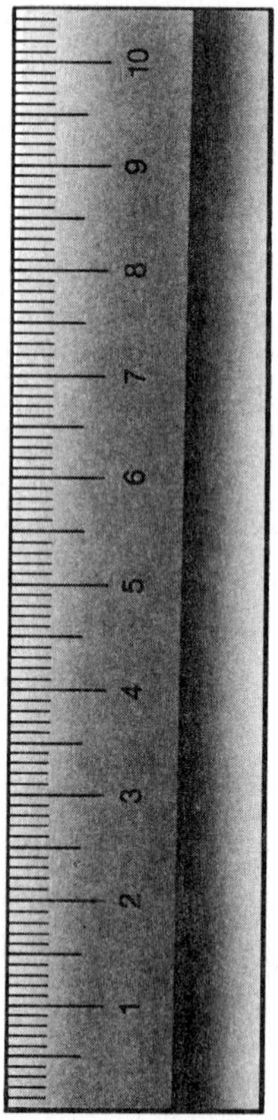

MASS VS. WEIGHT

Weight refers to the pull of gravity on an object. An astronaut on earth might "weigh" 50 kilograms. While in space, the

Fig. 1–22 The numbers on the ruler mark the centimeters.

Fig. 1–23 The weight of an astronaut in zero gravity is zero.

astronaut would be "weightless." See Fig. 1–23. Obviously, the amount of matter in his or her body doesn't change. But his or her weight *does* change. For this reason, scientists measure the amount of matter in an object. This is called its **mass.** **Grams** are units of *mass*. You would say the astronaut has a mass of 50 kilograms. A kilogram is equal to 1,000 *grams*. We will speak of mass instead of weight throughout this book.

To make accurate measurements, scientists use instruments. To find the mass of an object, a balance is used. See Fig. 1–24. There are many types of balances. All of them compare the object to another object whose mass is known.

Mass The measure of the amount of matter contained in an object.

Gram (g) The unit used to measure mass in the metric system.

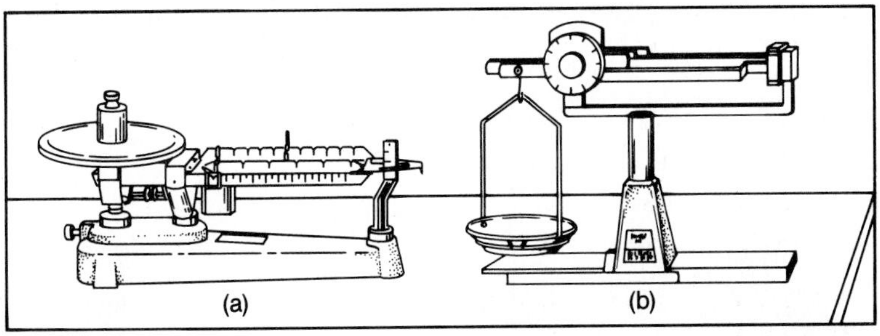

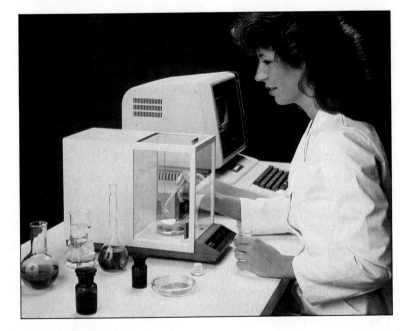

(a)   (b)

Fig. 1–24 (left) *A triple-beam balance;* (right) *a dialed, double-beam balance, sensitive to 0.01 g.*

Many of the instruments used by scientists have become more precise since the development of electronics. Electronic balances like the one in Fig. 1–25 make this task easier and more accurate.

Fig. 1–25 This expensive electronic balance is sensitive to .00001 g. It can be attached to a printer or computer.

MEASURING VOLUME

Volume is the amount of space a solid or liquid takes up.

Liter (L) A unit of volume used in the metric system.

The amount of space a solid or liquid takes up is called its **volume.** The *volume* of a liquid may be measured in **liters.** You may be familiar with this unit. Most soft drinks are now sold by the *liter*. The liter is slightly larger than a quart. We will use liters to measure the volumes of liquids in this book.

The SI unit for volume is the centimeter cubed. The cube drawn in Fig. 1–26 measures 1 cm in length, 1 cm in width, and 1 cm in depth. If the cube were filled with water, it would hold one milliliter of liquid. So the volume of the cube is either one cubic centimeter or 1 mL. Cubic centimeters are the units used for finding the volume of solids and gases. The units are equal.

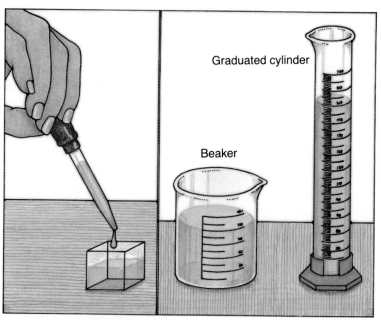

Graduated cylinder

Beaker

Fig. 1–26 Beakers and graduated cylinders are used to measure liquid volume in the lab. The graduated cylinder is the most accurate of the two.

Liquid volumes are measured with a *graduated cylinder* such as the one in Fig. 1–26. The cylinders are marked in milliliter units. The surface of most liquids in a glass cylinder is curved with the edges slightly higher than the center. It is customary always to read the level at the bottom of this curve. Be sure the cylinder is standing on a level surface. The curve of the liquid should be at eye level when you read it.

Many of the experiments in this book require you to use a beaker. Beakers also measure liquid volume. They are handy to use when measurements do not have to be very precise.

TEMPERATURE

Several units are used to measure temperature. You are probably most familiar with degrees Fahrenheit. If your body temperature is over 98.6°F, you know that you are sick. Weather reporters announce the air temperature in degrees F. In recent years weather reporters have also begun to report the temperature in degrees Celsius. The Celsius degree is the unit most used by the world's scientists. On the Celsius scale, the freezing point of water is 0°C. The boiling point of water is 100°C. A healthy body temperature is about 37°C.

Thermometers are the instruments used to measure temperature. See Fig. 1–27. The most common type is made from a glass tube. A liquid is added and then sealed. As the end of the tube is heated, the liquid expands. It fills a larger amount of the tube. The degrees are marked on the side of the tube.

Whatever the unit, whatever the instrument, accuracy is the watchword. No measurement is ever totally exact. No instrument is ever perfect. Measurements are made and remade to reduce possible errors.

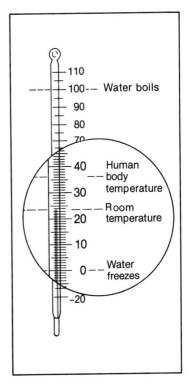

Fig. 1–27 The liquid sealed in the thermometer may be mercury, alcohol, or an organic compound called toluene.

SUMMARY

Many of the observations made by scientists are in the form of measurements. The use of instruments makes the measurements more accurate. The metric system is used throughout the scientific community. It includes the units meter, gram, liter, and degrees Celsius. A common system of prefixes makes it simple to convert each of these units.

QUESTIONS

Use complete sentences to write your answers.

1. List the prefixes used in the metric system, in order, from smallest to largest.
2. At what temperature does water boil on the Celsius scale? At what temperature does it freeze?
3. How are mass and weight different?
4. What are two units used for measuring volume? How are they related to each other?
5. A certain distance is measured at 16.8 m. How many cm would it be? How many km would it be?

INVESTIGATION

CAN YOU LOWER THE TEMPERATURE OF ICE WATER?

PURPOSE: To use the units for measuring mass, temperature, and time.

MATERIALS:

250-mL beaker	crushed ice
graduated cylinder	salt
lab thermometer	graph paper
stirrer	clock with
water	second hand

PROCEDURE:

A. On a separate piece of paper, set up a data table like the one in Table 1–4.

B. Fill the beaker half full of crushed ice. Add 25 mL of water.

C. Record the room temperature beside time 0 in your data table.

D. Position the thermometer so that the bulb is in the water. Record the temperature of the water every 30 seconds until it stops changing. Stir the mixture gently, *using the stirrer,* every minute.

E. Continue recording the temperature for two minutes after it stops changing.

F. Make a graph like the one below.

G. Plot the data you collected.

H. Empty the beaker. Half fill it with ice.

I. Dissolve one gram of salt in 20 mL of water. Add the mixture to the ice.

J. Repeat steps D, E, and F.

K. Plot the points for this data on the same graph. Connect the points with a line. Use a different colored pencil.

Time	Temperature	
(Seconds)	Ice/Water	Ice/Water/Salt
0		
+ 30		
+ 60		
+ 90		
+ 120		
+ 150		
etc.		

Table 1–4

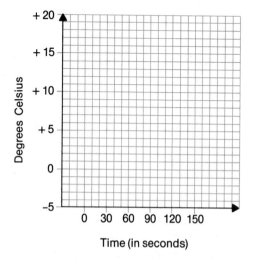

Fig. 1–28

CONCLUSIONS:

1. At what temperature did the ice/water mixture level off?

2. At what temperature did the ice/water/salt mixture level off?

3. Can ice water be colder than 0°C?

4. Why is salt spread on icy walks and roads in winter?

T E C H N O L O G Y

INTRODUCTION

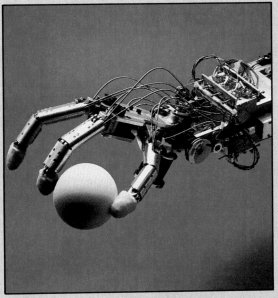

Welcome to the future, a world created by scientific research and development. The path of the scientific method and the scientist's curiosity have brought about astounding accomplishments and possibilities by creating a base of knowledge upon which new technologies have been built. Because of these developments, horses can give birth to zebras; sound waves can diagnose strokes; and genes can be spliced, diced, and changed. Topics such as these provide the content of the Technology sections.

We will inform you of some current research and technological developments that have extended the subject matter of science and/or changed the nature of what you must learn in order to live and work in modern society. The Special Features sections go beyond textbook material to encourage you to explore, on your own, events in the scientific world.

Some of the new research discussed does not involve strange or radically new techniques. It may, instead, be the result of traditional research methods, used under conditions that have changed dramatically owing to new technology. For example, developments in cryogenics, the study of the effects of low temperatures, now allow the freezing of living tissue so that it neither grows nor dies while in a frozen state. The success of embryo transplants depends, in part, on our well-developed ability to use liquid nitrogen to freeze living cells.

To find work in today's society, you have to understand how the marketplace has been transformed by technology. Many jobs that exist today, such as that of dialysis technician (see Chapter 16), simply couldn't exist if the technology wasn't available.

The computer, too, has changed the shape of our work. For example, a few years ago textbooks were set into print by a typographer using actual type—that is, pieces of metal with letters engraved on them. He or she would arrange these letters, line by line, in order to set up a page. Gigantic printing presses would then stamp out the pages. The words of the textbook you are now reading, however, were entered into a word processor, stored on a floppy disc, and became type when a compositor programmed a computer to translate the floppy disc into printed pages. Even the image of the owl in the Computer section was not drawn by a human artist; it was created by a computer. Changes such as these mean that people must continually learn new job skills.

A word to the wise: New developments often need new words to explain them. In each Special Features section, you will find words that may not be completely familiar to you, such as imaging, tomography, exoskeletal, and receptor. Explanations and descriptions are provided to help you understand these terms.

VOCABULARY

On a separate piece of paper, list the numbers of the sentences in the paragraph below. Then match the number of each sentence with the term that best completes it. Use each term only once.

grams   atmosphere   variable   control
laws   organisms   experiments   observations
data   scientific method   environment   hypotheses
mass

At the end of 1983, the space shuttle *Columbia* lifted the Spacelab through the earth's __1__ and into orbit. Six astronauts spent the next nine days conducting scientific __2__. They worked in the weightless __3__ of space. The experiments involved making __4__ and collecting __5__ in the form of photographs or measurements. Some of the data will be used to test __6__ about how plants grow in a weightless environment. Other data will test accepted scientific __7__ about the nature of the universe. The biology experiments included tests of several living __8__, including the humans. In one test, an astronaut studied the location of several objects. Then he tried to point to them with his eyes closed. His performance in space is compared with the same performance on earth. The performance on earth is the __9__. In this way, the effect of the single __10__ of weightlessness can be studied. In another test, he picks up a pair of metal balls and tries to decide which has the greater __11__. The balls are the same size and only a few __12__ different in mass. Whatever the experiment, wherever it is done, scientists always use the thinking skills we call the __13__.

QUESTIONS

Give brief but complete answers to each of the following. Unless otherwise indicated, use complete sentences to write your answers.

1. What is the biosphere? Describe the limits of the biosphere as we now know them.
2. Name four environmental factors that would be different in locations such as the Alaskan mountains and the Texas prairie.
3. What is meant by the *scientific method?*
4. List the steps of the scientific method.
5. What part of an experiment is the "control"? Why is it necessary?
6. A television commercial states that soap X cleans clothes better than soap Y. How would you set up an experiment to test this statement?

7. You are to find the average number of chocolate chips per cookie in three different brands of cookies. Design a simple data table to organize the information you would gather.
8. How can the weight of an object change while its mass remains the same?
9. Match the letter of each unit to the number of what it measures:

    a. meter   1. volume
    b. gram   2. temperature
    c. liter   3. distance
    d. degrees Celsius   4. mass
10. What does each of these prefixes mean? deci-, deca-, kilo-, milli-

APPLYING SCIENCE

Scientists share their discoveries with other scientists in many ways. One way is by writing articles in scientific magazines. These articles are often difficult for nonscientists to understand. Recently, however, several new science magazines have been published. They are for the general public. Among these are *Science News, Science Digest, Discover, Science 86,* and others. The articles in these magazines are interesting and informative, but you have to know how to read them.

Visit your school or public library and locate an article on life science in one of these magazines. Use the following outline to write a report on the article.

A. Who is the author?
B. What is the title of the article?
C. What magazine was it in? (Include the issue date and page numbers.)
D. What was the main idea of the article?
E. Briefly summarize the facts the author uses to support the main idea.
F. Briefly summarize the conclusions that the author makes.

BIBLIOGRAPHY

Morris, Desmond. *Animal Days.* New York: Morrow, 1980.

Morris, Richard. *Dismantling the Universe: The Nature of Scientific Discovery.* New York: Simon & Schuster, 1983.

Schneider, Herman. *How Scientists Find Out.* New York: McGraw-Hill, 1976.

Shapiro, Stanley Jay. *Exploring Careers in Science.* New York: Richards Rosen Press, 1981.

Smith, Sandra. *Discovering the Sea.* New York: Time-Life, 1981.

UNIT

1

THE LIVING WORLD

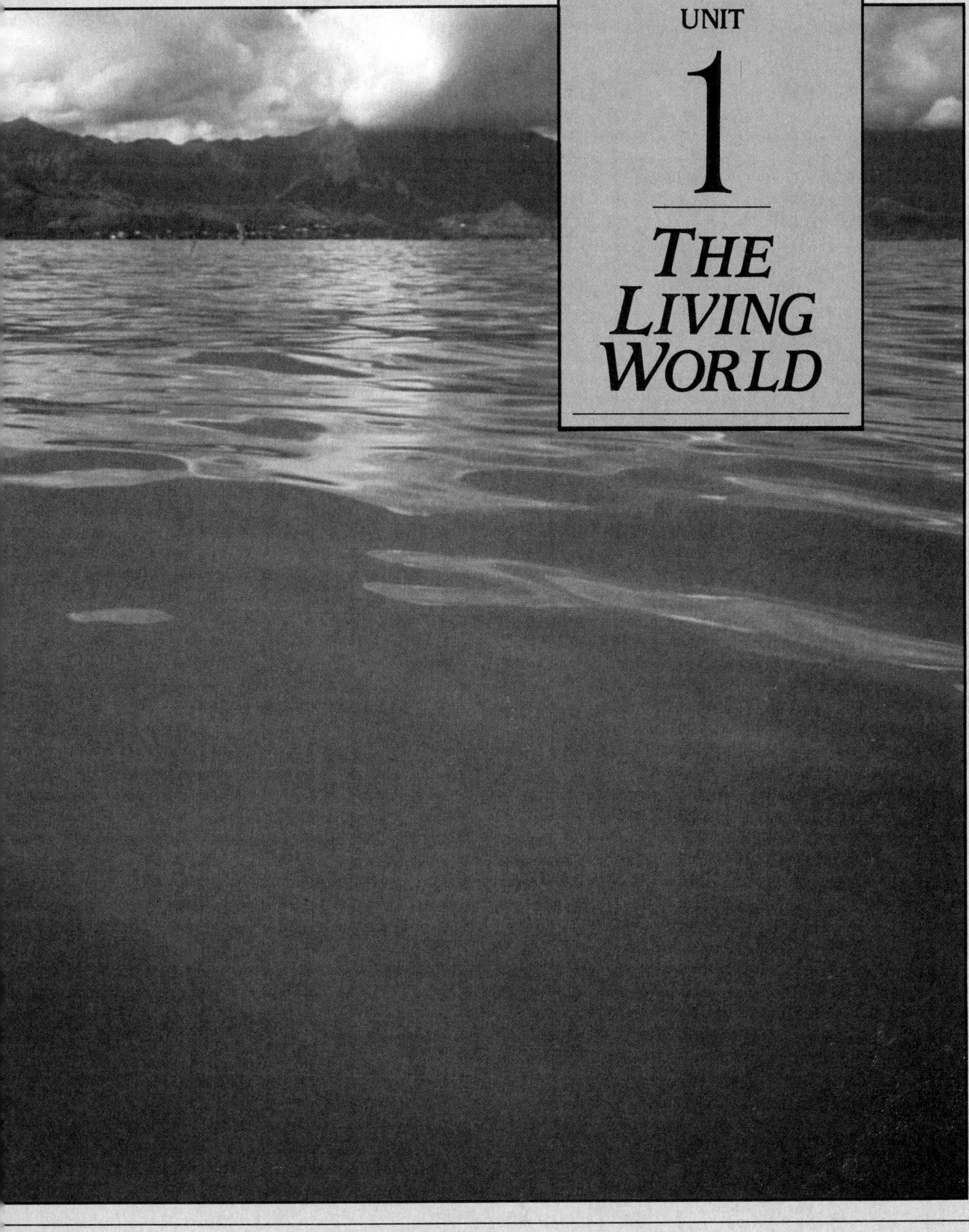

These day moths and the desert flower are living organisms. What can they do that nonliving things cannot? Both plants and animals are made of the same basic units of matter and life.

MATTER AND LIFE

CHAPTER GOALS

1. Identify the life processes and use them to distinguish between living and nonliving things.
2. Identify the basic particles of matter.
3. Compare and contrast living and nonliving matter.
4. State the cell theory and discuss how it developed.
5. Identify and explain the functions of the parts of animal and plant cells.
6. Describe the levels of organization in living things.

2-1. The Activities of Life

At the end of this section you will be able to:

- ☐ Describe the *life processes* common to all living things.
- ☐ Identify life processes shown by nonliving materials.
- ☐ Classify objects as either living or nonliving.

In 1986, the spacecraft *Voyager II* is scheduled to fly past the planet Uranus. If all goes well, the spacecraft will reach Neptune in 1989. It will then leave the solar system for outer space. Attached to *Voyager II* are recorded messages. They explain who made them, where they came from, and what life on our planet is like. Will any other form of life ever read them? Is there life anywhere else than on earth? If you were to search for life on another planet, what would you look for? How do we know what is and what is not alive on earth?

THE LIFE PROCESSES

Most biologists agree that living things can carry on certain activities that nonliving things cannot. These activities are necessary for life. What are some of the activities you carry on that a nonliving thing cannot? Living things can:

1. Take in food and useful gases.
2. Release energy from food.
3. Get rid of wastes.
4. Grow.
5. Move.

Fig. 2–1 A meadow vole is alive; a computer is not. What life processes does each show?

6. Respond to changes in the environment.

7. Give off useful chemical substances.

8. Reproduce.

Sometimes a nonliving object appears to carry on one or more of these activities. For example, a car moves. It uses gasoline to run. A car gives off wastes through the exhaust pipe. A car also releases energy from its fuel. A computer responds to changes in its environment. See Fig. 2–1. So does a smoke or heat detector. But only living things can carry on *all* of these **life processes.** Each of the *life processes* is explained below.

1. *Food getting:* Living things take in materials that provide energy and are needed for growth. For example, a frog eats insects, and a plant takes in gases and water to make food.

2. *Respiration:* Living things release energy from food. Most living things use oxygen to do this.

3. *Excretion:* Organisms remove wastes produced by the life processes. For example, you exhale and perspire to get rid of wastes.

4. *Growth:* The size of an organism increases. Living things also repair injuries. This includes a tree growing taller, or a cut on your finger healing.

5. *Movement:* The whole organism may move and materials inside the organism also move. The legs of a horse allow it to move. A system of tubes inside a tree moves food and water.

6. *Response:* Organisms react to changes in the environment. A plant responds to light by growing toward it.

7. *Secretion:* Secretions are useful chemicals that living things make and give off. Your mouth "waters" when you eat. Saliva is a secretion that helps you swallow food. Poison ivy secretes an oil. The oil protects it from some animals.

8. *Reproduction:* Reproduction means producing more of its own kind. Every organism has a limited life span. For some bacteria, it is a matter of minutes. For the bristlecone pine tree, it may be several thousand years. Each organism must produce young if its kind is to survive. Humans have babies. Trees produce seeds. See Fig. 2–2. Simple forms of pond life just divide to form two separate living organisms.

These life processes are common to living things. Some of these processes are quite obvious. Others are very difficult to observe. Biologists study how each organism carries on the life processes. But the ways in which they are carried on vary greatly. Living things are very complex. Even the smallest of living organisms carries on complicated physical and chemical activities. Living things are also similar in that they are made of the same basic chemical materials. These materials are found throughout the universe. Could life be there also?

Fig. 2–2 *Living things, such as this milkweed plant, reproduce to ensure survival of their kind.*

SUMMARY

All living things carry on eight basic life processes. Some nonliving things appear to carry on a few of these activities. Scientists apply this list of life processes to determine whether things they discover are living or nonliving.

QUESTIONS

Use complete sentences to write your answers.

1. Read the following list of organisms and objects: robot, molds, pond scum, polar bear, snowman, computer. Which ones are alive? Which ones are not?
2. What was the basis of your classification?
3. Which of the life processes did the nonliving objects seem to carry on?
4. Can you suggest other examples of nonliving materials showing some of the life processes?

SKILL-BUILDING ACTIVITY

PART I: USING THE MICROSCOPE

PURPOSE: To become familiar with the use and care of a light microscope.

MATERIALS:

microscope	salt
microscope slide	cane sugar

PROCEDURE:

A. Fig. 2–3 is a diagram of a compound microscope. Learn where each part is on your own compound microscope and what it does.

B. Carry a microscope to your desk using both hands. Hold the arm in one hand and place the other under the base.

C. Use lens paper to clean the lenses.

D. Turn the nosepiece so that the low-power objective clicks into position under the tube. Turn the microscope so that it is facing a light source. CAUTION: Do not place it in bright sunlight.

1. Why is this caution necessary?

E. While looking through the eyepiece, adjust the mirror so that light reflects upward through the lenses.

F. Open the diaphragm so maximum light comes through the lenses.

G. Place a few grains of table salt on a clean microscope slide. This is known as a "dry mount." Separate the crystals.

H. Place the slide on the stage. Center the salt over the hole in the stage. Use the clips to hold the slide in place.

I. Look at the microscope from the side. Use the coarse adjustment to lower the low-power objective until it *almost* touches the salt.

J. Now look through the eyepiece and slowly raise the tube, using the coarse

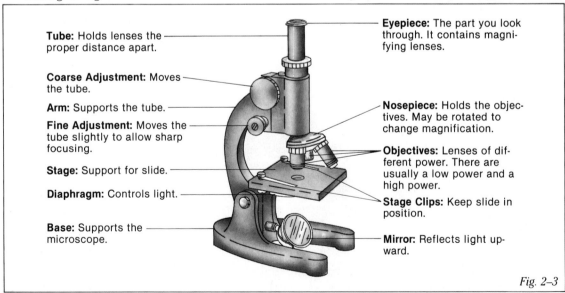

Tube: Holds lenses the proper distance apart.

Coarse Adjustment: Moves the tube.

Arm: Supports the tube.

Fine Adjustment: Moves the tube slightly to allow sharp focusing.

Stage: Support for slide.

Diaphragm: Controls light.

Base: Supports the microscope.

Eyepiece: The part you look through. It contains magnifying lenses.

Nosepiece: Holds the objectives. May be rotated to change magnification.

Objectives: Lenses of different power. There are usually a low power and a high power.

Stage Clips: Keep slide in position.

Mirror: Reflects light upward.

Fig. 2–3

adjustment, until the salt crystals are in focus. Use the fine adjustment to sharpen the focus. Sketch the appearance of the salt.

K. Adjust the diaphragm to change the amount of light.

    2. How does the salt change in appearance under different lighting?

L. To switch to high power, turn the coarse adjustment away from you to raise the tube. Turn the nosepiece until the high-power lens clicks into place.

M. Look at the microscope from the side. Lower the objective until it *almost* touches the slide. Use the fine adjustment for focusing.

N. Clean the stage and the lenses before putting the microscope away.

PART II: OBSERVING LIFE PROCESSES

PURPOSE: To observe the life processes in pond water organisms.

MATERIALS:

microscope	pond water samples
cover slips	cotton ball
microscope slides	forceps

PROCEDURE:

A. Place a drop of water from the sample on the center of a clean microscope slide. Include some of the material you see in the water.

B. Add a few strands of cotton to the water drop. This will slow down the organisms and allow them to be seen more easily.

C. Hold a cover slip with forceps as shown in Fig. 2–4.

D. Move the cover slip toward the drop until they touch.

E. Gently lower the cover slip onto the slide. Be careful not to trap any air bubbles under the cover slip. This type of slide is called a "wet mount."

F. Examine the water drop under the microscope, using low power.

G. Draw and describe in detail each of the organisms you find. Include the life processes you observe.

    1. Which of the life processes is easiest to observe?

    2. Which of the other life processes do you observe?

    3. Which life processes were you unable to observe?

CONCLUSIONS:

    1. Why can't all of the life processes be observed?

    2. How can you tell the living material from the nonliving material?

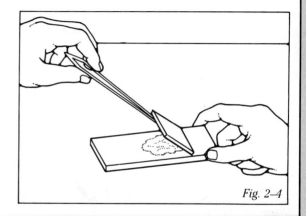

Fig. 2–4

2-2. The Chemistry of Life

At the end of this section you will be able to:

☐ Compare and contrast elements and compounds.

☐ Compare and contrast atoms and molecules.

☐ Recognize chemical symbols and formulas.

For years, the study of science was separated into areas such as biology, chemistry, and physics. Now, however, scientific research often overlaps these areas. Scientists seeking to answer a question such as "What is life?" must have a firm background in both biology and chemistry. These scientists are called *biochemists*. See Fig. 2–5.

A chemist is interested in the materials that make up the universe. A biochemist is interested in the materials that make up living things. The basic materials they study are the same. Both study how these materials behave and change.

Fig. 2–5 *This biochemist works in a medical laboratory.*

MATTER AND ITS CHANGES

Matter Anything that takes up space and has mass.

Everything in the universe may be classified as either **matter** or energy. *Matter* is the scientist's term for whatever takes up space. These objects can be in the form of solids, liquids, or gases. *Energy* is the term used for all the forces that act on the matter. These forces often cause changes in the makeup or appearance of the matter.

ELEMENTS

Elements Substances composed of only one kind of atom.

Chemists have discovered 92 basic natural substances. All the matter in the universe is made from them. These substances are called **elements.** The list of *elements* contains many with which you are familiar. Gold, silver, iron, and oxygen are elements. Biochemists have found that living organisms are made from the same elements as all other matter.

Atom The smallest particle of an element that has the characteristics of that element.

Think of the iron that is used to build a skyscraper or a bridge. The same kind of iron is used by your body, only in tiny amounts. The smallest possible particle of iron that can exist is called an **atom** of iron. Each element is made up of only one kind of *atom*. If there are 92 different kinds of elements, there must be 92 different kinds of atoms. Table 2–1 lists the elements that are the most common in living matter.

COMPOUNDS

Look around your classroom and at the world outside your windows. If you were to list the different kinds of things you see, how long would your list be? Fifty items? Ninety-two? Or more? Obviously, there are more than 92 different kinds of materials in the world. How can that be? Chemists have learned that the 92 basic elements can join with each other to make a tremendous number of new substances. Substances that are made of two or more different kinds of atoms are called **compounds.** Water is a *compound*. It is formed when atoms of the elements hydrogen and oxygen join together. Table salt is another compound. It is formed from atoms of the elements sodium and chlorine. Some compounds are simple. Others are very complicated. Sugars, proteins, fats, and vitamins are a few of the complex compounds found in living matter.

Compounds are formed when two or more elements are combined. Therefore, the smallest particle of a compound contains two or more different kinds of atoms. A new term is needed to describe this grouping of atoms. The smallest particle of a compound is called a **molecule.** See Fig. 2–6.

How do atoms join together to form *molecules*? What holds the molecules together? The answer is energy. A form of energy called chemical energy joins, or "bonds," the atoms together. When these bonds are broken, this energy is released and can be used for other purposes. The energy we get from food is the energy holding the food molecules together. When you eat food, some of this energy is released.

THE MOST COMMON ELEMENTS IN LIVING THINGS	
Element's Name	Symbol
Oxygen	(O)
Carbon	(C)
Hydrogen	(H)
Nitrogen	(N)
Phosphorus	(P)
Sulfur	(S)
Potassium	(K)
Magnesium	(Mg)
Calcium	(Ca)
Iron	(Fe)
Sodium	(Na)
Chlorine	(Cl)
Iodine	(I)

Table 2–1

Compounds Substances composed of two or more different kinds of atoms.

Molecule The smallest unit of a compound. Usually it is two or more atoms held together by energy.

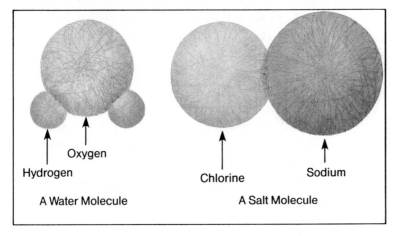

Oxygen
Hydrogen
A Water Molecule

Chlorine
Sodium
A Salt Molecule

Fig. 2–6 These are models of compounds. What are their smallest particles called?

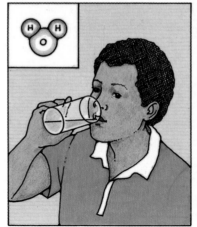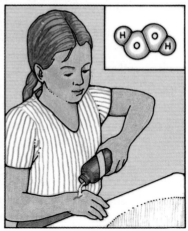

Fig. 2–7 Small differences in the formulas of compounds create very different materials.

Formula A combination of symbols and numbers that shows the kind and amount of atoms in a molecule.

FORMULAS AND SYMBOLS

Scientists use a one- or two-letter symbol to represent each different kind of atom. Several atoms and their symbols are shown in Table 2–1. The makeup of a molecule can be shown by combining these symbols. This is called a **formula.** For example, the *formula* for water is H_2O. The number 2 just after and below the H means that this molecule contains two atoms of hydrogen. The O shows that there is one atom of oxygen present. The formula H_2O_2 is not water. The second atom of oxygen makes the molecule behave very differently. This compound is hydrogen peroxide. It is used as a bleach or to clean wounds. See Fig. 2–7.

SUMMARY

The elements that make up living matter are the same as those that form the rest of the universe. Atoms join to form molecules to create the great variety of matter we see around us. Scientists use symbols and formulas to identify them.

QUESTIONS

Use complete sentences to write your answers.

1. If there are only 92 elements, how can there be so many different materials in the world?
2. Compare and contrast an element, a compound, a molecule, and an atom. Explain how they are related.
3. What information can you get from this formula: $C_6H_{12}O_6$?

INVESTIGATION

MAKING ATOMIC MODELS

PURPOSE: To construct models of simple molecules.

MATERIALS:
8 large spheres, 8 small spheres, toothpicks

PROCEDURE:

A. Separate the spheres into two piles according to size. You now have models of two different elements. The pile of large spheres represents atoms of the element oxygen. The pile of small spheres represents atoms of the element carbon.

B. Use the toothpicks to attach two atoms of oxygen to one atom of carbon. The atoms should be in a straight line with carbon in the middle. The new model is carbon dioxide. Your model of carbon dioxide is made up of two or more different atoms.

1. Does the model represent an element or a compound? Why?

2. How many molecules of carbon dioxide are in your model?

3. What is the formula for the molecule?

4. What do the toothpicks represent?

5. If you break apart a molecule, what can happen to the energy that held it?

C. Take the molecule of carbon dioxide apart and separate the different atoms again. The large spheres will now represent the element chlorine and the small ones the element sodium.

6. Why are the piles of atoms no longer compounds?

D. Attach a large atom to a small atom.

7. Is the new model an element or a compound? Why?

E. The model you just made is sodium chloride, or common table salt. Now join all the spheres into pairs of sodium and chlorine atoms. Take two pairs and join the sodium of one to the chlorine of the other. They should form a straight line. Make four of these lines by joining the other pairs in the same way. Place two of these chains side by side so that the sodium atoms are always beside chlorine atoms. Add a third and fourth chain.

F. Join with another student team. Place their model on top of yours so that sodium is on top of chlorine and vice versa. Add several layers. See Fig. 2–8. You have just built a crystal of salt.

8. Judging from this model, what is the shape of a salt crystal?

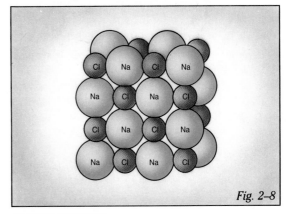

Fig. 2–8

CONCLUSIONS:

1. Explain how atoms are different from molecules?

2. Explain how elements are different from compounds?

SCIENCE INPUT

Cells are the smallest organized units of living protoplasm. They are important to study and understand because, as you learned in Chapter 2, only cells can produce other living cells. The materials composing cells can be grouped into three divisions: cell boundary, cytoplasm, and nucleus. Each section of the cell helps it to carry out its functions. Knowing how the various parts of the cell function has helped life scientists in their research, as, for example, with cloning (which you will read about in the next chapter's technology feature).

COMPUTER INPUT

Learning new concepts, while important to the subject area you're studying, can sometimes be difficult and frustrating. Being able to repeat definitions over and over is often helpful. A computer can be programmed to function as a tutor, asking you for the correct answer, telling you when you are right, and giving you the correct answer if yours was wrong. Unlike a human tutor, however, a computer will repeat the questions endlessly, without getting either angry or frustrated. In Program Cell Vocabulary, the computer will be used to help you learn the new words and concepts associated with cells used in this text. Enter the program carefully, and consult your computer manual. Commands are not the same for every computer.

WHAT TO DO

On a separate piece of paper, make a chart similar to the one at right, listing terms and their definitions. The definitions should be short and simple. Study the chart before entering the program. When you run the program, it will produce a quiz of four questions, using the definitions of cell structure you wrote.

Lines 220 to 310 are for data statements. Your terms and definitions are data. Be sure to type the statement in the correct form. An example is given in line 220.

The program will tell you the correct answer if your answer is wrong. After four questions, it will give you a score and tell you how many questions out of the four you've answered correctly. Run the program at least five times, recording the terms you've missed, so you will know which you need to study.

Sample Data Chart

Cell Term	Definition
Wall	A rigid layer around a plant cell
Membrane	
Vacuole	
Mitochondria	
Ribosome	
Chloroplast	
Plastid	
Nucleoplasm	
Nucleolus	
Chromosome	
etc.	

LIVING MATTER: PROGRAM CELL VOCABULARY

GLOSSARY

REM   short for REMark. These statements are for the user's information and are ignored by the computer. They help make the steps in the program clear.

DATA   a command in BASIC that instructs the computer to store information for use in the program.

READ   a command in BASIC that instructs the computer to read a value from a data statement and give that value to a variable in the program. You *must* have a READ statement, or the data will not be included in the working of the program.

RUN   a command in BASIC that instructs the computer to execute the program you have entered.

PROGRAM

```
100   REM CELL VOCABULARY
110   DIM T$(10), D$(10)
115   REM READING IN OF DATA
120   FOR X = 1 to 10: READ T$(X),D$(X):
      NEXT X:K = 0
130   FOR Q = 1 TO 4: LET N = INT (10 *
      RND (1) + 1)
135   REM BEGIN PRINTING QUESTIONS
140   PRINT : PRINT "WHICH 'CELL'
      TERM HAS THIS MEANING?"
150   PRINT "-> ";D$(N): INPUT "-> ";A$
160   IF A$ = T$(N) THEN PRINT "YOU'RE
      CORRECT!"
170   IF A$ = T$(N) THEN K = K + 1
180   IF A$ < > T$(N) THEN PRINT "THE
      CORRECT ANSWER IS:    ";T$(N)
190   FOR T = 1 TO 2000: NEXT: NEXT Q
200   PRINT "-> END OF QUIZ"
210   PRINT "-> ";K;"   OF 4 ANSWERS
      WERE CORRECT"
220   DATA WALL,    A RIGID LAYER
      AROUND A PLANT CELL
310   DATA
330   END
```

PROGRAM NOTES

This program is written to be used with ten data statements. See line 110. To change the number of data statements, replace the number 10 wherever it appears. New words and definitions can, however, replace the old ones at any time. The program can also be used for more than ten cell vocabulary terms or for any other set of words and their definitions. If you did use it for another subject, you would have to change the kinds of data statements as well as lines 100 and 140.

BITS OF INFORMATION

Learning through computers, as well as about computers, is beginning to change the character of our educational system. Can you imagine a 24-hour-a-day university with no campus? A company in California has developed a network of 170 university-level courses that are available to students 24 hours a day through their personal computers. Communications with teachers, homework assignments, and projects are all done via computer! If our entire school system were designed this way, what might some effects of these methods be?

2-3. The Cell Theory

At the end of this section you will be able to:

- ☐ Explain the two parts of the cell theory.
- ☐ Contrast organic and inorganic compounds.
- ☐ Describe the basic structures found in cells.

"Fireflies rise from the morning dew,

fish and frogs from a muddy stew,

maggot worms from rotting meat,

while mice shall come from sweat and wheat."

Does this rhyme sound like a witch's chant? Actually, each of these ideas was once accepted as how these organisms came into being.

LIFE FROM LIFE

That living things could form directly from nonliving things was an idea that was held by many people as recently as the 1800's. It wasn't until the methods of scientific problem solving and controlled experiments were developed that the idea was seriously challenged. It was finally proved wrong.

Maggot worms are a stage in the development of flies. People had seen the worms on decaying meat. Not having seen the flies laying eggs, they assumed that the worms developed *from* the meat. Francesco Redi, an Italian doctor, didn't believe this idea. Like many educated people of his time, he was interested in how life developed. In 1668, he carried out a series of controlled experiments that disproved this idea about the origin of maggot worms. See Fig. 2–9.

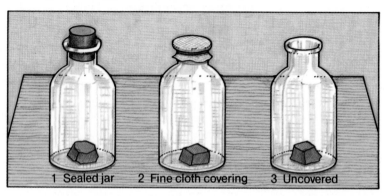

1 Sealed jar   2 Fine cloth covering   3 Uncovered

Fig. 2–9. Which jar was the control in Redi's experiment?

1 No maggots 2 Maggots on cover 3 Maggots on meat

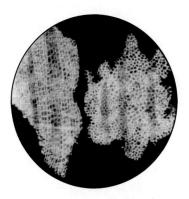

Fig. 2–10 What did the results of Redi's experiment prove?

Redi placed pieces of meat into several jars. Some jars he sealed tightly. Some he covered with a fine cloth. Some he left open to the air. This last group acted as the control in the experiment. They would show what usually happened when meat is left uncovered.

No maggots developed in the sealed jars. See Fig. 2–10. The meat in the open jars attracted flies and soon maggots appeared on the meat. Flies were attracted to the cloth-covered jars too. But the maggots hatched from eggs laid on top of the cloth. The flies could not reach the meat. This experiment proved that the maggots did not come out of the decaying meat. However, it took another 200 years and many more scientific experiments to prove that, in all cases, life comes only from life.

THE CELL THEORY

Also in the late 1600's, people first observed things smaller than the naked eye could see. An Englishman named Robert Hooke examined many materials with a simple microscope that he designed. One of these materials was a piece of cork. He cut a very thin slice of cork and examined it with his microscope. Hooke observed tiny, orderly, empty spaces that reminded him of the spaces in a honeycomb. He called these spaces **cells.** See Fig. 2–11.

Ten years later, a Dutchman named Anton van Leeuwenhoek focused his microscope on a drop of water. He saw a new world of life that no one ever knew existed. He called the tiny creatures he saw "wee beasties." Many of these organisms were made of a single *cell.* As the years went by, more and more observations of living things were made. Investigators found that all living things were made up of the tiny units

Cell The smallest organized unit of living protoplasm.

Fig. 2–11 Robert Hooke drew these cork cells.

called cells. Gradually, the *cell theory* was developed. This theory may be stated:

1. The cell is the basic unit of all living things.
2. Only living cells can produce new living cells.

LIVING MATTER

Scientists later discovered that cells contained a thick liquid they called **protoplasm.** *Protoplasm* is sometimes called the "stuff of life." It is a mixture of compounds in which the basic life activities occur.

The compounds that make up protoplasm may be grouped as **organic** or **inorganic.** All *organic* compounds contain the element carbon. Any compound that doesn't contain carbon is called *inorganic.* Many inorganic compounds are basic to life. Water is one important inorganic compound. It makes up about 65 to 95 percent of living material. Certain minerals and the gases oxygen and carbon dioxide are other inorganic compounds necessary for life. Carbon dioxide is inorganic, even though it contains carbon. This is because carbon dioxide is much simpler than most organic compounds.

Organic compounds include sugars and starches, which are energy sources; fats, which store energy; proteins, which serve as the building blocks of living matter; and other compounds with specialized functions. Each of these organic compounds will be discussed in later chapters.

BASIC CELL STRUCTURE

Cells are made of protoplasm and its products. Cells are not all the same size and shape. Many cells have special structures that have special purposes. But all cells are similar in some respects.

Protoplasm Living material, found in a cell, capable of carrying on all the life processes. It is made up of organic and inorganic compounds.

Organic Refers to compounds that contain carbon.

Inorganic Refers to compounds that do not contain carbon.

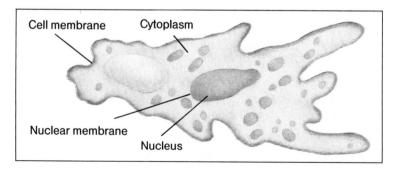

Fig. 2–12 The major parts of all cells are identified on this ameba.

Surrounding the cell is a covering called the **cell membrane.** See Fig. 2–12. It controls what materials enter or leave the cell. Most of the cell is made of a type of protoplasm called **cytoplasm.** Many of the cell's activities are carried on in the *cytoplasm*. Near the center of the cell is a structure called the **nucleus.** The *nucleus* is the "control center" that directs all the cell's activities. It is surrounded by a **nuclear membrane.** Inside the nucleus is a type of protoplasm called *nucleoplasm*.

All these structures can be seen with a standard laboratory microscope. These microscopes usually magnify objects from 100 to 450 times. In the past fifty years, newer, more powerful types of microscopes have been developed. The electron microscope can enlarge things 300,000 times or more. The scanning electron microscope gives a three-dimensional picture of objects. Both have helped scientists discover more about the structure of the cell. See Fig. 2–13.

Cell membrane The part of the cell that determines what enters and leaves the cell.

Cytoplasm The protoplasm surrounding the nucleus of the cell. The cell's activities are carried on here.

Nucleus The "control center" of the cell that directs all the cell's activities.

Nuclear membrane Boundary separating the nucleus from the cytoplasm.

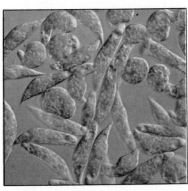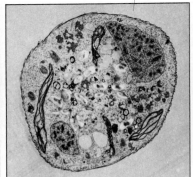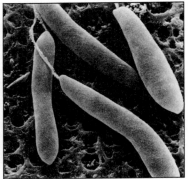

Fig. 2–13 Euglena *as seen under a light microscope* (left), *electron microscope* (middle), *and scanning electron microscope* (right).

SUMMARY

The cell theory states that the cell is the basic unit of life and that all life comes from living cells. Cells are made of protoplasm, a mixture of organic and inorganic compounds. Although cells vary, they all have certain basic parts in common.

QUESTIONS

Use complete sentences to write your answers.

1. State the two parts of the cell theory.
2. How are organic and inorganic compounds different?
3. Why is protoplasm called "the stuff of life"?
4. Name three basic cell structures and give their functions.

LIFE FROM LIFE

PURPOSE: To repeat Redi's experiment.

MATERIALS:

small plastic aquarium	2 rubber bands
3 baby food jars with lids	container of fruit flies
cloth, 10 cm square	3 slices of ripe banana
cloth cover for aquarium	1/4 package of yeast
wide masking tape	100 mL water
	paper towels

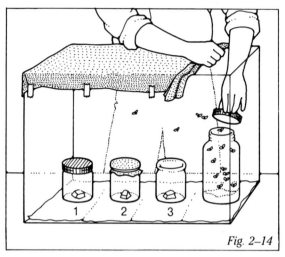

Fig. 2–14

PROCEDURE:

A. For better visibility, line the bottom of the aquarium with the paper towels.

B. Dissolve 1/4 package of yeast in 100 mL of water.

C. Dip a slice of banana (about 2 cm thick) into the yeast solution.

D. Place the banana into a jar and seal tightly with the lid.

E. Dip a second slice of banana into the yeast solution.

F. Place the banana in a second jar. Cover with the cloth square. Secure the cloth with the rubber bands.

G. Dip the third piece of banana into the yeast solution.

H. Place the banana into the third jar. Leave this jar uncovered.

　1. Which of these jars will be the control?

I. Place all three jars into the aquarium.

J. Cover the aquarium with the large cloth. Secure all but one end of the cloth tightly to the aquarium with the masking tape.

K. Place the container of fruit flies in the aquarium. Open the container and quickly seal the rest of the cover with tape. See Fig. 2–14.

　2. Write a hypothesis. State which jars you think will contain eggs or worms.

L. Each day for about a week, observe the jars for evidence of eggs and worms.

　3. Do any eggs or worms appear in or on the sealed jar? If so, where?

　4. Do any eggs or worms appear in or on the cloth-covered jar? If so, where?

　5. Do any eggs or worms appear in or on the open jar? If so, where?

CONCLUSIONS:

　1. Did the eggs and worms come from the bananas or the fruit flies? How do you know?

　2. How does this activity support the second part of the cell theory?

　3. How does this experiment differ from Redi's experiment?

2-4. Living Cells

At the end of this section you will be able to:

- ☐ Compare the functions of the parts of the cell to the workings of a factory.
- ☐ Explain the function of each cell part.
- ☐ Compare the structure of a typical animal cell and plant cell.

Living organisms, including you, are made up of cells. These cells carry on the activities necessary for life. Cells have special parts that perform these activities. It may help you to understand the function of each part if we compare the cell to a factory.

A factory is a place of great activity. See Fig. 2–15. Fuel and raw materials are delivered to the factory through its gates. The factory workers follow a set of directions from the main office as they do their jobs. Fuel is burned in generators to provide energy. Energy is used to put the raw materials together into finished products. During the manufacturing process, wastes are produced and need to be removed. The finished product is packed. It is stored until it is used or shipped out of the factory. These manufacturing processes may be compared to the life processes carried on in a cell. The finished products of the cell are the compounds that form the parts of this cell and other cells.

Fig. 2–15 What part of a cell is the factory fence similar to?

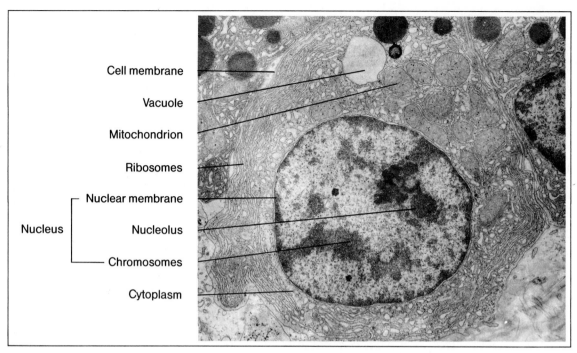

Cell membrane

Vacuole

Mitochondrion

Ribosomes

Nuclear membrane

Nucleus

Nucleolus

Chromosomes

Cytoplasm

Fig. 2–16 Compare the electron micrograph (magnified 5,000 times) and drawing of a typical animal cell (right).

Chromosomes Rod-shaped structures found in the nucleus of the cell.

THE NUCLEUS

The main office and planning department of our factory cell is the nucleus. The nucleus is the control center of the cell. It controls everything that goes on inside the cell.

The microscopes in most schools are not powerful enough to show many details inside the nucleus. You may see a dark spot called the *nucleolus* inside the nucleus. The nucleolus has a function related to making proteins. See Fig. 2–16.

Throughout the nucleus are structures called **chromosomes.** The *chromosomes* contain the instructions for the manufacture of the finished products of the cell. The chromosomes can be compared to the files of blueprints in the factory's main office. Blueprints contain the plans for what the finished product will look like. In a factory, the original blueprints never leave the office. Copies are made and sent from the office into the work area. In the cell, copies of the instructions are made and sent out of the nucleus.

THE CYTOPLASM

The cell's manufacturing processes are carried on in the cytoplasm. This is the "shop area" of the cell. Here are located

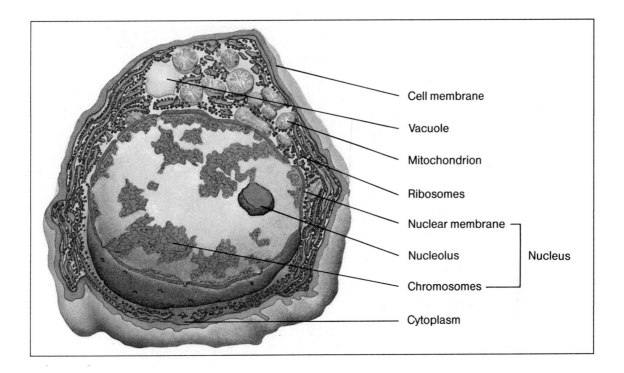

Cell membrane

Vacuole

Mitochondrion

Ribosomes

Nuclear membrane ⎤

Nucleolus   Nucleus

Chromosomes ⎦

Cytoplasm

the structures that do the cell's work. These structures include storage areas called **vacuoles.** Some of these *vacuoles* store food for future use. Some store waste matter until it is removed from the cell. Other vacuoles store fluids such as water and chemicals. Vacuoles are sometimes large enough to see. They resemble tiny air bubbles.

The power generators, called **mitochondria**, are also located in the cytoplasm. The *mitochondria* release the energy needed to run the cell. The fuels of a cell are sugars and starches.

Some of this energy is used by structures called **ribosomes.** The *ribosomes* are attached to a long, winding network in the cytoplasm. They help build proteins. The ribosomes can be compared to the machines of our factory model. The proteins are the final products made from the raw materials brought into the cell. Scientists use electron microscopes to magnify these cell parts over 200,000 times so they can be studied.

The *cell membrane* that surrounds the cytoplasm acts like the factory wall. The cell membrane allows only certain substances to enter and leave the cell. If the cell membrane is damaged in any way, the cell may die.

Vacuoles Storage areas located in the cytoplasm.

Mitochondria Structures in the cytoplasm that release energy from food.

Ribosomes Structures in the cytoplasm that make proteins.

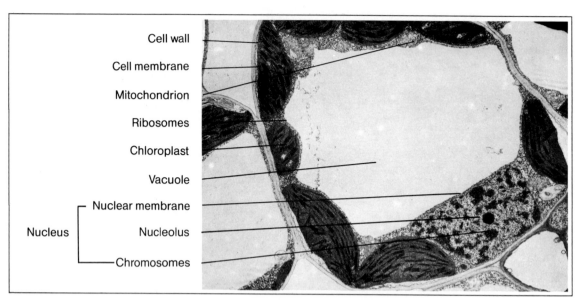

Cell wall

Cell membrane

Mitochondrion

Ribosomes

Chloroplast

Vacuole

Nuclear membrane

Nucleus — Nucleolus

Chromosomes

Fig. 2–17 Compare the photo (above) and drawing (page 51) of a plant cell to the animal cell in Fig. 2–16.

Cell wall The rigid protective layer that surrounds the plant cell.

Chloroplasts Oval-shaped structures in plant cells containing chlorophyll.

Chlorophyll Green material in green plants that uses sunlight to make food.

THE PLANT CELL

Plant cells contain all the structures that are found in animal cells. See Fig. 2–17. One of the most obvious differences between plant and animal cells is the size of the vacuoles. Vacuoles in plant cells are very large. The plant cell vacuole also acts as a storage area. It is filled with a clear fluid. This fluid is mostly water but also contains sugar, starch, and protein molecules.

Plant cells also have a thick, firm, outer boundary called the **cell wall.** The *cell wall* supports and protects the cell. Cell walls formed the little "cells" that Robert Hooke saw in the thin slice of cork. Animal cells do not have a cell wall. Plant cells also have a cell membrane like animal cells. The cell membrane in plant cells is pressed up against the inside of the cell wall by a large vacuole.

These thick and rigid cell walls will remain long after the plant cell dies. In trees, many of these cell walls are crushed together into layers. Eventually, these layers form the rigid material that we call wood.

The ability to make food is the major difference between plant and animal cells. In the cytoplasm of some green plant cells are many small, green structures called **chloroplasts.** These *chloroplasts* contain molecules of **chlorophyll.** The *chlorophyll* allows a plant cell to make its own food. Animal

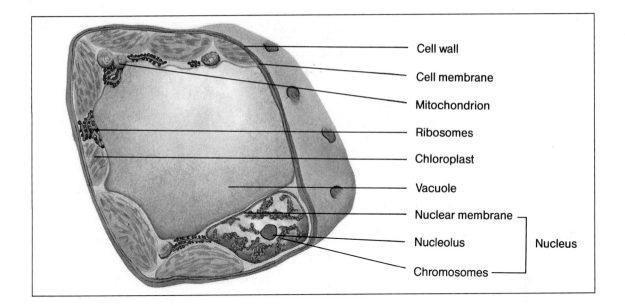

Cell wall

Cell membrane

Mitochondrion

Ribosomes

Chloroplast

Vacuole

Nuclear membrane ⎱
Nucleolus   ⎰ Nucleus
Chromosomes

cells do not have chlorophyll. Therefore, they cannot make their own food. Not all of the cells in a green plant have chlorophyll, however.

SUMMARY

A factory is a place of great activity. A living cell is, too. The "machinery" of both uses power to turn raw materials into finished products. Both have a "main office" that controls what happens in the "shop areas." Although there are differences between plant and animal cells, the basic functions are the same.

QUESTIONS

Use complete sentences to write your answers.

1. Which of the parts of an animal cell is represented by each of these parts of a factory? (a) power generators, (b) blueprints, (c) storage areas, (d) main office, (e) machinery, (f) factory wall, and (g) shop area.

2. What structures do plant cells have that animal cells do not have?

3. List the cell part(s) most involved with the following life processes: (a) food getting, (b) respiration, (c) response, and (d) growth. Explain the reasons for your choices.

INVESTIGATION

IDENTIFYING CELL STRUCTURES

PURPOSE: To compare and contrast animal and plant cells.

MATERIALS:

microscope	medicine dropper
cover slips	iodine stain
2 microscope slides	*Elodea* leaf
toothpicks, flat	scissors

Fig. 2–18

PROCEDURE:

A. Place a drop of iodine stain on a clean microscope slide. The stain is used to make colorless protoplasm visible.

B. Gently scrape the inside lining of your cheek with the flat end of a *clean* toothpick. See Fig. 2–18. Stir the cheek material into the drop of stain in order to separate the cells.

C. Cover the stained cheek material with a cover slip. Using low power, examine it under the microscope.

D. Locate some of the separated cells. Sketch several cells and any cell parts you can see.

   1. Do the cells all have similar shapes? Are they all about the same size?

E. Focus on one cell with the highest power of your microscope. The dark brown spot in your cheek cell is the nucleus. Label the nucleus on your sketch.

   2. Where is the nucleus located?

   3. Does every cell on your slide have a nucleus?

   4. Are they all located in the same position in the cell?

F. Now carefully examine the cytoplasm around the nucleus of the cheek cell. Label the cytoplasm on your sketch of the cheek cell.

G. Locate the cell membrane surrounding the cytoplasm. Label the cell membrane on your sketch.

H. Cut off the tip of an *Elodea* leaf. Place the tip of the leaf in a drop of water on a clean slide. Add a clean cover slip. Observe under low power. Sketch a group of cells. Label any parts you can identify.

   5. Does the *Elodea* cell have a nucleus?

   6. What fills in the area between the nucleus and the outer edge of the cell?

   7. Are there chloroplasts present? Describe them.

   8. What is the function of chloroplasts?

   9. How are the edges of *Elodea* and cheek cells different?

CONCLUSIONS:

   1. How does a plant cell differ from an animal cell?

   2. How does the size of the vacuoles differ in plant and animal cells?

2-5. Levels of Organization

At the end of this section you will be able to:

☐ Explain why cells specialize.
☐ Describe each level of organization.
☐ Name examples of each level in animals and plants.

Two hundred years ago, our nation began to expand westward into the Central Plains. Each family, such as the one in Fig. 2–19, settled far from others. Raising food, finding water, building shelter, and making tools were up to each family unit.

As time went on, more and more families settled in the wilderness areas. Some of these people were better farmers than others. Some were skilled blacksmiths. Some could make and repair tools. Others had skills in carpentry.

A kind of exchange system developed. The skilled blacksmith would shoe the farmer's horse. The farmer would provide the blacksmith with food. The carpenter would build an office for the doctor. The doctor would provide medical care for the carpenter's family. As years passed, each person became a specialist.

Fig. 2–19 This pioneer family settled on land in the state of Washington.

CELLS ORGANIZE

Single-celled organisms are similar to those early families of settlers. Each cell must perform all the life functions by itself. Each part of the cell has a special function.

Most organisms are not made of one cell. They are made of many cells. These organisms are called **multicellular** (mul-tih-**sel**-you-lahr). See Fig. 2–20. Each cell in a *multicellular*

Multicellular Made up of more than one cell.

Fig. 2–20 A cardinal flower (left) and red fox (right) are multicellular organisms.

Specialized Refers to the fact that each cell has a particular function, and that different kinds of cells have different functions.

organism has all of the basic cell parts. However, each of these cells has also become **specialized.** Each has a special function that benefits the other cells. In this way, the cells of a multicellular organism depend on one another for survival.

This division of labor is similar to our society today. Each working person adds to the benefit of the community. In turn, each individual benefits from the work of others.

The *specialized* cells of multicellular organisms are organized into different levels. The cell itself is at the first level of organization. Most cells have a certain size and shape related to their purpose. See Fig. 2–21. For example, in humans there are two types of blood cells and three types of muscle cells. There are also covering cells, nerve cells, bone cells, and fat cells. Each cell has a special shape that allows it to perform its task. In plants there are cells specialized for absorbing water from the soil. There are also cells in plants for covering and protecting, transporting, and growing.

Fig. 2–21 The shapes of cells vary with their functions: Muscle cells (left), *blood cells* (middle), *and plant transport cells* (right).

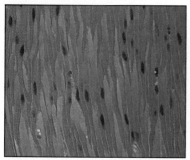 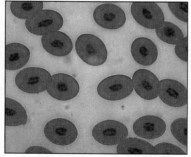 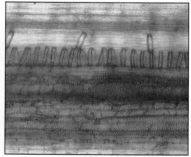

TISSUES, ORGANS, AND SYSTEMS

Tissues Groups of similar cells with similar functions.

The second level of organization is **tissues.** *Tissues* are groups of similar cells with similar functions. Blood cells form blood tissue. One of the functions of blood tissue is to carry oxygen to the body cells. Muscle cells form muscle tissue. Muscle tissue is responsible for movement.

Plant cells also form specialized tissues. Covering cells on leaves are called *epidermal* tissue. Its function is to protect the cells beneath it.

Organs Groups of different tissues working together to perform a specific function.

Organs represent the third level of organization. An *organ* is a group of different tissues working together to perform a special task. The heart is one such organ. The heart is made up of muscle, nerve, covering, and blood tissues. Together, these tissues have a special purpose. They pump blood

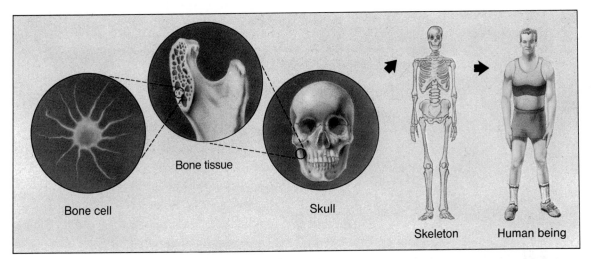

Bone cell

Bone tissue

Skull

Skeleton

Human being

through the body. Other examples of animal organs are the eye, the stomach, and the lung. Leaves are an example of a plant organ. The leaf contains several tissues. Each performs a special task. Together, as an organ, they function to make food. Roots and stems are other plant organs.

A group of organs working together is known as a **system.** *Systems* are the fourth level of organization. In humans, there are many systems of organs. These include the digestive, respiratory, nervous, circulatory, and reproductive systems. In plants, systems include transport and reproductive systems.

The highest level of organization is the *organism.* A single living organism is the combination of all its systems, organs, tissues, and cells. See Fig. 2–22.

Fig. 2–22 The different levels of organization, from bone cell to organism.

System A group of related organs performing a major function for an organism.

SUMMARY

The cells in multicellular organisms are specialized for certain jobs. Specialization allows each job to be performed better. There are five levels of organization of these cells.

QUESTIONS

Use complete sentences to write your answers.

1. Arrange the following in order from simple to more complex: organ, cell, organism, tissue, system.
2. What is the advantage of specialized cells?
3. What level of organization is each of the following: tree, heart, blood, skeleton?

INVESTIGATION

LEVELS OF ORGANIZATION

PURPOSE: To use a model of how cells are organized in living things.

MATERIALS:

(Teams of 4 students)	metric ruler
set of activity pieces	pencil
	paper

PROCEDURE:

A. Follow steps B through F by yourself, using one set of 16 pieces of the same color. At step G, the four students in your group will join the four different sets of colors to complete the activity.

B. Spread out the pieces of one color on your desk. These pieces are models of the first level of organization. Each piece represents the basic unit of living things.

   1. What is the basic unit of life called?

   2. How are these models different from each other? These differences represent the different functions of these model cells.

C. Group the model cells with identical sizes and shapes together. You should have six different groups. These groups represent "similar cells with similar functions."

   3. What level of organization do these "similar cells" represent? What are these groups called?

D. Draw a 9-cm square on your paper. Note that there are three basic shapes of models (rectangles, triangles, and trapezoids). Separate the shapes into these three groups. Arrange the two different sizes of rectangles so that they fill the square you have drawn. This new model represents level III organization.

   4. What are structures at level III called? Why does this model fit the description of this level?

E. Arrange the other two groups to form two 9-cm squares. Remember to keep the basic shapes together.

   5. Do these new structures also represent organs? Why?

F. Arrange your three "organs" in a row to form a model of level IV organization.

   6. What are structures at level IV called? Why does this model fit the description of this level?

G. Now arrange your system with those of your teammates to form a model of level V organization.

   7. What are structures at this level called? Why does this model fit this description?

   8. How many cells are there in your organism?

   9. How many different tissues are in the organism?

   10. How many different organs?

   11. How many different systems?

CONCLUSIONS:

   1. How is a model similar to the real thing?

   2. How is it different?

   3. Why do we sometimes use models instead of the real objects?

CAREERS IN SCIENCE

LABORATORY TECHNICIAN

The laboratory technician is responsible for most of the lab's routine work. Since laboratories are usually divided into different departments, most lab technicians specialize in a single medical area. Technicians working in the hematology department might collect, process, and store blood for the blood bank. In the microbiology department, technicians help locate and identify bacteria or viruses that cause infection or disease.

Technicians must be high school graduates who have completed two years of AMA-approved clinical laboratory training. Students who pass the training may take a national exam given by the Board of Registry of the American Society of Clinical Pathology. Certain states require successful completion of this test for employment.

For further information, write: American Medical Association, 535 North Dearborn Street, Chicago, IL 60610.

NEONATOLOGIST

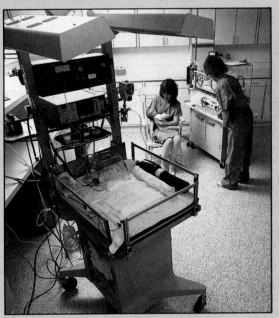

Neonatologists are physicians who specialize in the care of newborn infants up to two months old. Many of these children are under intensive care. Some have life-threatening infections. Others, born to mothers with chronic illnesses or drug addictions, suffer the effects of their mothers' illnesses or addictions when they are born. Still other infants are in intensive care because they were born prematurely and cannot survive without life-support systems.

Preparation for a career as a neonatologist demands a great deal of commitment. After completing college and medical school, it is necessary to intern and do residency before being certified as a doctor. Then, one usually does a year or so of special study in neonatology itself, combining classroom work with supervised clinical experience.

For more information, write: Director, Dept. of Neonatology, Columbia Presbyterian Baby Hospital, Room 1201, New York, NY 10032.

CHAPTER REVIEW

VOCABULARY

On a separate piece of paper, match each term with the number of the statement that best explains it. Use each term only once.

protoplasm   element   matter   specialized

formula   organic   multicellular   compound

inorganic   cell

1. Different kinds of cells have different functions.
2. Living material found in a cell capable of carrying on all the life processes.
3. Anything that takes up space.
4. Compounds that do not contain carbon.
5. A substance composed of two or more different atoms.
6. The smallest organized unit of living protoplasm.
7. A type of organism composed of more than one cell.
8. A combination of symbols and numbers that shows the kind and number of atoms in a molecule.
9. Compounds that contain carbon.
10. A substance composed of only one type of atom.

QUESTIONS

Give brief but complete answers to each of the following questions. Unless otherwise indicated, use complete sentences to write your answers.

1. What is meant by the life processes? List and briefly explain three life processes.
2. How do scientists determine if an object is living or nonliving?
3. What are the differences between the following pairs of terms? (a) Matter and energy, (b) element and compound, and (c) atom and molecule.
4. What does each of the following symbols and formulas tell you about these molecules?

    CH_4 H_2O_2 H_2SO_4 SO_2 $C_{12}H_{22}O_{11}$
5. Which of the molecules in question 4 are organic? Why?
6. State the two parts of the cell theory.
7. Construct a chart that will show how plant and animal cells are alike and how they are different.
8. Identify the numbered parts of the plant and animal cells in Fig. 2–23. In your own words, write a brief statement explaining the function of each cell part.
9. Give an example, from both plant and animal organisms, of a tissue, an organ, and a system.

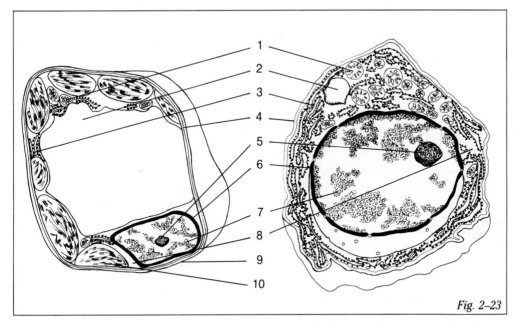

Fig. 2–23

10. Why would you expect to find a lot of mitochondria in a muscle cell?

11. Many living things are made up of just one cell. Some scientists place these living things in both level I and level V organization. Explain why this is possible.

APPLYING SCIENCE

1. Construct a three-dimensional model of a typical animal or plant cell. Use materials such as a plastic bag, gelatin, marbles, and pipe cleaners to represent cell parts.

2. This chapter describes how pioneer communities and living organisms both have levels of organization. Describe how your school or school district is organized. Identify several levels of organization. Create a chart to illustrate how the levels are related.

BIBLIOGRAPHY

Asimov, Isaac. *How Did We Find Out About Atoms?* New York: Walker, 1976.

Langone, John. "Meteorites and the Stuff of Life." *Discover,* November 1983.

McKean, Kevin. "Life on a Young Planet." *Discover,* March 1983.

Hidden Worlds. Washington National Geographic Society, 1981.

CHAPTER
3

When magnified 125 times, you can see the individual cells in this plant root tip. All cells, in plants or animals, carry on certain similar activities.

LIFE PROCESSES OF CELLS

CHAPTER GOALS

1. Explain why and how materials move into and out of cells.
2. Explain why cells need energy, and compare two ways that they release it.
3. Explain why new cells are needed, and describe the sequence of events by which cells reproduce.

3-1. Cells and Diffusion

At the end of this section you will be able to:

☐ Define *diffusion*.

☐ Describe how diffusion takes place through the cell membrane.

☐ Explain why water is needed to move materials into and out of cells.

☐ Compare and contrast diffusion with active transport.

What do solid air fresheners, mothballs, perfume, and cell membranes have in common? Here's a hint: What allows you to smell the odors of the first three?

You have learned that all substances are made up of molecules. These molecules are in constant motion. They move in a nonorderly, random way. In all substances, some molecules move faster than others.

Molecules are too small to see. But the use of models can help you understand how molecules move. Scientists often use models to explain how particles of matter behave. Have you ever seen or been on bumper cars at an amusement park? If so, think of them now. Suppose the cars are bunched together in one corner. They are facing in all directions. As the cars begin to move, they bounce off one another and then move in all directions. Each time a car collides with another, it is likely to move away from the corner in which it started. Although some cars may move back toward the starting corner, most of the movement is away from it. The cars will continue to bump into each other. After a while, they will be fairly evenly spread out.

Fig. 3–1 Molecules of red food coloring diffuse among the molecules of blue water.

Diffusion The spreading out of molecules from a crowded area to a less crowded area.

DIFFUSION

Molecules act in much the same way as bumper cars. Molecules even bounce off each other as the cars do. This constant movement of molecules causes **diffusion.** *Diffusion* is the process by which molecules are spread out. This movement occurs from a crowded area to a less crowded area.

A drop of food coloring in water is an example of diffusion. Look at Fig. 3–1. At first the molecules of food coloring are crowded together. Since molecules are constantly moving, some food-coloring molecules move away from the others. Eventually, the molecules of food coloring become evenly spread out in between the water molecules.

Solid air fresheners, mothballs, and perfume all work by diffusion. In the solid air freshener, the scented molecules of the stick are crowded together. See Fig. 3–2. The molecules leave the stick and spread out into the air in the room. This occurs very slowly.

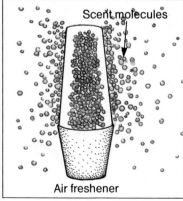

Fig. 3–2 Molecules from a solid air freshener diffuse into the air quite slowly.

THE CELL MEMBRANE AND DIFFUSION

A cell is surrounded and protected by a thin covering called the cell membrane. The cell membrane also gives the cell strength and shape. In this way it is like a plastic bag. However, unlike a plastic bag, the cell membrane allows some materials to pass through it.

Materials that move into and out of the cell must pass through the cell membrane. This membrane acts as a "traffic controller." Materials must enter and leave the cell in order for the cell's life processes to continue. The cell needs to take in food and get rid of wastes.

As the "traffic controller," the cell membrane is very selective. It allows some materials to pass through quickly, and others to pass through slowly. Some materials are not allowed to pass through at all. The cell membrane is described as **selectively permeable.** *Permeable* means that substances can pass through it. The cell membrane will not let many large molecules pass through. This allows the cell to store large molecules such as starches, fats, and proteins, since they cannot escape through the cell membrane.

Selectively permeable Refers to a membrane that will allow some substances to pass through, but not others.

To diffuse through the cell membrane, a material must be *dissolved* in water. See Fig. 3–3 (left). When a substance dissolves, it separates into individual molecules. For example, when salt crystals dissolve, they separate into individual salt molecules. One substance dissolved in another is called a solution. Salt dissolved in water is an example of a solution. See Fig. 3–3 (right). The salt molecules are spread evenly throughout the water. They are too small to be seen. Another example of a solution is sugar dissolved in water.

Fig. 3–3 When salt dissolves in water, the salt particles are too small to be seen.

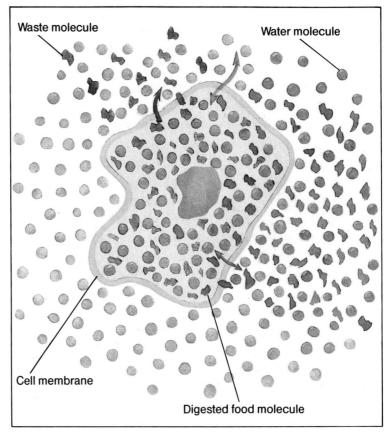

Fig. 3–4 Diffusion of food into the cell and waste out of the cell take place at the same time.

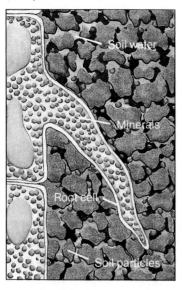

Fig. 3–5 Why do plant root cells have to use active transport?

Most cells are bathed by some type of water solution. Water is needed for materials to move into and out of a cell. A material that does not dissolve in water cannot diffuse through the cell membrane.

Different materials can be moving in and out of a cell at the same time. Food may be moving into the cell while wastes are moving out. See Fig. 3–4. Diffusion will continue until the materials are evenly spread out.

ACTIVE TRANSPORT

Sometimes, a cell must take in materials through the cell membrane even from areas where these materials are not crowded. Even though the materials inside the cell are more crowded than the materials outside it, the materials outside are still able to enter the cell. For example, the root cells of land plants take in minerals from water in the soil. There are fewer minerals in soil water than in the root cell. See

Fig. 3–5. This type of movement of materials through the cell membrane is called **active transport.** The cell must use energy to do this.

Let's use our imaginations to compare diffusion and *active transport*. Imagine you are riding in a bicycle race and you come to a hill. Riding down a hill is a relief because all you have to do is steer. You don't have to pedal very hard to keep the same speed. You don't get tired because you don't use much energy. Riding up a hill is very tiring. You have to stand up and use all your energy to pedal.

During diffusion, the cell uses little or no energy to move the materials through its membrane. Thus, diffusion can be compared to riding a bicycle down a hill. Look at the racer on the left in Fig. 3–6.

During active transport, materials are moved opposite to the way they are moved by diffusion. Materials move from where they are less crowded to where they are more crowded. The cell must use energy to do this. Active transport can be compared to riding a bicycle up a hill. The racers going uphill must also use a lot of energy.

How does active transport take place in cells? Scientists have developed several theories to explain it. They think that active transport may take place in several ways. One idea is that the membrane has openings in it. These openings, or pores, are lined with "squeezing" molecules. The transported

Active transport The process by which a cell must use energy to move materials into or out of the cell.

Fig. 3–6 How is diffusion like riding downhill (left)*? How is active transport like riding uphill* (right)*?*

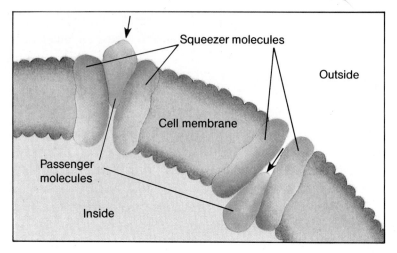

Fig. 3–7 The current theory to explain active transport is that molecules are squeezed through pores in the cell membrane.

Squeezer molecules

Outside

Cell membrane

Passenger molecules

Inside

molecule is pushed through the membrane and into the cell. See Fig. 3–7.

Imagine a crowded subway train. Each car is like one cell. Workers stand at each doorway. Their job is to push more people into the cars. See Fig. 3–8. Which part of the cell is the subway worker similar to? This is a model of how active transport may work. Like the subway workers, the cell must use energy to do this.

SUMMARY

The cell membrane controls what gets into and out of the cell. Many materials move into or out of the cell by diffusion. In order to do this, they must be dissolved in water. The cell membrane also plays an active role in transporting material into and out of the cell. Unlike diffusion, active transport requires energy.

QUESTIONS

Use complete sentences to write your answers.

1. Why must materials enter and leave the cell?
2. What is diffusion?
3. Explain why the cell membrane is called the "traffic controller" of the cell.
4. Why must the cell membrane be selectively permeable?
5. What is the difference between diffusion and active transport in the cell?

Fig. 3–8 A Japanese subway worker pushes passengers into a crowded train.

A DIFFUSION MODEL

PURPOSE: To observe the process of diffusion of materials through a selectively permeable membrane.

MATERIALS:

glass container	starch solution
1 piece of dialysis	water
tubing, about 25	iodine solution
cm long	grease pencil
2 rubber bands	

PROCEDURE:

A. The tubing looks like a strip of stiff plastic when dry. When wet, however, it acts like a selectively permeable membrane. Wet the strip of tubing thoroughly. Open it up by rubbing it between your fingers. Now knot one end tightly.

B. Fill the tube with water, making sure it doesn't leak. Empty the tube and fill it with starch solution. It should be half-filled. Tie a knot in the open end. Rinse off the tubing to remove any excess starch solution.

C. Label a glass container with your initials. Fill the container three-quarters full of water. Add two droppers of iodine to the water. The water should change to a rusty color. (CAUTION: Iodine stains! Be careful not to get it on your hands or clothing.) The iodine is used to test for the presence of starch. Iodine turns black when it comes in contact with starch.

D. Set the tubing in the container. Use a rubber band to secure the ends to the rim of the glass container. Most of the tubing

should be in the iodine solution, as in Fig. 3–9.

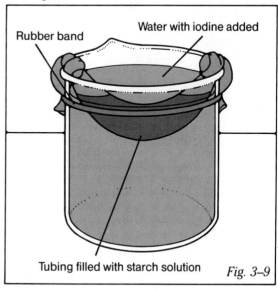

Rubber band

Water with iodine added

Tubing filled with starch solution

Fig. 3–9

E. Leave your container in a safe place overnight. Now answer these questions.

1. How is the wet dialysis tubing like a cell membrane?

2. What do you predict the setup will look like the next day? Why?

F. The next day, get your container and make careful observations.

3. What happened to the starch solution?

G. Read over the prediction you made for question 2.

4. How does it compare with what actually happened?

CONCLUSION:

1. Based on what you know about a selectively permeable membrane and diffusion, explain what happened in this activity.

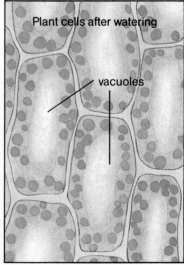

3-2. Osmosis and Water Balance

When you finish this section you will be able to:
- ☐ Describe the process of osmosis.
- ☐ Explain why osmosis is important to the water balance in a living thing.

Look at Fig. 3–10. Can you give a possible explanation for the condition of these garden plants? A simple explanation might be a lack of water. Water is very important to living things. Organisms can live much longer without food than they can without water.

Fig. 3–10 What substance do the cells of these plants need?

HOW CELLS GET WATER

You have learned that materials dissolved in water diffuse into and out of the cell. In this section, we will explore the diffusion of *only* water into and out of the cell. The diffusion of water through a selectively permeable membrane is called **osmosis** (oss-**moe**-sis).

Osmosis The diffusion of water through a selectively permeable membrane.

Imagine that you watered the plants pictured in Fig. 3–10. Water is added to the soil around the plant. There are more water molecules outside the plant cells than inside them. As a result, water from outside the cells will move into the cells. See Fig. 3–11. This is an example of *osmosis.* This added water inside the cells will cause them to swell.

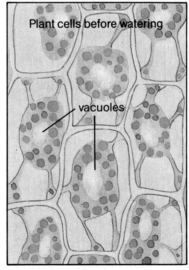

Fig. 3–11 The cell of a wilted plant (left), and the cell after it has taken in enough water.

Plant cells have a cell wall that does not stretch. As water moves into a plant cell, the cell wall stops the cell from getting bigger. Therefore, no more water can enter the cell. When this happens, the cell membrane and cytoplasm are pressed against the cell wall. This causes the cell to become stiff. A limp piece of celery placed in water becomes stiff again. This is due to the water moving into the cells by osmosis.

Animal cells do not have cell walls. An animal could have a problem if too much water moved into its cells. The added water might cause the cells to burst. For this reason, many animals have special structures to remove excess water.

HOW CELLS LOSE WATER

Did you know that watering a plant with salt water can kill it? Let's see why. A normal plant cell has many dissolved materials in it. However, in Fig. 3–12 only the salt and water particles are shown. Outside the cell is salt water. The cell membrane and cell wall separate the contents of the cell from the salt solution. The cell membrane is permeable to water molecules but not to salt particles. This means that only water, not salt can move across the membrane.

The salt particles are more crowded outside the cell than inside the cell. This means that there are fewer water molecules in the salt water than inside the cells. Diffusion will move water from a crowded area to an area less crowded with water. In which direction will the water in Fig. 3–12 move?

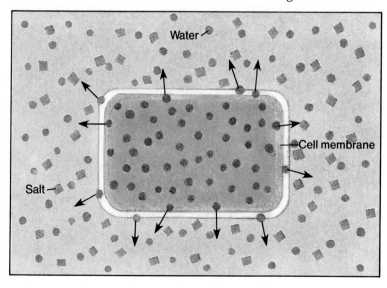

Fig. 3–12 A cell surrounded by salt water will lose much of its own water by osmosis.

Fig. 3–13 Where do sailors get their drinking water on a long ocean voyage?

There is less water outside the cell. As a result, the water inside the cell will move out into the salt water. When this happens, the cell membrane shrinks from the cell wall. If this condition continues, the cell will die.

What would happen if you drank only salt water? Your cells would be surrounded by salt water only. The water inside your cells would move out to where there was less water. The loss of water would cause the cells to shrink. Thus a person stranded at sea would truly die of thirst. See Fig. 3–13.

SUMMARY

Osmosis is the process that maintains water balance in a living cell. Too much water either entering or leaving a cell may eventually cause the cell to die.

QUESTIONS

Use complete sentences to write your answers.

1. What is osmosis? How does it affect the water balance of a cell?
2. What would happen to the cells of a shipwrecked person who drinks salty sea water? Explain.
3. Why do some animals need special structures to get rid of excess water? Why don't plants need these structures?
4. Why do plants become limp and wilted when they don't receive enough water?

WHAT EFFECTS DO SALT SOLUTIONS HAVE ON PLANTS?

PURPOSE: To observe evidence of osmosis in plant cells.

MATERIALS:

3 containers   10% salt solution
potato slices   pure water
celery pieces   graduated cylinder
1% salt solution

PROCEDURE:

A. Label the three containers A, B, and C. Add a potato slice and a celery piece to each container.

B. To container A, add 150 mL of pure water. To container B, add 150 mL of 1 percent salt solution. To container C, add 150 mL of 10 percent salt solution. See Fig. 3–14.

C. Let the containers stand for at least 20 minutes.

D. After 20 minutes, remove the pieces of potato and celery and feel them. Place each beside the container it came from.

1. From which container are the potato and celery pieces most spongy and flexible?

2. From which container are the potato and celery pieces most rigid and stiff?

3. Compare the potato and celery pieces from containers B and C. Is there any difference?

CONCLUSIONS:

1. Which vegetable pieces have lost water from their cells?

2. Does the amount of salt in the water seem to make a difference?

3. Why did the cells of these vegetables lose water? Explain.

4. Could the cells of the vegetables in container A have gained water? Explain your answer.

5. How does the percentage of salt in a solution affect plants?

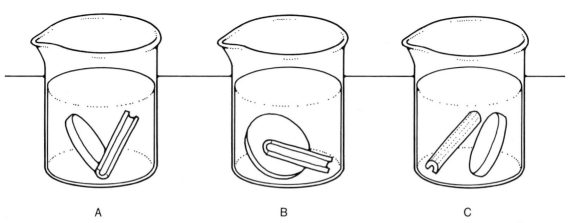

A   B   C

Fig. 3–14

¡COMPUTE!

SCIENCE INPUT

The average human being's body is composed of one hundred trillion cells. These cells perform the life processes necessary for a human to continue to live in a state of good health. However, these cells do need to be replaced continuously, as some are damaged through injury or disease, while others die. When needed, or upon reaching a certain age, the cells reproduce themselves by the cell division process called MITOSIS. The speed at which mitosis occurs is determined by environmental factors that include: AVAILABLE NUTRIENTS and WATER, TEMPERATURE, and POLLUTION FACTORS.

COMPUTER INPUT

Because of the way in which computer hardware is engineered, a computer can immediately organize data, complete the necessary mathematical functions, and produce the resulting information in a meaningful and organized form. The speed at which computers can do all of this varies according to the computer's power and the nature of its electronic circuitry.

The more functions a computer must perform, the greater its power needs to be. Computers store and use information in a special machine language called "bits." Computers can translate words and numbers to "bits" for later use. A computer that can store more bits than another has a greater memory and, therefore, greater power. The speed at which this information goes through the computer depends on the distance it has to travel. You might have read about the efforts to produce the smallest silicon chip for use in computers. The reason behind this effort is that the smaller the chip, the less distance the information must travel. Then, the data can be run through the computer faster.

In this exercise, you will see the speed with which the computer can calculate the number of cells produced during mitosis.

WHAT TO DO

On a separate piece of paper, copy the data chart below. Then, using the format and commands needed for your computer, enter Program Mitosis. Save the program, then run it. The program will ask you to ENTER data from the chart. The computer will then compute the number of cells produced for a given cell type in the time you indicated. Use this OUTPUT to fill in your data chart. Study the chart. With the data given, what conclusions might you make in comparing the cell division of plant and animal cells?

Data Chart

Cell Mitosis Completed in One Day			
Cell Type	Mitosis Time (hr/min)	Temp. °C	Number of Cells
Animal Group			
Fibroblast[1]	17/00	37	
Neuroblast[2]	3.5/00	38	
Neuroblast	8/00	26	
Plant Group			
Root tip	22/00	20	
Tradescantia[3]	00/135	10	
Tradescantia	00/30	45	

1. A fibroblast is one of the six types of cells that make up connective tissue. Connective tissue supports or joins and bends various parts of the body (see Chapter 13). When a fibroblast grows older and becomes inactive, it is sometimes called a fibrocyte.
2. Neuroblast—one of the cells of the nervous system.
3. *Tradescantia* is a plant genus.

GLOSSARY

INPUT   data entered by the user.

OUTPUT   data produced by the computer.

PROGRAM

```
100   REM MITOSIS
110   DIM CT$(6),HR(6),MI(6),CN(6): FOR X
      = 1 TO 6
120   PRINT : PRINT : PRINT : PRINT :
      PRINT : PRINT : PRINT : PRINT :
      PRINT : PRINT : PRINT : PRINT :
      PRINT : PRINT : PRINT : PRINT :
      PRINT : PRINT : PRINT : PRINT :
      PRINT : PRINT
130   PRINT X;"-": INPUT "WHAT CELL
      TYPE? ";CT$(X)
140   INPUT "HOW MANY CELL DIVISION
      'HOURS'? ";HR$(X)
150   HR(X) = VAL (HR$(X))
160   IF HR(X) < .0 THEN GOTO 140
165   IF HR(X) = 0 THEN GOTO 180
170   IF HR(X) > 0 THEN GOTO 210
180   INPUT "HOW MANY CELL DIVISION
      'MINUTES'? ";MI$(X)
190   MI(X) = VAL (MI$(X))
200   IF MI(X) < = 0 THEN GOTO 120
205   IF MI(X) > 0 THEN GOTO 220
210   MI(X) = HR(X) * 60
220   CN(X) = INT (2 ^ (1440 / MI(X))):
      NEXT X
230   PRINT : PRINT : PRINT : PRINT :
      PRINT : PRINT : PRINT : PRINT :
      PRINT : PRINT : PRINT : PRINT :
      PRINT : PRINT : PRINT : PRINT :
      PRINT : PRINT : PRINT : PRINT :
      PRINT : PRINT
240   PRINT "CELL TYPE","# CELLS"
245   FOR X = 1 TO 6
250   PRINT : PRINT : PRINT
      X;"-";CT$(X),CN(X): NEXT X
260   END
```

PROGRAM NOTES

If you enter mitosis time in minutes, type in "00" when the program asks you for the number of cell division hours. It will automatically go on to ask you for minutes.

BITS OF INFORMATION

Thinking of buying a computer? First, read a bit about computers and do some window-shopping. For a 50-page illustrated guide to home computers, write to:

"How to Buy a Home Computer"
Electronics Industries Association
PO Box 19100
Washington, DC 20036

Enclose a stamped, self-addressed envelope that is 6″ × 9″ or larger, with $0.54 postage on it.

3-3. Cells Release Energy

At the end of this section you will be able to:

- ☐ Describe the process of respiration.
- ☐ Explain why cells need energy.
- ☐ Compare the energy-releasing processes of fermentation and respiration.

Do you know what happens in a power plant? Fuel such as oil or coal is burned. To burn, the fuel must combine with oxygen. When this happens, energy is given off. Water and carbon dioxide are also formed. The energy from the burning fuel runs special generators that make a more usable energy—electricity. Electrical energy is used to run our hair dryers, radios, lights, etc. See Fig. 3–15.

Each of your cells is like a little power plant. In this section you will see why and how.

Fig. 3–15 A power plant releases energy from coal. How is this energy used?

RESPIRATION

Power plants do not create energy. They release the energy from fuel. Then it is changed to electricity. Cells must also release energy from fuel. The fuel that cells use is digested food. When you eat, the food is broken down into molecules that can diffuse into your cells. Once inside the cells, the food

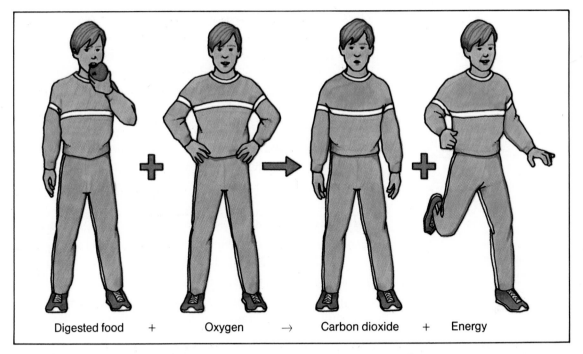

| Digested food | + | Oxygen | → | Carbon dioxide | + | Energy |

is "burned" in special generators called mitochondria. In order to burn, a substance must combine with oxygen. You take in this oxygen when you breathe in air.

When oxygen combines with digested food, energy is given off. This energy is released when the chemical bonds that hold the atoms together are broken. The atoms will form different molecules. Carbon dioxide and water are two new molecules that are formed. Organisms give off this carbon dioxide and water when they breathe out. This energy-releasing process is called **respiration.** *Respiration* in a cell can be shown as an equation, as in Fig. 3–16.

CELLS NEED ENERGY

Cells need energy in order to survive. The energy released from food is used by cells to do all types of work. Energy is needed to carry on all the life processes. Energy is used for moving, growing, and reproducing.

A power plant cannot change all its fuel energy into usable energy. Some of it is lost as heat. Some is lost as light. Some is lost in ashes. Likewise, cells cannot change all the energy from food into usable stored energy. What happens to the rest

Fig. 3–16 All organisms need energy. Respiration releases energy by combining fuel with oxygen.

Respiration The process in which oxygen combines with food to release energy for life activities.

of this energy? A lot of it is lost as heat. Some organisms lose this heat to the environment. Others, like yourself, use this heat to help keep a constant body temperature.

Burning in a power plant occurs as one simple step. The release of energy by burning is rapid. It is not a controlled energy release. In contrast, respiration is a series of many steps. Each step is controlled by a separate **enzyme.** The *enzymes* slow down the release of energy. Enzymes are special compounds made by cells. They cause chemical changes to occur. They can also affect how fast chemical changes occur.

Enzyme A substance that helps to cause or to control the speed of chemical changes in living things.

FERMENTATION

Some organisms can release energy from food without oxygen. This process is called **fermentation.** These organisms can survive without oxygen. Inside canned foods is an oxygen-free environment. Some bacteria thrive in an oxygen-free environment. For example, if food containing the botulism bacteria is canned, the bacteria will then be able to grow. See Fig. 3–17.

Fermentation The release of useful energy from food without the use of oxygen.

Fermentation releases less energy than respiration. For this reason, organisms with greater energy needs must get their energy by respiration. Some organisms can carry out both fermentation and respiration depending on their environmental conditions.

Yeast, a single-celled organism, carries on fermentation. This process is called *alcoholic* fermentation. It can be shown by the following equation:

Fig. 3–17 Canned foods must be rid of organisms that can survive in an environment with no oxygen.

Digested Food → Alcohol + Carbon Dioxide + Energy

Compare this equation to the one for respiration in Fig. 3–16. What is missing? If you said oxygen, you are right.

There are three common types of fermentation. Each type gets its name from what it makes. You have already learned about one type—alcoholic fermentation. Alcoholic fermentation is important in the brewing industry.

Another type of fermentation is carried on by some kinds of bacteria. In this type, an acid, carbon dioxide, and energy are produced from digested food. This is called acid fermentation. The equation below shows what happens.

Digested Food → Acid + Carbon Dioxide + Energy

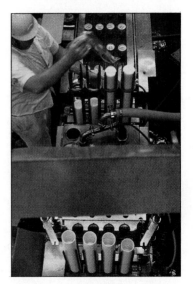

Fig. 3–18 Bacteria ferment milk, changing it to yogurt.

Dairy products, such as cottage cheese and yogurt, are made by this type of fermentaton. Yogurt is made by stirring the bacteria *Lactobacillus bulgaricus* into milk. See Fig. 3–18. The milk must be between 38°C and 49°C. This temperature allows the yogurt bacteria to multiply rapidly. As they grow, the bacteria feed on milk sugar. The acid produced thickens the milk to a puddinglike product. It also gives the yogurt a tangy taste. Acid fermentation is also important in the making of sauerkraut and pickles.

SUMMARY

Organisms get energy by changing digested food either by respiration or fermentaton. They use energy to carry out all of the life processes. Respiration and fermentation differ both in the kinds of materials needed to release energy and in their end products.

QUESTIONS

Use complete sentences to write your answers.

1. What happens during respiration?
2. What do respiration and fermentation do for living things?
3. How are the cells in your body like small power plants?
4. What substance makes respiration different from fermentation? How does this affect the amount of energy released?

SKILL-BUILDING ACTIVITY

INFERRING

PURPOSE: To make observations and gather evidence of processes that cannot be observed directly.

MATERIALS:
mirror

PROCEDURE:
A. Look at the photo in Fig. 3–19. Make careful observations.
   1. What do you see in this photo? List your observations.

Fig. 3–19

B. Look at your list of observations. Are they really all things that you can see in the picture?
   2. Did you say that an argument is taking place?
C. If you did, you made an *inference.* An observation must be information obtained from your senses. An inference is a conclusion based on your observations.
   3. Look at your list from question 1. Which items are really inferences?

Fig. 3–20

D. Look at Fig. 3–20.
   4. What happens to the windows when you take a hot bath or shower?
   5. What substance on the windows causes this to happen?
E. Use the mirror, glass, or a shiny surface provided by your teacher. Breathing through your mouth, exhale strongly on that surface.
   6. What do you observe?
   7. What substance on the mirror do you think caused this to happen?
   8. What process in the body would explain the source of this substance? Whenever you make a decision based on what you already know and what you observe with your senses, you are inferring.
   9. What did you infer in this part of the activity?

CONCLUSIONS:
   1. What is the difference between an observation and an inference?
   2. Is the process of respiration one that can be observed or inferred?

3-4. Cell Division

At the end of this section you will be able to:

- ☐ Explain why a cell can only grow to a certain size.
- ☐ Describe the sequence of events in cell division.
- ☐ Contrast cell division in animal cells and plant cells.

Which single cells would hold the record for size? An ostrich egg cell is 12 to 18 centimeters long, and its yolk is about the size of an orange. See Fig. 3–21. An elephant's nerve cell may be over two meters long. However, it is extremely thin. Most cells can only be seen with a microscope. Why are most cells so small?

Fig. 3–21 *An ostrich egg is the biggest single cell known.*

CELL SIZE

Cells must be able to get enough food. They must also be able to get rid of all their wastes. As you learned earlier in this chapter, these materials must enter and leave through the surface of the cell. There must be enough membrane surface to meet the needs of the cell's contents. If a cell grew too large, there would be more cell content than there would be membrane surface for the exchange of materials. Thus cells can grow to only a certain size.

CELL DIVISION

If cells grow only to a certain size, how do organisms grow larger? The answer is **cell division.** Cells divide to make more cells. The size of most organisms depends on the number of cells they have. Thus, a tree is larger than a flower because it has more cells. As you grow from an infant to an adult, your total number of cells increases.

An injury can also cause *cell division* to occur. If an organism is injured, the cells around the wound will divide. New cells will replace the damaged ones.

Cell division also replaces cells that die or are lost. Some of your body cells, such as red blood cells, only live for a short time. Did you know that skin cells are constantly being worn away? Three billion of your body cells die or are lost every minute. That same number is replaced by cell division in the same minute.

Cell division The process by which a cell divides to form two new cells.

THE PHASES OF MITOSIS

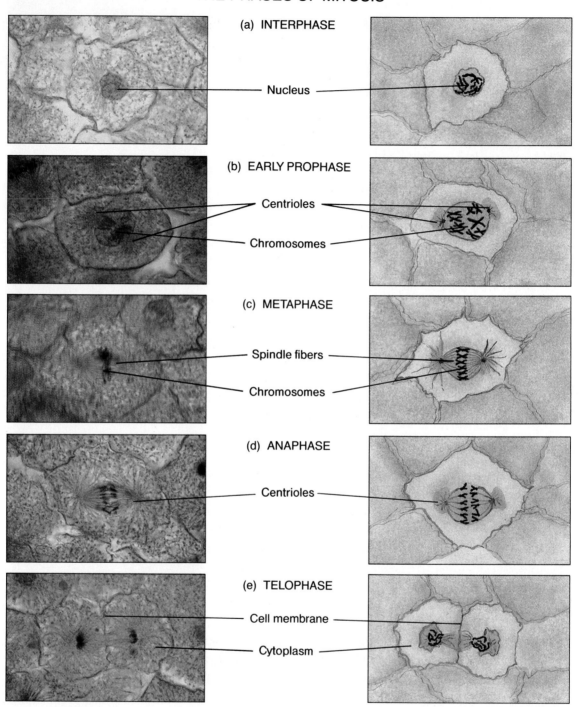

(a) INTERPHASE

Nucleus

(b) EARLY PROPHASE

Centrioles

Chromosomes

(c) METAPHASE

Spindle fibers

Chromosomes

(d) ANAPHASE

Centrioles

(e) TELOPHASE

Cell membrane

Cytoplasm

Fig. 3–22 The animal cells (left) *are magnified 450 times. Use the matching drawings* (right) *to find all the cell parts.*

MITOSIS

Cell division is a series of events in which two cells form from one cell. The division of the nucleus takes place first. This is called **mitosis** (mie-**toe**-sis). The events of *mitosis* are described in a series of steps called *phases*.

The nucleus contains chromosomes. Chromosomes are the blueprints of the cell. They carry all the information needed to direct and control the cell's activities. Suppose you wanted to have two identical houses built at the same time. You wouldn't cut the blueprints in two and give one half to each builder. You would make an exact copy or duplicate of the blueprints. Then each builder would have a full set of plans. The same is true for the blueprints of a cell: When the nucleus divides, each division carries a full set of chromosomes. Chromosomes must be duplicated before the cell divides.

A cell that is not dividing is still very active. It is performing all its life functions. During this nondividing stage, the cell is also growing. In this stage, chromosomes are not visible in the nucleus. See Fig. 3–22 (a).

In animal cells, there is a special area just outside the nucleus. This special area contains two small structures called *centrioles* (**sen**-tree-oles). The centrioles begin moving apart during the first phase of mitosis. In the beginning of this phase, the doubled chromosomes appear as long, thin threads. As this phase continues, each chromosome gets shorter and thicker. The chromosome looks like two separate strands that are joined in the middle. See Fig. 3–22 (b). Toward the end of this phase, the centrioles are at opposite sides of the nucleus. Fibers appear to go out in all directions from them. Some fibers connect one centriole to the other. These fibers form a football-shaped structure called a *spindle*. The chromosomes begin to move toward the middle of the spindle. By the end of this phase, the nuclear membrane has disappeared.

In the next phase, the chromosomes line up across the middle of the spindle. Each doubled chromosome attaches to a spindle fiber. At the end of this phase, the doubled chromosome splits apart. The two halves of the doubled chromosome are no longer connected. See Fig. 3–22 (c). Each half is a separate and complete chromosome. The cell now contains two complete sets of identical chromosomes.

Mitosis The process in cell division in which the nuclear material divides.

During the next phase, the two sets of chromosomes move apart. One set moves toward one end of the cell. The other moves toward the opposite end of the cell. The spindle fibers seem to guide this movement. This phase ends when each set of chromosomes reaches its centriole. See Fig. 3–22 (d).

The last phase of mitosis is nearly the reverse of the phase in Fig. 3–22 (b). The fibers disappear. Each set of chromosomes becomes enclosed in a nuclear membrane. The chromosomes lengthen and look like long, thin threads again.

Also during the final phase, the cytoplasm begins to divide. The cell membrane appears to be pinching inward. Eventually, the cytoplasm pinches apart to form two cells. See Fig. 3–22 (e). By the end of this last phase, the two cells separate. Each of their nuclei look the same as the nucleus of a nondividing cell.

Daughter cells Cells that result from the division of a parent cell. They have the exact same chromosomes that the parent cell had.

The two new cells are called **daughter cells.** The original parent cell no longer exists. *Daughter cells* are identical to each other. They are also identical to the original parent cell. Each daughter cell has the same number and kind of chromosomes as the parent cell. See Fig. 3–23. These newly formed daughter cells will grow for a while before they, too, divide.

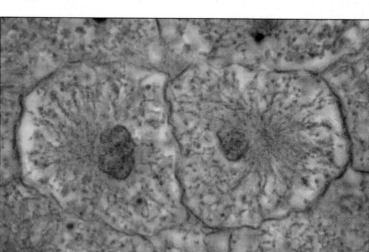

Fig. 3–23 Daughter cells are the result of cell division.

CELL DIVISION IN PLANTS

As in animals, plants need a constant supply of new cells for growth and repair. The most cell division takes place at the tips of the roots and stems. In plants, cell division is slightly

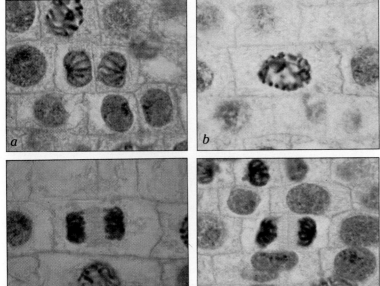

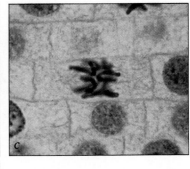

Fig. 3–24 Mitosis in a plant cell: (a) interphase; (b) prophase; (c) metaphase; (d) anaphase; (e) telophase. What differences can you see between these and the animal cells in Fig. 3–22?

different. Plant cells cannot pinch in half. Their cell walls are too stiff. Instead, a new wall and membrane, called a cell plate, forms across the middle of the cell. The cell plate grows outward until the cell is fully divided into two new cells. See Fig. 3–24.

SUMMARY

The size of cells is limited by their need to move materials through the cell membrane. Cell division allows organisms to grow in size and replace worn-out, damaged, or lost cells. Cell division begins in the nucleus. It ends as the cell splits into two cells, which occurs differently in plants and animals.

QUESTIONS

Use complete sentences to write your answers.

1. Why do cells divide?
2. List the sequence of events in cell division.
3. Describe how the cytoplasm divides in an animal cell and in a plant cell.
4. Why are so many cells dividing at the tips of roots and stems?

INVESTIGATION

OBSERVING THE PHASES OF MITOSIS

PURPOSE: To observe the sequence of phases of mitosis.

MATERIALS:

prepared Type 1 slide   microscope
prepared Type 2 slide   blank, unlined paper

PROCEDURE:

A. Take a prepared Type 1 slide. Fig. 3–25 is an example of a Type 1 slide. Using low power, examine it under the microscope. Find an area with many cells.

B. Now focus the slide under the highest power of your microscope. You should see cells at different phases of mitosis.

   1. Are these plant or animal cells? How can you tell?

C. Title one side of your unlined sheet of paper Type 1 slide and draw as many cells in different phases of mitosis as you can find. Label each cell that you draw with its phase of mitosis. For assistance, refer to the appropriate Figures in Section 3–4.

D. Take a prepared Type 2 slide and repeat steps A through C. Fig. 3–26 may be similar to your Type 2 slide. Be sure to answer question 1 again. For step C, use the other side of the unlined paper and title it Type 2 slide. Be sure to label your cell drawings appropriately.

E. When you have finished making and recording your observations, answer the following questions.

2. Why can't you draw chromosomes in a cell that is not dividing?

3. If eight chromosomes are visible in the fourth phase, how many chromosomes will each daughter cell have? How many chromosomes did the parent cell phase have?

CONCLUSIONS:

1. What major, observable difference can you see between a cell that is dividing and one that is not?

2. What differences are there between animal and plant cells in the last phase of cell division?

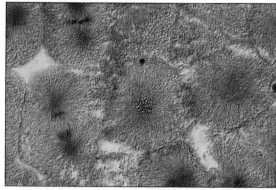

Fig. 3–25

Fig. 3–26

SEEING DOUBLE: NO TRICK WITH MIRRORS

What is a clone? If you do any gardening, you may have already produced a clone. A clone is a genetically identical copy—that is, a cell or group of cells that all come from the same original cell. If you take a cutting from a plant—for instance, cut off a part of the stem and let it grow—the plant that results is a clone.

Common types of cloning research have been done with mice. For example, an embryo is taken from a gray donor mouse at a very early stage of cell division. The inner cell mass is removed and separated into individual cells. The nuclei of each of these cells is alike, and they each have chromosomes that contain the genetic information needed to develop the whole organism. The nucleus is then taken from one of these cells and injected into the fertilized egg of a black mouse. This fertilized egg, which has been removed from the black mouse, is cultivated in a nutrient solution until the cell multiplies and becomes an embryo. This new mouse embryo—grown from a gray nucleus in a black mouse's fertilized egg—is then implanted into the uterus of a white mouse. The offspring of the white mouse mother is a gray mouse, genetically identical to the embryo of the original gray donor mouse.

In more recent experiments with frogs, the cell used to start a new individual does not come from either an egg or sperm. The nucleus of a frog egg cell is removed and replaced with a nucleus that comes from the intestinal cell of the donor frog. The egg cell with the new nucleus divides by mitosis. Then those two cells divide, and so on. From that one cell, an entire frog that looks exactly like the donor frog is formed. However, a great deal of genetic manipulation is necessary in order for this to occur.

Most research in cell biology involves molecular cloning and not the cloning of complex organisms such as plants or animals. Usually, scientists select one particular type of cell to clone—in order to study it. The cells are rinsed in salt water and separated. Then they are placed into dishes containing material in which, through mitosis, the cells begin to divide. The cell population may double in a day or two. After two weeks, a large enough group of cells is formed so that experiments can be done.

Many of these experiments are designed to find out how different tissue cells produce antigens and fight disease, particularly cancer. Other research in molecular cloning has been exceptionally important to our understanding of the body's production of insulin.

A great many questions are raised by the technology that enables scientists to clone cells and animals. What guidelines should there be for research and development in cloning, and who should set these guidelines? So far, there are more questions than answers.

CHAPTER REVIEW

VOCABULARY

On a separate piece of paper, match each term with the number of the statement that best explains it. Use each term only once.

active transport   mitosis   osmosis
respiration   fermentation   enzyme
diffusion   daughter cells   cell division
selectively permeable

1. The movement of molecules from a more crowded area to a less crowded area.
2. Diffusion of water through a selectively permeable membrane.
3. Cells that result from cell division.
4. The cell using energy to move materials into and out of the cell.
5. A membrane that allows only some materials to pass through it.
6. The process by which cells form two new cells.
7. A substance that helps to cause or to speed up chemical changes in living things.
8. The process by which the nucleus of a cell divides.
9. A process that uses oxygen to release useful energy from digested food.
10. A process that does not use oxygen to release energy from digested food.

QUESTIONS

Give brief but complete answers to each of the following questions. Unless otherwise indicated, use complete sentences to write your answers.

1. Why do the cells of living things need energy?
2. What is a cell's source of energy?
3. What are three reasons why cells divide?
4. List, in order, the steps of cell division in an animal cell.
5. How does each daughter cell come to have the same kind and number of chromosomes as the parent cell?
6. What material or materials diffuse(s) through a selectively permeable membrane during osmosis?
7. Each statement below fits one of these terms: respiration, fermentation, both respiration and fermentation. Match one of these to each statement.

(a) Process that needs oxygen to release energy from digested food. (b) Digested food is the source of energy in this process. (c) Alcohol and carbon dioxide are the products of this process. (d) Process that releases large amounts of energy. (e) Carbon dioxide is a product of this process. (f) This process takes place in the mitochondria of your cells.

8. How is cell division in a plant cell different from that in an animal cell?

9. If the cell membrane loses its selective permeability, the cell will most likely die. Briefly explain why.

APPLYING SCIENCE

1. Describe how a tea bag is like a selectively permeable membrane. To determine the answer to this, use two cups and two tea bags that are identical. Fill one cup three-quarters full of hot water. Fill the other three-quarters full of cold water. At the same time, gently place a tea bag in each cup. Let each stand undisturbed. Do not move the tea bag once it has been placed in the water. Record your observations and draw your own conclusions.

2. Explain why eating salty popcorn or potato chips makes you thirsty. (Hint: Think of the shipwrecked person drinking salt water.)

3. Many products sold today rely on the process of diffusion for their use. Name at least two such products and explain how they show diffusion. (Example: In solid air fresheners, the scent diffuses into the surrounding air.)

4. A new person begins with a single cell. When this cell divides, two cells result. When the two cells divide, four cells are formed. Create a chart to show how many cells would result after each of ten cell divisions.

BIBLIOGRAPHY

Cyganiewicz, Paul F. *The Cell Theory.* West Haven, CT: Pendulum Press, 1975.

Gore, Rick, with photos by Bruce Dale and paintings by Davis Meltzer. "The Awesome World Within a Cell." *National Geographic Society,* September 1976.

McNamara, Louise Greep. *Your Growing Cells.* Boston: Little, Brown, 1973.

Pfeiffer, John, and The Editors of *Life. The Cell.* New York: Time-Life Books, 1964.

Tufty, Barbara. *Cells, Units of Life.* New York: Putnam's, 1973.

SIMPLE LIVING THINGS

Scientists have already discovered over one million different types of insects. Imagine how difficult it is for insect biologists to keep track of them all.

CLASSIFYING LIFE

CHAPTER GOALS

1. Discuss why and how scientists classify living things.
2. List the seven levels used in classification systems.
3. Compare and contrast the different structures of bacteria and viruses.
4. Explain why viruses are not classed as living things by many scientists.

4-1. Keeping Order

At the end of this section you will be able to:

☐ Explain the need for classifying living things.
☐ List the seven levels of scientific classification in their proper order.
☐ Explain how living things are named.

Do you like things to be in order? Do you have a place for everything? Do you keep your socks, shirts, and slacks in different places in your room? Do you keep clothes that "go together" in one place? People seem to need to **classify** things in order to understand or use them better. For example, when you visit a supermarket or library, you see things in order. Scientists also like to see things in order. They *classify* many things, from rocks and minerals to stars, chemicals, and living things. Classifying helps scientists understand how objects are related to each other. It also helps them see how living things have changed over long periods of time.

Classify To arrange things into groups according to ways in which they are alike and not alike.

WAYS OF CLASSIFYING

Putting living things into different classes is not a new idea. The earliest known system was developed over 2,000 years ago. A Greek scientist named Aristotle (**air**-ih-staw-tul) tried to explain how things in nature are related. As a child, he had been trained in medicine by his father. During the years he lived on an island, Aristotle studied the sea life around him. He divided living things into two large groups—plants and animals. Then he divided these into smaller groups. Plants

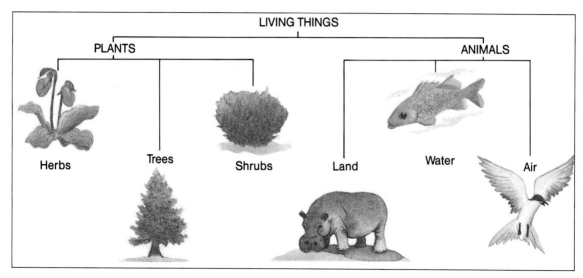

Fig. 4–1 What characteristics did Aristotle use to classify plants and animals?

were herbs, shrubs, or trees. The animals were land, water, or air animals. See Fig. 4–1.

Later scientists tried other systems. At one time animals were classified as useful, harmful, or unnecessary. Plants were once grouped as to whether they produced fruits, vegetables, fibers, or wood. Today we classify all sorts of things, including baseball teams. See Table 4–1.

Over 200 years ago, a Swedish biologist developed a new way of classifying plants and animals. His name was Carolus Linnaeus. His system was based on likenesses in form. Which animals have backbones? Which have lungs? Which plants have roots, stems, and leaves, and so on. Linnaeus looked at

MAJOR LEAGUE BASEBALL			
American League		National League	
Eastern Division	Western Division	Eastern Division	Western Division
Baltimore Orioles	California Angels	Chicago Cubs	Atlanta Braves
Boston Red Sox	Chicago White Sox	Montreal Expos	Cincinnati Reds
Cleveland Indians	Kansas City Royals	New York Mets	Houston Astros
Detroit Tigers	Minnesota Twins	Philadelphia Phillies	Los Angeles Dodgers
Milwaukee Brewers	Oakland A's	Pittsburgh Pirates	San Diego Padres
New York Yankees	Seattle Mariners	St. Louis Cardinals	San Francisco Giants
Toronto Blue Jays	Texas Rangers		

Table 4–1

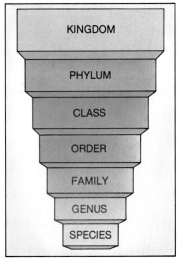

thousands of living things. He described each one in detail. Linnaeus found that there were certain traits that were alike in different kinds of living things. All the living things that were alike in many ways he put into the same group. He called this group a **species** (**spee**-sheez).

Many characteristics are used to classify organisms today. Many early systems used likenesses and differences in form alone. Scientists often still use the form of the organisms. Now, however, differences and likenesses in the chemicals found in the organisms' cells are also used. Scientists try to decide which organisms are related to each other by gathering as much information as they can.

Organisms of the same species have the same basic characteristics. However, they may vary in size, shape, color, and in other ways. People, for example, are all the same species, even though they may be different in appearance. Look at Fig. 4–2 (left). All dogs are in the same species. But all dogs do not look alike. This variety within a species will be discussed again in a later section.

Many species differ only in a few characteristics. Linnaeus placed those similar species into a larger group called a **genus** (**jee**-nus).

CLASSIFICATION LEVELS

The classification systems in use today have seven groupings or levels. Look at Fig. 4–2. The highest level is called

Fig. 4–2 All people belong to the same species (left). The seven levels of classification are on the right.

Species A group of related organisms that are alike in many ways. They can mate and produce living young.

Genus A group of related species differing in only a few ways.

Kingdom The level of the classification system with the largest number of living things that are alike in some ways.

a **kingdom**. A *kingdom* includes the largest number of different organisms. Each of the other levels contains a smaller number of living things than the one above it.

THE KINGDOMS

There are many ways to classsify living things. There is no one right way. One reason for this is that scientists are always finding new information. Another is that scientists do not all agree on what is the best system. Also, each year thousands of new species are discovered. One example is the tube worms recently found near vents in the ocean floor. They may be totally unlike any known species of living things. New families or even new *phyla* may be needed to classify these tube worms.

As new ways of studying living things are found, our ideas about classification can change. For example, for over 2,000 years all living things were grouped as either plants or animals. After the invention of the microscope, new organisms were

A FIVE-KINGDOM CLASSIFICATION SYSTEM		
Kingdom	Description	Examples
Monera	*Monerans* are all single-celled organisms that do not have nuclei.	bacteria; blue-green algae
Protista	*Protists* are one-celled or many-celled organisms that do have nuclei.	algae; protozoa
Fungi	Mostly many-celled. *Fungi* absorb food directly from living or dead organisms.	yeasts; molds; mushrooms
Plantae	Many-celled. *Plants* make their own food using chlorophyll and sunlight.	ferns; conifers; seed plants
Animalia	Many-celled. *Animals* cannot make their own food. They eat other organisms and digest them.	insects; reptiles; birds; mammals

Table 4–2

found. It was hard to determine whether they were plants or animals. In recent years, systems with three, four, or even five kingdoms have been proposed. There is still much discussion as to what differences should be used to establish these kingdoms. The five-kingdom system shown in Table 4–2 seems to be gaining the most support among biologists. Therefore, it is the system that will be used in this text.

NAMING LIVING THINGS

People have always given common names to the living things they can see. A common name is one that is used in one part of the world for a living thing, such as a wildflower. One problem with common names is that all people do not use the same name. For example, the mountain lion is also called a cougar, a puma, and a panther. Sometimes the common names of organisms are misleading. Jellyfish and starfish are not fish. Horseshoe crabs are more like spiders than crabs. Look at Fig. 4–3. Early biologists saw the need to develop a less confusing naming system.

Fig. 4–3 A horseshoe crab (left) *is not a crab at all. By looking at its body form, can you see why it is related to spiders?*

The Swedish biologist Carolus Linnaeus devised a new naming system that is still used today. He gave each different type of plant and animal a two-part name. He used the Latin language because at that time it was understood by educated people all over the world.

The scientific name of an organism has two parts. First comes the *genus* name, then the *species* name. These names are always given in Latin. With this system, each species of living thing has its own two-part name. This means that scientists all over the world can and do use the same name for the same organism. When they discuss this organism, they use this name. In this way, other scientists know what organism they are talking about. For example, humans are placed in the genus *Homo* and in the species *sapiens*. The correct scientific name for a human is *Homo sapiens* (**hoe**-moe **say**-pee-ens). Notice that only the genus name is capitalized.

You may have a dog or a cat as a pet. To the scientist, a dog is named *Canis familiaris* (**kahn**-iss fah-mil-ee-**ar**-iss). You also know that there are many different breeds of dogs. Collies, poodles, and shepherds are a few examples. See Fig. 4–4. Sometimes the different breeds of dogs are referred to as different *varieties*.

Other animals such as wolves and coyotes are very similar to dogs. They are similar enough to dogs to be in the same genus, but different enough to be in different species. Thus wolves are named *Canis lupis* (**kahn**-iss **loop**-iss), and coyotes are named *Canis latrans* (**kahn**-iss **lay**-truns). Members of different species usually cannot mate and produce young. Scientists have studied the differences between lions, tigers, and house cats. Many feel that there are enough differences to place lions and tigers in a different genus. See Fig. 4–5.

Lions and tigers were placed in the genus *Panthera*. Lions are called *Panthera leo*. Tigers are named *Panthera tigris*. House cats are named *Felis domesticus* (**fee**-liss doe-**mes**-

Fig. 4–4 A wolf (left) *and Husky dogs* (right) *are in the same genus, but they are different species.*

tih-kus). The mountain lion, regardless of its common names, is in still another genus; it is named *Profelis concolor*. Because of their likenesses, all of these cats are placed together in the next larger group. This group is the family *Felidae*, which means *cat family* in Latin.

Fig. 4–5 Lions and tigers are in family Felidae. *What other animals might be in this family?*

SUMMARY

Scientists classify living things. It helps them find out how organisms are related. Living things are classified by how they are alike and how they are different. Living organisms are classified by structure and by chemical makeup. There are seven levels of classification of organisms. Each kind of organism has a two-part name made up of its genus and species.

QUESTIONS

Use complete sentences to write your answers.

1. What are some of the reasons why scientists try to classify living things?
2. Why is a scientific system of naming living things necessary? How would it be without one?
3. Arrange the following classification levels in order, beginning with kingdom: phylum, genus, species, order, kingdom, family, class.
4. List some of the traits that scientists would use to help them classify a newly discovered form of life.

SKILL-BUILDING ACTIVITY

CLASSIFYING

PURPOSE: You are the manager of a new food store. Your task is to classify, or sort, the items you have ordered and have them arranged in the store.

MATERIALS:
activity cards   large piece of paper

PROCEDURE:
A. Copy Table 4–3 on a large piece of paper. Make the column below each letter as long as your page. Each item on the cards belongs to only one of the four major areas of the store. The items that are most alike will be located in the same part of the store.
B. Separate your cards into the four main areas. Complete the classification of these items with the help of the following questions.
   1. Which items would you place in the grocery area?
C. Separate the grocery items into three sections. The items in each section should be related to each other.
   2. What would you title sections A, B, and C of the grocery area? Add the section and item names to your chart on the large piece of paper.

D. Separate the dairy products into three groups.
   3. What would you label sections D, E, and F of the dairy products area? Add the section and item names to your chart.
E. Separate the produce items into two sections. Produce includes any freshly picked plant materials.
   4. What would you call sections G and H? Label them.
F. Now we come to the meats and poultry area. Poultry is another name for birds raised for food. Separate your item cards into the three sections.
   5. What would you label sections I, J, and K?
   6. Which items would you put in each section? Label them on the chart.
G. Add each of the following items to your table: vegetable oil, paper towels, applesauce, tomatoes, tea and coffee, lobsters, pizza dough, and cake mixes.

CONCLUSIONS:
1. Why are the cheese and butter placed in the same area as the milk?
2. Why are they in different subgroups?
3. How is this type of reasoning the basis for all classification systems?

Groceries			Dairy Products			Produce		Meat and Poultry		
A	B	C	D	E	F	G	H	I	J	K

Table 4–3

4-2. Kingdom Monera: Bacteria

At the end of this section you will be able to:

☐ Describe the major characteristics of bacteria.

☐ Describe the ways bacteria obtain food.

☐ List some ways in which bacteria are harmful and other ways in which they are helpful to humans.

Bacteria, like the ones in Fig. 4–6, are everywhere. They were among the "wee beasties" found by van Leeuwenhoek as he looked at drops of pond water. *Bacteria* have been found in many places where other life forms cannot exist. They have been discovered eight to ten kilometers up into the atmosphere. Some thrive in the hot-water vents on the ocean floor. They even live inside other living things. Some bacteria have ways to survive very unfavorable conditions. They surround themselves with a hard coat and become inactive. When conditions are better, they can become active again. Some of these bacteria can survive for over 50 years in this way. Bacteria have been on the earth for a very long time. See Fig. 4–7.

Bacteria Very small organisms with a simple cell structure and no nucleus.

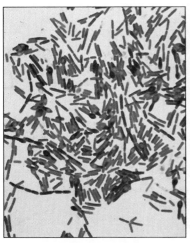 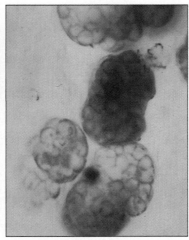

Fig. 4–6 (left) *This bacteria, magnified 400 times, causes botulism (food poisoning).*
Fig. 4–7 (right) *Fossilized bacteria from ancient times.*

KINGDOM MONERA

Bacteria are very small organisms. They were once thought to be like plants because they have cell walls. However, it was found that their cell walls are different from the cell walls of plants. Many scientists now place the bacteria in a separate kingdom, the Monera.

Bacterial cells do not have a nucleus. This is a major difference between members of the Kingdom Monera and other organisms. Reproduction is a simple splitting of one cell into two cells. Under ideal conditions, bacterial cells divide very rapidly. Look at Fig. 4–8. It shows a bacterial cell with flagella and a capsule. The flagella are whiplike and help the bacterium move. The capsule protects the cell. Not all bacteria have these structures, however.

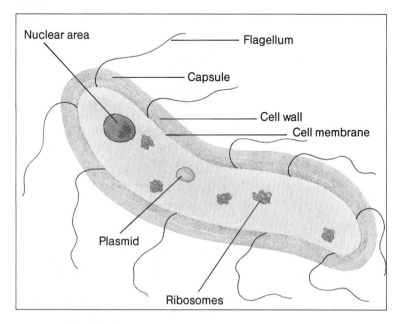

Fig. 4–8 The structure of a typical bacterial cell. Most bacteria have a rigid cell wall, covered by the capsule. A plasmid contains genetic material.

HOW BACTERIA OBTAIN FOOD

Most bacteria cannot make their own food. They get their food from other living or dead organisms. Those bacteria that get their food from dead organisms are called **saprophytes** (**sap**-roe-fites). Many of these types of bacteria are very useful. *Saprophytes* break down the tissues of dead organisms. Then, the released chemicals can go back into the air and the soil.

Some bacteria are **parasites**. *Parasites* live in or on another living organism. They obtain their food from it. Parasites are usually harmful to the "host." These bacteria often cause diseases. However, in some cases bacteria live in other living things without harming them. They may even help. For example, termites cannot digest the wood they eat. Bacteria in their bodies help with this process. Even cows need bacteria to help them digest the grasses they eat.

Saprophyte An organism that gets its food from dead or decaying organisms.

Parasite An organism that lives in or on another living organism and usually causes harm to its host.

The blue-green bacteria can produce their own food. They contain chlorophyll, which allows them to use the sun's energy to make food. Most of these bacteria live in fresh water. Some contain gas bubbles, so they can float near the surface and receive more sunlight. They produce oxygen and also serve as food for other organisms.

Blue-green bacteria can reproduce very rapidly in water that is polluted with chemical fertilizers. They can make the water pea-soup green and smelly. When they die, other bacteria feed on their remains. This may use up all the oxygen in the water. The fish in the pond may die from lack of oxygen while the bacteria thrive.

Other bacteria can also make their own food. However, they do not use the energy of the sun. These bacteria get their energy from chemical reactions. The bacteria in the hot-water vents on the ocean floor are in this group. So are some types of bacteria found in the soil.

CLASSIFYING BACTERIA

Bacteria are among the smallest living things. Some 50,000 bacteria could fit on the head of a pin. Even at a magnification of 1,000 times, only their general shape can be seen. Bacteria have many different shapes and sizes. Simple bacteria are one-celled. They are shaped like rods, spheres, or spirals. See Fig. 4–9. A rod-shaped bacterium is called a *bacillus.* The bacteria in your intestines are this shape. The bacteria that cause strep throat are round, or spherical. A round bacterium is called a *coccus.* The spiral, or corkscrew, shape is called a *spirillum.*

Fig. 4–9 (left) *rod-shaped bacteria;* (middle) *coccus;* (right) *spirillum. All are magnified about 200 times.*

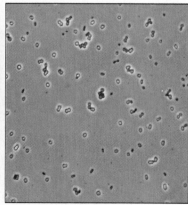
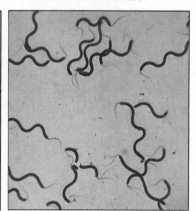

Fig. 4–10 Cheddar cheese begins as a soupy mixture, including milk and bacteria.

The rod-shaped bacilli may stick together after they divide and form long chains. The round cocci may form long chains, or clusters, or groups of two or four. The simple bacteria are classified according to their shapes and whether or not they form chains or clusters. Differences in the way they make or obtain food may also be used to classify them.

BACTERIA AND HUMANS

Bacteria play a large part in the lives of humans. Your own body is a place where bacteria grow. Many bacteria live on your skin, in your nostrils, mouth, and large intestine. Many of these bacteria are not harmful.

Some bacteria do cause diseases, but others are very useful. For example, they can be added to milk to produce yogurt, sour cream, cottage cheese, and hard cheeses. Look at Fig. 4–10. Sauerkraut, sour pickles, and vinegar are also made by the action of bacteria. One type of bacteria found in soil is used to make medicines. These medicines help our bodies fight other, attacking bacteria.

Bacteria also cause heavy losses in stored foods. Bacteria are in all the foods we eat. They cause rotting and spoilage as they break down these foods for themselves to consume. Refrigeration slows down the growth of bacteria. The heat of cooking reduces the number of bacteria. This is one reason why we cook food.

SUMMARY

Monerans are single-celled organisms that lack nuclei. The different ways they obtain food make them both harmful and helpful to other organisms.

QUESTIONS

Use complete sentences to write your answers.

1. What are two major differences between bacteria and plants?
2. Describe three ways bacteria obtain food.
3. Describe how bacteria that live in other organisms can be both harmful and helpful to them.
4. What are the three basic shapes of simple bacteria?

OBSERVING BACTERIA COLONIES

PURPOSE: To culture and observe common bacteria.

MATERIALS:

disposable petri dish with food supply
cotton ball
toothpick
rubbing alcohol
sticky tape
glass-marking pen or pencil
magnifier

PROCEDURE:

A. Turn the petri dish upside down on your desk. Use the marking pen to divide and label the bottom of the dish into quarters, as shown in Fig. 4–11.

B. Turn the dish right side up. Press a short piece of sticky tape onto a surface such as a doorknob or light switch. Peel the tape off. Gently press the tape onto the dish section marked "Surface." Slowly peel the tape off and close the cover.

C. Use the blunt end of a toothpick to remove some material from beneath a fingernail. Lift the cover of the dish slightly. Wipe the material on the section marked "Nail." Close the cover.

D. Touch the circle marked "D" with the tip of a finger. Close the cover. Wipe the fingertip with alcohol to clean it. Touch the circle marked "C" with the cleaned fingertip. Close the cover.

E. Snip a 1-cm length from a single hair. Place the hair in the section marked "Hair." Close the cover. Do not shake or tip the dish. Label the cover with your initials. Place the dish in a dark, warm place.

F. Examine the dish each day for the next five days. CAUTION: Do not open the dish. Always treat bacterial cultures as possible causes of disease. Use a magnifier to examine any colonies that develop. Keep a careful record of what you observe. Describe the size, shape, and color of any colonies that develop.

1. Write a report on the results of this activity. Include day-by-day sketches of the dish.

G. After five days, return the dish to your teacher for disposal. CAUTION: Do not open the dish. Wash your hands with soap and water after handling the closed dish.

CONCLUSIONS:

1. From your observations, make a general statement about where bacteria can be found.

2. Reread the instructions given. What three environmental factors are needed for the growth of bacteria?

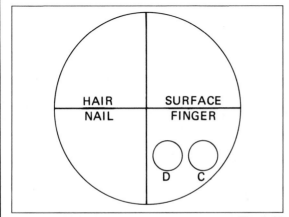

HAIR	SURFACE
NAIL	FINGER

D C

Fig. 4–11

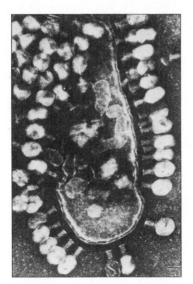

Fig. 4–12 What is this? How big do you think these objects are?

Virus A particle that is not a cell but can reproduce in a cell of a living organism.

4-3. Viruses

At the end of this section you will be able to:

- ☐ Describe the major characteristics of *viruses.*
- ☐ Describe how viruses reproduce.
- ☐ Discuss reason why scientists think that viruses are not living.

What do you see in the picture in Fig. 4–12? Is it an abstract design printed on a piece of fabric? Is it a group of animals gathered around a watering-hole? Is it a squadron of alien spacecraft about to land on earth?

Actually, it is the beginning of an invasion. But it is the invasion of a single bacteria cell by a particle called a **virus**.

WHAT ARE VIRUSES?

Viruses are very small. Even the largest virus can just barely be seen with a light microscope. This is one of the reasons why viruses were not discovered until the 1930's. However, scientists have been treating diseases caused by viruses for over 200 years. Viruses cause smallpox, mumps, measles, warts, influenza ("flu"), cold sores, and the common cold.

Modern equipment makes it easier to study viruses. Photographs of viruses enlarged over 300,000 times have been made with the help of the electron microscope. See Fig. 4–13. These show that viruses come in many shapes and sizes. They may be round or shaped like rods, needles, or cubes. They may have many-sided shapes.

Viruses are not really cells. They do not have a nucleus,

Fig. 4–13 These viruses, photographed with an electron microscope are magnified 80,000 times (left) *and 27,000 times* (right). *The virus at right causes measles.*

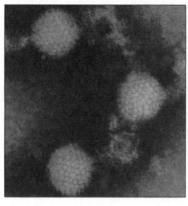
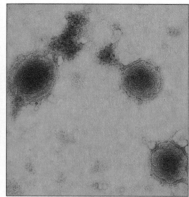

cytoplasm, or a cell membrane. The simplest viruses seem to be made of a **nucleic** (nyoo-**clay**-ik) **acid** with a coating of protein. See Fig. 4–14. *Nucleic acids* control activities in a cell. They are the means by which traits are passed on to new cells.

Nucleic acid A chemical that controls activities in a cell and passes on traits to new cells.

ARE VIRUSES ALIVE?

Viruses do not carry on the life processes in the way that the cells of living things do. They do not take in food or give off wastes. This makes it difficult to decide if viruses are alive or not. They can only reproduce in the cells of living things. The mumps virus, for example, only becomes active when it comes in contact with the gland cells of a person's jaw. The polio virus multiplies only in one kind of nerve cell in the brain and spinal cord.

Viruses can be grouped by the kinds of organisms they affect. One modern classification system uses the type of nucleic acid, the shape, and the size of the virus. Viruses are not included in the five-kingdom system of classification.

Scientists have learned much about viruses by studying a type that attacks bacteria. When a virus attacks a bacteria cell, the protein coat of the virus attaches to the cell surface. The nucleic acid of the virus is injected into the cell. Sometimes the whole virus enters the cell. Once this happens, the virus may become active, or it may instead enter into an inactive state.

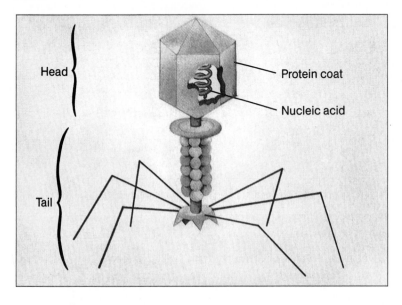

Fig. 4–14 The structures of a virus like the ones shown in Fig. 4–13 (right).

HOW DO VIRUSES FUNCTION?

If the nucleic acid remains inactive, it can be passed from cell to cell as the host cell divides. Look at Fig. 4–15. Later, a change in conditions may cause the virus' nucleic acid to become active.

When the nucleic acid is active, it may cause changes in the cell. These changes may harm the cell and cause it to die. Some viruses, such as the ones that may cause cancers, change the cell but do not kill it. The cells lose control over their rate of division and divide rapidly. They cannot do the jobs they did before the virus entered them.

The virus' nucleic acid may take over the cell's nucleus and cause it to make hundreds of new viruses. Soon the cell bursts, and the new viruses are released. These, in turn, attack other cells and repeat the process. In this way, new viruses are formed and spread from cell to cell.

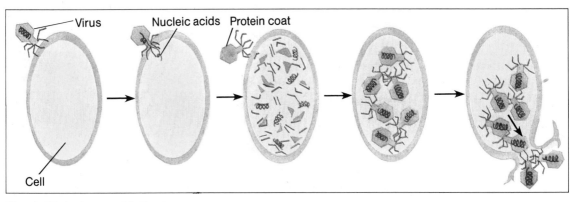

Fig. 4–15 A virus multiplies by using the host cell's energy and chemicals.

SUMMARY

Viruses seem to be made up of a nucleic acid covered with protein. Viruses do not appear to carry on all the life processes. Therefore, many scientists do not classify viruses as living. Viruses reproduce only in the cells of living things.

QUESTIONS

Use complete sentences to write your answers.

1. What is the structure of viruses?
2. How do viruses reproduce?
3. How are viruses different from cells of living things?
4. Describe how viruses can affect living cells.

CAREERS IN SCIENCE

POLLSTER

At one time, the word "poll" meant head. Eventually, polling came to mean the counting of heads for one purpose or another. Every ten years the government takes a poll to collect information about our society. This poll is called the census. Companies use polls, often by telephoning consumers at home to check what brands are used. And social scientists are continually polling people with questionnaires they have designed to measure and analyze the attitudes, opinions, and behavior patterns of different segments of society.

Polling takes many forms and requires a variety of skills. To be a pollster, you have to be good at meeting and talking to strangers. You must be friendly, articulate, and skilled at getting a response. A background in such subjects as social studies, human behavior, and math, especially statistics, is helpful.

For further information, write to: U.S. Bureau of the Census, Washington, DC 20233.

NUMISMATIST

A numismatist is a specialist in the science of coins, tokens, medals, paper money, and objects that closely resemble them. A numismatist is a classifier who is also part historian. He or she must know what the coins or medals are made of, where they were used, and at what point in history they circulated. Some knowledge of chemical metallurgy is helpful in order to be able to detect fraudulent copies of valuable objects. A professional numismatist is usually a collector who is also a dealer in the items collected. Some people, however, do not buy or sell; instead, they look over the items brought to them by others and determine whether the objects are genuine or not.

An interest in numismatics often begins as a hobby and becomes a profession as expertise increases. Experience in business is helpful.

For further information, contact: American Numismatic Society, Broadway at 155th Street, New York, NY 10032.

CHAPTER REVIEW

VOCABULARY

On a separate piece of paper, match each term with the number of the phrase that best explains it. Use each term only once.

species   nucleic acids   saprophyte   classify
parasite   bacteria   kingdom   genus
virus

1. An organism that gets its food from dead or decaying organisms.
2. A particle that is not a cell but which can reproduce in the cells of living organisms.
3. To group things according to their likenesses or differences.
4. A group of organisms that has many of the same traits and can mate and produce young.
5. Very small organisms with simple cell structure and no nucleus.
6. Chemicals that control activities in a cell and the means by which traits are passed to new cells.
7. A group of related species differing only in a few characteristics.
8. The highest classification level.
9. An organism that lives in or on another living organism and is usually harmful to its host.

QUESTIONS

Give brief but complete answers to each of the following questions. Unless otherwise indicated, use complete sentences to write your answers.

1. What is meant by the phrase "to classify organisms"?
2. List the seven levels of classification from the smallest grouping to the broadest grouping.
3. Name the five kingdoms of living things used in this text. Briefly describe each.
4. In which kingdom are bacteria placed? Why?
5. Why are organisms given a two-part scientific name?
6. How do bacterial cells differ in structure from other cells?
7. How are viruses different in structure from living cells?
8. What are some of the characteristics used to classify living things?
9. Why are viruses not considered to be living organisms?
10. Describe what may happen when a virus particle invades a living cell.

11. Explain two methods used by bacteria to obtain food.
12. Describe the characteristics that are used to classify bacteria.

APPLYING SCIENCE

1. In what ways can classifying help scientists show how living things have changed over long periods of time?
2. Why is it difficult to store food products for any length of time?
3. Why have scientists who use a five-kingdom classification system not made a separate kingdom for viruses?
4. Review the Investigation on classification on page 98. Develop a key of the same type to identify the organisms in Fig. 4–16. You must first decide how they are alike and how they are different. Then develop a question that will divide the organisms into two groups. Each question should be answered by either YES or NO. Each subgroup should be divided by the answer to a second question, and so on. When an organism has been placed in a group by itself, give its name. Test your key on your friends.

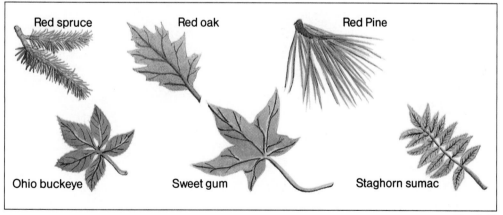

Fig. 4–16

BIBLIOGRAPHY

Angier, Natalie. "Bugs That Won't Die." *Discover,* August 1982.

Dixon, Bernard. "Attack of the Phages." *Science 84,* June 1984.

Nourse, Alan E. *Viruses,* rev. ed. New York: Franklin Watts, 1983.

Rose, Kenneth Jon. *Classification of the Animal Kingdom: An Introduction to Evolution.* New York: McKay, 1981.

Diatoms are neither plants nor animals. They are single-celled members of Kingdom Protista. Their markings are actually openings in their rigid cell walls. (Magnification: 450 times)

PROTISTS AND FUNGI

CHAPTER GOALS

1. Explain why the Protista and Fungi kingdoms were created.
2. Compare the forms and life processes of the protozoans and algae.
3. Discuss the characteristics that are used to classify the organisms in each of these kingdoms.
4. Identify some ways that protists and fungi affect other living things.

5-1. Protists That Obtain Food

At the end of this section you will be able to:

- ☐ Explain why the five-kingdom classification system was developed.
- ☐ Describe the characteristics of protists.
- ☐ Explain how different *protozoans* carry on the basic life processes.
- ☐ Compare and contrast *sexual reproduction* and *asexual reproduction.*

Anton van Leeuwenhoek discovered his "wee beasties" in drops of water in the late 1600's. He had no idea he was starting an argument that would continue even today.

KINGDOM PROTISTA

For over 150 years, biologists argued over these small organisms. They didn't know if the organisms should be classified as plants or as animals. One problem was that they were mostly single-celled, whereas both animals and plants are multicellular. Also, some of these organisms behaved like animals, and still others had characteristics not typical of either plants or animals. Where did these organisms belong? Some scientists classified them as plants at the same time as other scientists classified them as animals.

Over a hundred years ago, a third kingdom was proposed. It was to include all these "simple" organisms. It was called Kingdom **Protista.** Kingdom *Protista* was meant to include

Protista A kingdom of living organisms that are neither plants nor animals.

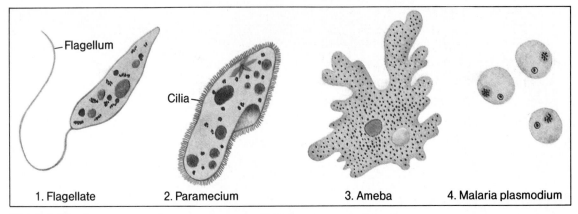

| 1. Flagellate | 2. Paramecium | 3. Ameba | 4. Malaria plasmodium |

Fig. 5–1 The four groups of protozoans each have a different method of movement.

Protozoans Single-celled, animal-like protists.

all single-celled organisms. It also included multicellular organisms that lack specialized tissues and organs.

A short time later, a separate kingdom was proposed for the bacteria. Bacteria cells are different in that they do not have a definite nucleus. Recently, a fifth kingdom has been proposed for the fungi. This is because fungi do not make or eat food. Instead, they absorb food molecules.

Not all biologists accept the five-kingdom system. Even among those that do, there is disagreement as to where certain organisms should be classified.

There are two large groups in Kingdom *Protista*, the **protozoans** and the *algae*. *Protozoans* must find their food. Algae can make their own food. Algae will be discussed in the next section.

Protozoans are single-celled, animal-like protists. They move about freely. They can be seen eating food. They do not have cell walls. They do not have chloroplasts and therefore cannot make food.

There are several thousand species of protozoans. They are found in fresh and salt water, in the soil, and even in other plants and animals.

Protozoans are classified by the way they move. They can be divided into four groups. See Fig. 5–1.

1. Protozoans that use a whiplike tail, called a *flagellum*, for moving. They are called flagellates. They may have one, two, or more of these tails.

2. Protozoans that move by means of hairlike structures called *cilia* (**sill**-ee-uh). The paramecium (par-uh-**mee**-see-um) is an example of this group.

3. Protozoans that change their shape as they move. The *ameba* (uh-**mee**-buh) is an example of a protozoan that constantly changes shape.

4. Protozoans that cannot move by themselves. The protozoan that causes malaria belongs to this group.

HOW DO PROTOZOANS LIVE?

Let's see how these tiny organism move, get food, and respond to changes around them. *Movement* is one of the life processes. Materials within an organism must be moved from place to place. Many organisms also must move to find food. *Flagellates* use their whiplike tails as propellers. They can pull or push themselves through the water.

The *paramecium* is covered with tiny, hairlike structures called cilia. The cilia act like thousands of tiny oars. The beating of the cilia can move the paramecium backward, forward, or even allow it to turn.

The movement of an ameba is fascinating to watch. It doesn't have any special structures to help it move. Instead, long, thin projections called *pseudopods* stretch in the direction of movement. Pseudopod means "false foot." The cell changes shape as the cytoplasm flows into the pseudopods. Old pseudopods are pulled back into the cell as the organism flows along.

To *obtain food*, the ameba wraps its pseudopods around another protist. See Fig. 5–2. The pseudopods meet and join. The food is trapped inside. A new food vacuole, or storage

Fig. 5–2 An ameba, from left to right, uses pseudopods to approach and capture its food.

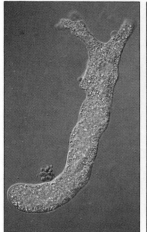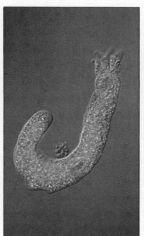

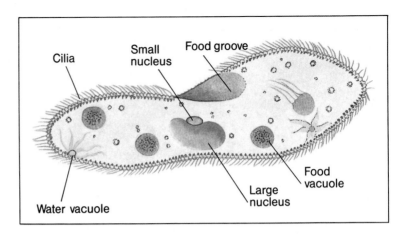

Fig. 5–3 *The structures of a paramecium.*

area, is formed. Enzymes from the cytoplasm enter the vacuole and break down the food. Useful materials are taken into the cytoplasm. The cell membrane then opens to get rid of wastes.

The paramecium also moves to get food. See Fig. 5–3. The beating cilia set up currents in the water around the cell. These currents draw food into a deep groove on the side of the cell. Then the food is enclosed in a food vacuole. The vacuole drifts through the cytoplasm. The food is broken down by enzymes. It is absorbed into the cytoplasm. Finally, the vacuole attaches to the cell membrane. Then wastes are released out of the cell.

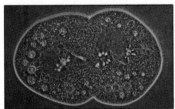

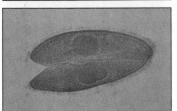

Fig. 5–4 *A paramecium can reproduce by fission* (above) *or sexual reproduction* (below).

LIFE PROCESS	PROTOZOAN		
	Ameba	Paramecium	Flagellates
Movement	Flowing	Cilia	Flagella
Food getting	Pseudopods	Mouth groove	Both
Reproduction	Fission	Fission and sexual	Fission
Excretion	All excrete wastes through cell membrane		
Response	All respond to light, touch, chemicals, heat		

Table 5–1

All protists *respond* to changes in their environments. The ameba responds to light by moving away from it. Some flagellates move toward light. Amebas respond to touch by moving away from all objects except food. Yet these organisms have no eyes or nerves. Paramecia behave like bumper cars in an amusement park. They bang into objects. Then they back up, turn, and move away. Protists also react to heat, cold, and chemicals in the water.

REPRODUCTION

Another life process of all living things is *reproduction*. Most single-celled organisms reproduce by **fission.** *Fission* means splitting in half. It is a form of **asexual reproduction.** *Asexual reproduction* requires only one parent cell. In fission, one parent cell produces two identical daughter cells. This is the same as cell division. During fission, the nucleus and cytoplasm divide equally. If conditions are right, a paramecium can reproduce by fission several times in the same day. See Fig. 5–4 (top).

Most protists can also reproduce by **sexual reproduction.** *Sexual reproduction* involves the joining of two parent cells. Sometimes two paramecia connect. See Fig. 5–4 (bottom). They exchange some of their nuclear material. Then they separate and divide. The new cells formed have characteristics of both parent cells.

Fission When a single-celled organism divides to form two new cells.

Asexual reproduction Reproduction that requires only one parent.

Sexual reproduction Reproduction that requires two parents.

SUMMARY

Protozoans are a large group of microscopic organisms. They were once thought of as the least complex types of animals. Now they are placed in the Kingdom Protista. Like all organisms, protozoans must carry on the life processes. They do this in a variety of interesting ways.

QUESTIONS

Use complete sentences to write your answers.
1. Why did biologists feel that more than two classification kingdoms were necessary?
2. Describe three ways protozoans move.
3. How do the means of food getting differ between amebas and paramecia?
4. Describe two ways paramecia reproduce.
5. Define reproduction, then compare and contrast asexual and sexual reproduction.

INVESTIGATION

OBSERVING PROTOZOANS

PURPOSE: To compare and contrast how two protozoans move and get food.

MATERIALS:

microscope	ameba and parame-
2 microscope slides	cium samples
2 cover slips	stained yeast cells
medicine dropper	cotton fibers or
forceps	slowing agent
paper towels	

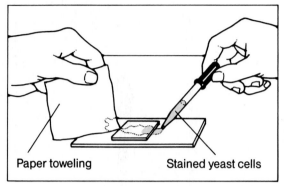

Paper toweling   Stained yeast cells

Fig. 5–5

PROCEDURE:

A. Place a drop of the ameba sample on a glass slide. Add a few cotton fibers or a drop of slowing agent to slow down their movement. Cover with a cover slip. Focus on the specimen with low power. Then move to high power. NOTE: Many protozoans move away from bright light. Use the *diaphragm* to reduce the brightness. This will also prevent rapid drying of the water and the specimen.

   1. Describe the ameba's shape.

   2. Is the ameba moving? If so, describe how it moves.

B. Make a diagram of the ameba. Label all the cell parts you can identify.

   3. Does it have a cell membrane?

   4. Does it have a nucleus? If so, where is it located?

   5. Do you see any clear, circular structures in the cytoplasm? What do you think these structures are?

C. The ameba cannot make its own food. It must eat other organisms. Place a drop of stained yeast cells along one edge of the cover slip. Touch a piece of paper towel to the opposite edge of the cover slip. This will pull the yeast under the cover slip by pulling water out the other side. See Fig. 5–5. Observe the ameba.

   6. How does the ameba take in food?

D. Place a drop of paramecium sample on a clean slide. Cover with a cover slip. Observe it under low power and then under high power. Draw and label its parts.

   7. Does the paramecium change shape?

   8. How does the paramecium move?

E. Carefully observe the vacuoles located at each end of the paramecium.

   9. What are the vacuoles doing?

F. Add a drop of stained yeast cells under the cover slip.

   10. How does the paramecium feed?

CONCLUSIONS:

   1. How do the movements of the ameba and the paramecium differ?

   2. How does food getting in the ameba and the paramecium differ?

5-2. Algae

At the end of this section you will be able to:

- ☐ Describe the characteristics of *algae*.
- ☐ Distinguish between several groups of algae.
- ☐ Explain the ecological and economic importance of algae.

Each year, thousands of people visit the giant redwood forests of California. Few of these visitors know that just offshore stands another forest of giants. This forest is beneath the waters of the Pacific Ocean. The giants in this forest are almost as tall as the redwoods. Some reach a height of 100 meters. But they are not true plants. They lack the complexity of higher plants. These giants are the giant kelp, more commonly called giant seaweeds. See Fig. 5–6. They are **algae** and belong to the Kingdom Protista.

Algae are the plantlike protists. They all contain chlorophyll. They are able to make their own food. Their tissues are not specialized. They lack roots, stems, leaves, and tissues for carrying water and food. For this reason they are not in the Plant Kingdom.

Fig. 5–6 Imagine swimming through a giant algae forest.

Algae Plantlike protists that contain chlorophyll and make their own food.

SINGLE-CELLED ALGAE

Many algae are microscopic, single-celled organisms. Single-celled algae include *Euglena* (yoo-**glee**-nah) and *diatoms*. Euglena and its relatives are one of the reasons that Kingdom Protista was created. See Fig. 5–7. Under the microscope, Euglena resembles many protozoans. It moves freely using its

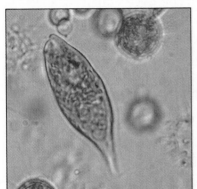

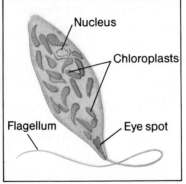

Nucleus
Chloroplasts
Flagellum
Eye spot

Fig. 5–7 Euglena can take in food or use its chlorophyll to make food.

flagellum. It doesn't have a cell wall. It can take in food in much the same way as a paramecium. However, Euglena does have chlorophyll. If conditions are favorable, it can make its own food. It has a special mass of protoplasm, called an *eyespot*, that is sensitive to light. The eyespot allows Euglena to locate areas where there is enough light for making food. Which is Euglena, a plant or an animal? Does the whole class agree? Biologists couldn't agree either.

Diatoms are single-celled algae that add *silica* to their cell walls for strength. Silica is the same material that makes sand and glass. Often these protective "shells" are very complex. See Fig. 5–8. When the organism dies, the "shells" pile up on the ocean bottom. We use these tiny shells in polishes and toothpaste because they are so hard. They are also used in water and air filters because their tiny openings trap particles.

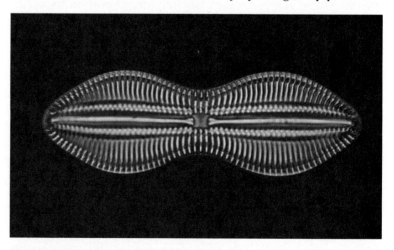

Fig. 5–8 Diatoms may live in fresh or salt water.

MULTICELLULAR ALGAE

Multicellular algae may form colonies, strands, or sheets of cells. They are grouped by their color *pigments*. Pigments are molecules that give living cells their color. There are green, red, and brown groups as well as several smaller color groups.

Green algae are the most common type. They are mostly found in fresh water, although some live in the oceans. They are also found growing on rocks in streams. Green algae make up much of the green scum on ponds. They also develop in your aquarium. Some are found living on wet tree trunks and wood. Most are single-celled. But even those that are multicellular don't get very large. A common green alga found in

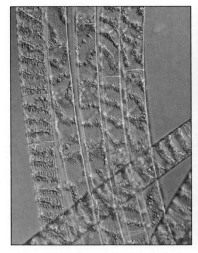

ponds is *Spirogyra* (spie-roe-**jie**-ruh). See Fig. 5–9 (left). Spirogyra is made of a threadlike chain of cells. Each cell has one or more ribbonlike chloroplasts.

Red algae seldom reach more than one meter in length. There are strand, sheet, and branching forms. All red algae live attached to a rock or some other surface. See Fig. 5–9 (middle). Most species are found in the sea. Some are found as deep as 100 meters below the surface. This is the limit of usable light for photosynthesis. Other species are found at or even above the high tide mark. These are kept wet by the splashing waves at high tide.

Brown algae include the giant kelp and the common brown seaweed found along the shore. Most brown algae have special structures to anchor them to rocks or the ocean bottom. They also have *air bladders,* which help keep the free ends floating near the surface so they can receive sunlight. See Fig. 5–9 (right). The forests of giant kelp, like those off the western coast of North America, are brown algae. They provide food and shelter for many organisms, from shrimp to seals.

Fig. 5–9 Spirogyra (left) *is a green alga; Corallina* (middle) *is a red alga; Fucus* (right) *is a brown alga.*

ALGAE AND THE ENVIRONMENT

Algae play an important role in the environments in which they live. In the oceans and lakes, they are the basic food source for most organisms. Algae are eaten by protozoans and other small organisms. These, in turn, are food for the larger fish. Some whales, the largest ocean organisms, feed directly on algae, the smallest. Where algae are abundant, so are other

 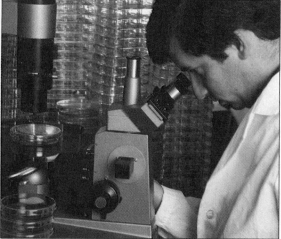

Fig. 5–10 In one Japanese dish, rice is rolled in sheets of algae (left). *A red alga is used to make the culture material called agar* (right).

forms of life. Parts of the oceans are barren of life because algae cannot grow there.

All life on earth depends on algae. Algae are the source of over half of the oxygen released into the atmosphere. The oxygen is released as a waste gas during photosynthesis.

Many forms of algae have an economic importance, too. Some algae are used as human and livestock food. See Fig. 5–10. Algae are also used in making fertilizers, cosmetics, and drugs. Even ice cream and marshmallows owe their smooth, creamy texture to chemicals obtained from algae.

SUMMARY

Algae are members of Kingdom Protista. They are plantlike, but not as complex as true plants. Algae are grouped by their characteristic color pigments. Algae serve as a source of food for many organisms. They also produce much of the oxygen we breathe.

QUESTIONS

Use complete sentences to write your answers.

1. How are algae similar to plants? How are they different from plants?
2. What is one characteristic used to group the multicellular algae?
3. Describe two contributions of algae to the biosphere.
4. Identify several ways algae are useful to humans.

SKILL-BUILDING ACTIVITY

OBSERVING

PURPOSE: To identify algae found in fresh water.

MATERIALS:

pond algae samples	cover slips
microscope	medicine dropper
microscope slides	toothpicks, round

PROCEDURE:

A. Prepare a wet mount of the pond algae sample. Be sure to include some of the green material.

B. If the green material is in long strands, use the toothpick to separate out a few.

C. Add a cover slip and observe under low power.

   1. In general, how can you tell the algae from any protozoans that might be present?

D. Sketch several different types of algae from the material. Compare your sketches with those in Fig. 5–11.

   2. What types of freshwater algae are in your sample?

E. Using high power, examine the algae samples carefully. Label any cell parts you recognize.

   3. Do all the cells have a nucleus?

   4. Do all the cells have a cell wall?

F. Observe the algae cells carefully.

   5. Do you see any bubbles forming?

   6. What could they be, and how were they formed?

G. Make several more slides of the pond material. Add details to your sketches and labels where possible. Also sketch any new types of algae you find.

H. If any protozoans are present, sketch and try to identify them.

   7. If protozoans are present, what life processes do you observe?

CONCLUSIONS:

   1. List the types of algae you observed in the pond material.

   2. Are there any organisms that move like protozoans but also have chlorophyll?

   3. If so, how would you classify them?

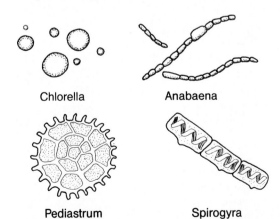

Chlorella   Anabaena

Pediastrum   Spirogyra

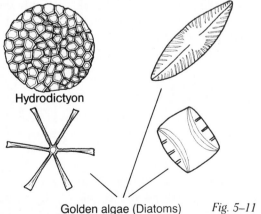

Hydrodictyon

Golden algae (Diatoms)   *Fig. 5–11*

SCIENCE INPUT

By comparing graphed data of their populations, you can determine whether one species in a community is an important food source for another. The graph below compares (2) protozoa, the DIDINIUM (predator) and PARAMECIUM (prey). The *Didinium* population change closely follows that of the paramecium, although there is a slight lag shown in the *Didinium* cycle. A food source must be available if a species or group is to survive and grow. The growth of the *Didinium* population after the peak of the paramecium growth suggests a predator–prey relationship.

COMPUTER INPUT

Computers are very useful in showing the relationship between a number of factors. Since research in life science must almost always deal with situations that are very complex, the computer's abilities to manipulate many variables at once allow calculations to be more accurate while still including as much information as possible. Program Predator will use data that you collect by reading the graph. It will then calculate how many predators and prey there are on any given day of their life cycles and tell you the average populations of each species per day.

WHAT TO DO

After studying the graph of the life cycles of a paramecium and a *Didinium* population, try to answer the questions asked on a separate piece of paper. Next, enter Program Predator into your computer. (Remember to save it on disc or tape before you RUN it.) Where you are asked to ENTER data, use the data from the graph. Record the OUTPUT from Program Predator on a separate piece of paper.

Consider the questions a second time, now answering them using the computer data. Are your answers the same? Was one set of data easier to work with than another? Or were they both pretty much the same?

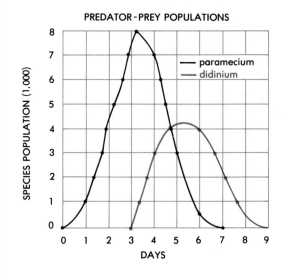

PREDATOR-PREY POPULATIONS

QUESTIONS

1. On which day did the paramecium population begin to grow? On what day did growth stop?
2. How many days after the beginning of the paramecium population's growth did the *Didinium* population start to grow? On which day did it reach its "peak"?
3. Was the paramecium population on the rise, at its peak, or on the decline for the *Didinium* peak day?

4. For how many days did both the paramecium and the *Didinium* population exist at the same time? For how many days did each exist alone?

GLOSSARY

IF-THEN STATEMENTS	directions for the computer's path based on certain conditions being met. "If-then" statements represent decision points in the program.
GOTO	a command in BASIC that instructs the computer to return to a previous statement or to go to another part of the program and follow the directions there.

PROGRAM

```
100   REM PREDATOR
110   FOR CS = 1 TO 24: PRINT : NEXT CS
115   PRINT : PRINT "        WHICH TIME UNIT
      DESCRIBES THE": PRINT "PREDATOR'S
      LIFESPAN?"
117   PRINT : PRINT "          1. DAYS": PRINT :
      PRINT "          2.   YEARS"
120   PRINT : PRINT : PRINT : INPUT
      "WHICH, 1 OR 2? ";T
123   IF T < 1 OR T > 2 THEN GOTO 110
125   IF T = 1 THEN GOSUB 400
127   IF T = 2 THEN GOSUB 500
130   FOR CS = 1 TO 24: PRINT : NEXT CS
135   PRINT : PRINT : INPUT "WHAT SPECIES IS THE
      PREDATOR? ";PR$
140   PRINT : PRINT "HOW MANY ";T$;" DID THE
      PREDATOR LIVE?"
145   PRINT : INPUT "   ";L
150   IF L < 0 OR L > 100 THEN GOTO 140: IF L > 0
      OR L < = 100 THEN PRINT : PRINT
160   DIM PR(L),PY(L):A = 0:B = 0
170   FOR X = 1 TO L
180   FOR CS = 1 TO 24: PRINT : NEXT CS
185   PRINT : PRINT T$;"     ";X
```

```
190   PRINT : PRINT : INPUT "HOW MANY
      PREDATORS? ";PR(X)
193   IF PR(X) < 0 THEN GOTO 190
197   A = A + PR(X): PRINT : PRINT : NEXT X
200   FOR CS = 1 TO 24: PRINT : NEXT CS
210   PRINT : PRINT : INPUT "WHAT SPECIES IS THE
      PREY? "PY$
220   FOR Z = 1 TO L
223   FOR CS = 1 TO 24: PRINT : NEXT CS
227   PRINT : PRINT T$;"        ";Z
230   PRINT : PRINT : INPUT "HOW MANY PREY?
      ";PY(Z)
233   IF PY(Z) < 0 THEN GOTO 230
237   B = B + PY(Z): PRINT : PRINT : NEXT Z
240   AP = A / L:AY = B / L
250   FOR CS = 1 TO 24: PRINT : NEXT CS
260   PRINT : PRINT : PRINT "PREDATOR ",
      "        PREY"
265   PRINT : PRINT PR$,"         ";PY$
270   PRINT : PRINT : PRINT "TOTAL − ";A,
      "     TOTAL − ";B
280   PRINT : PRINT : PRINT "AVERAGE − "; INT
      (AP),"         AVERAGE − "; INT (AY)
290   END
300   REM 400 & 500 GOSUB ROUTINES
400   T$ = "DAY(S)": RETURN
500   T$ = "YEAR(S)": RETURN
```

PROGRAM NOTES

Program Predator can be used with other predator–prey populations. Lifespan may be calculated in either days or years, depending on which is most appropriate. As the program is now written, lifespan can be as long as 100 days or years. If you wish to use the program with a population whose lifespan is greater than 100, change that number in line 150 of your program.

5-3. Fungi

At the end of this section you will be able to:

☐ List the characteristics of Kingdom *Fungi*.

☐ Distinguish between the several groups of fungi.

☐ Describe asexual reproduction methods found in fungi.

☐ Discuss the importance of fungi as *decomposers*.

Fungi A kingdom of organisms that have no chlorophyll and absorb food molecules.

Have you ever opened the refrigerator and found food that has developed mold? Or suffered from athlete's foot? Or seen mushrooms growing in decaying plant matter? If so, you are already acquainted with the Kingdom **Fungi.**

Fig. 5–12 Woodland mushrooms are members of Kingdom Fungi.

CHARACTERISTICS OF FUNGI

Fungi were once thought to be part of the Plant Kingdom. They do have cell walls. Some of their structures do resemble plant organs. But they do not have true roots, stems, or leaves, as plants do. Fungi also do not have chlorophyll. They do not reproduce in the same ways as plants. These differences were used to establish a separate kingdom for the fungi.

Fungi cannot make food because they lack chlorophyll. They also cannot eat food as animals do. Some fungi live as parasites on other living organisms. Others are saprophytes. They take their food from dead or decaying organisms. To obtain food, fungi secrete enzymes. The enzymes break organic matter into small molecules. These molecules are absorbed into the cells of the fungi.

Spore A special reproductive cell that grows into a new organism.

Fungi can reproduce both sexually and asexually. Spore formation is one method of asexual reproduction. A **spore** is a complete cell with a hard, protective covering. See Fig. 5–13. This covering allows it to survive unfavorable conditions. A single fungus can produce billions of *spores*. The spores are spread everywhere by the wind. When the spore lands and conditions are favorable, it may begin to develop. The shape of the structure on the fungus that contains the spores, or the lack of this structure, helps to classify fungi in groups. Some of these groups are *threadlike fungi, sac fungi, club fungi*, and *imperfect fungi*.

Fig. 5–13 These mushroom spores are magnified 800 times.

THREADLIKE FUNGI

The *threadlike fungi* include many types of molds. Molds grow well in warm, moist, dark areas. They will grow more slowly in cool areas. These are the organisms that spoil food left in the refrigerator for a long time. They also include the mold that grows on bread.

Bread molds develop when spores land on moist bread. See Fig. 5–14. The spores begin to send out thin, threadlike structures called *hyphae*. These hyphae penetrate the bread and release enzymes. The enzymes break sugars and starches into molecules that can be absorbed into the hyphae. The fungus obtains its food and water in this way. As more hyphae develop, they form a thick mass called the *mycelium*, which is the body of the fungus. After a few days, black spots develop at the tips of some of the hyphae. These are the *spore cases*. Thousands of spores are produced in them. Eventually, the cases break open and the spores are released, ready to continue their life cycle.

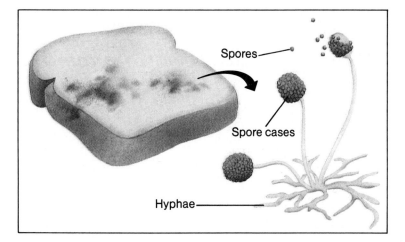

Spores

Spore cases

Hyphae

Fig. 5–14 Bread molds reproduce by producing spores.

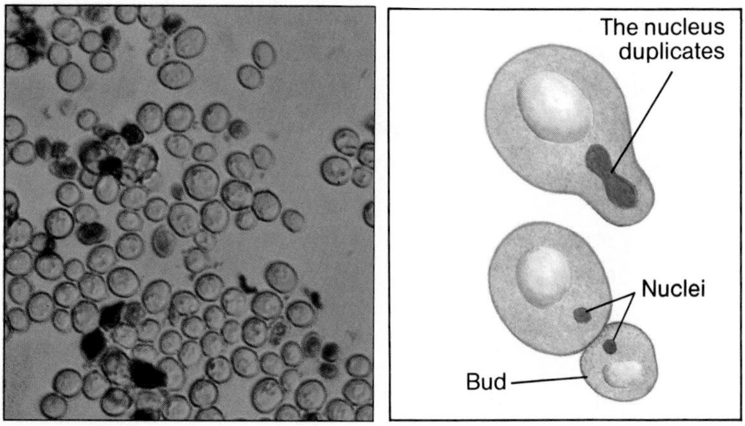

Fig. 5–15 Budding yeast cells (left) *are magnified 450 times. Art* (right) *shows how a bud forms.*

Budding A type of asexual reproduction in which an outgrowth on the organism develops into an entirely new organism.

SAC FUNGI

The *sac fungi* include mildews, some molds on cheese and fruits, some edible fungi, and yeast. Sac fungi attack trees such as elm, chestnut, apple, and pear. The hyphae penetrate into the wood, absorbing food and killing the tree. The saclike spore cases are found in a structure on the outside of the tree.

Yeast is a single-celled sac fungus. See Fig. 5–15. It is used in baking to make the dough rise. As yeast grows, it gives off carbon dioxide gas as a waste. As these gas bubbles expand, they cause the dough to rise. Yeast is also used to produce the alcohol in beer and wine.

Yeast reproduces asexually in two ways. One method is by spore formation. The other method is by **budding.** A new yeast cell grows out of the parent and is called a bud. See Fig. 5–15 (right). *Budding* is similar to fission. In both, the nucleus is divided equally by mitosis. However, in budding, the cytoplasm is not divided equally. A bud has a smaller amount of protoplasm. The bud grows in size and in time breaks away from the parent. Sometimes, the bud stays attached to the parent and then develops a bud of its own.

CLUB FUNGI

The *club fungi* include mushrooms, rusts, smuts, puffballs, and bracket fungi. See Fig. 5–16. The mushrooms you buy in the supermarket are club fungi. Some wild mushrooms are edible. However, others are very poisonous. Only an expert can tell them apart in the woods. Do not ever pick and eat wild mushrooms.

The bodies of the club fungi are made up of closely packed hyphae. Spores are formed in club-shaped structures on the underside of the mushroom cap.

The rusts and smuts do tremendous damage each year to crops such as wheat and corn. Millions of dollars are spent on fungicides to be sprayed on crops to control these fungi.

IMPERFECT FUNGI

The *imperfect fungi* include species that reproduce only asexually by spore formation. These fungi are grouped together because they don't fit in any of the other groups. They are the cause of many diseases that affect grains, fruits, and vegetables. Human diseases such as ringworm and athlete's foot are caused by imperfect fungi.

The valuable drug penicillin is made by a mold from this group. Penicillin kills bacteria cells without harming the other cells in our bodies.

FUNGI AS DECOMPOSERS

So far, we have mentioned mostly the bad effects of fungi. However, fungi also play a helpful role in all environments. Along with certain bacteria, fungi are **decomposers** of dead tissues. They seek food by breaking down the tissues of dead and decaying organisms. This action releases the minerals and other compounds into the environment. These materials can now be recycled. They will be used by other growing organisms. In this role, fungi are vital to all life.

Fig. 5–16 When a puffball is touched, its spores are pushed out.

Decomposer An organism that obtains its food by breaking down dead and decaying organisms.

SUMMARY

Organisms in Kingdom Fungi lack chlorophyll and cannot make food. They reproduce asexually. Fungi are grouped by their reproductive structures. They are important in recycling materials in the environment.

QUESTIONS

Use complete sentences to write your answers.

1. Why have fungi been placed in their own kingdom?
2. Fungi lack chlorophyll. How do they obtain food?
3. List several ways fungi are useful and harmful to humans.
4. Explain two ways fungi reproduce asexually.

INVESTIGATION

GROWING MOLDS

PURPOSE: To grow and observe several examples of mold type fungi.

MATERIALS:

Hand magnifier or dissecting microscope	bread
	cheese, Roquefort or blue
plastic containers	pond water
plastic wrap	dust
pieces of blotter or paper towels	dead insect or small fish
rubber bands	wheat or rice grains

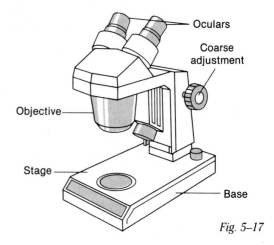

Fig. 5–17

PROCEDURE:

A. Line two plastic containers with pieces of blotter or paper towels. Soak the paper. Squeeze out the excess water.

B. Place the bread and cheese in different containers. Sprinkle dust on the bread. Cover the containers with plastic wrap. Secure the covers with rubber bands. CAUTION: Students who are allergic to dust or molds should perform an alternate activity.

   1. Why was the dust sprinkled on the bread?

   2. Why wasn't it sprinkled on the cheese?

   3. Why were the containers covered?

C. Label the containers with your name, and place them in a warm, dark place.

D. Examine the containers each day for a week or more. The mold *Rhizopus* should grow on the bread. The mold *Penicillium* should grow on the cheese.

   4. What does each mold look like? Make sketches as it grows. Use the magnifier to examine the mold.

E. Grow a water mold as follows: Place a dead insect or a dead tropical fish in a container of pond water. Water mold will develop.

F. Add wheat grains or rice grains to pond water. Change the water when it becomes cloudy. A mold called *Dictyuchus* will develop.

G. Use a dissecting microscope or magnifier to examine the structures of these molds. If you are using the dissecting microscope, first review its parts by looking at Fig. 5–17.

   5. Can the hyphae and spore cases of these molds be seen? If so, sketch the different types that are visible.

CONCLUSIONS:

   1. What conditions do molds need for good growth?

   2. How can foods be protected from molds?

TECHNOLOGY

MICROBE POWER

Scientists are developing ways in which microbes can make our food, produce our fuel, and clean up our environment. How can tiny microorganisms take on such large responsibilities? One reason why is that all microbes—bacteria, viruses, fungi, algae, and protozoa—reproduce very rapidly. If there is a job for a microbe to do, there can be an army of them in no time. Also, microbes are adapted to just about every environment on earth. This means they can use a wide variety of materials as food sources. In soil, bacteria, fungi, and protozoa feed on molecules that come from dead and decaying plants and animals. In a pond, photosynthetic algae produce their own food and are often eaten by protozoa. In turn, the decay bacteria feed on the dead protists, keeping a balance in the pond, which would otherwise be choking with dead algae.

As microbes feed, they produce new cells containing protein and waste products. This protein could be harvested for useful purposes. In various parts of the world, bacteria, cultured in tanks, are already used to produce protein, which is added to cattle feed. The "bacteria-burger" may not be here yet, but there are microorganisms that do provide food for human consumption. For example, for thousands of years, algae gathered from African lakes have been dried into cakes that are then used in a nutritious stew. More recently, the alga Spirulina (spih-roo-**lee**-nah) has become a common item on the shelves of health food stores along with many types of edible seaweed. The idea of algae cultivation tanks on spacecraft has long been part of a plan for helping humans grow their own food in space.

Algae may also be a new source of fuel. Over 30 species of algae produce oil that could be used for gasoline or in the manufacturing of products such as plastics and polyesters.

Perhaps the most significant area in which microbes can play a starring role is in the cleanup of the environment. Bacteria seem to have an appetite for just about anything. Some bacteria are able to ingest and metabolize chemicals found in poisonous industrial wastes. Scientists have, in fact, found two bacteria that seem to have a taste for a widespread pollutant called polyethylene glycol, or PEG. Billions of pounds of PEG are produced each year when antifreeze, detergents, explosives, plastics, and cosmetics are made. One type of bacteria gobbles up the long chains of the PEG molecules and produces short chains that are then eaten by another type. The final waste products are acetate, ethanol, and methane. All of these chemicals have industrial uses. In the future, an industry could manufacture chemicals, use microbes to clean up their operation, and recover a few more chemicals as a bonus, courtesy of microbe power.

CHAPTER REVIEW

VOCABULARY

On a separate piece of paper, match each term with the number of the phrase that best explains it. Use each term only once.

decomposers   sexual   budding   asexual
protozoans   reproduction   fission   reproduction
spore   Protista   algae   fungi

1. When a single-celled organism divides to form two new cells.
2. A kingdom of living organisms that were once called "simple animals" and "simple plants."
3. A special reproductive cell of some organisms that becomes a new organism under favorable conditions.
4. Plantlike organisms that have no chlorophyll and absorb food molecules.
5. A type of asexual reproduction in which an outgrowth on an organism becomes a new organism.
6. Plantlike protists that contain chlorophyll and make their own food.
7. Reproduction that requires two parents.
8. Organisms that obtain food by breaking down dead and decaying organisms.
9. Reproduction that requires only one parent.
10. Single-celled, animal-like protists that must find their food.

QUESTIONS

Give brief but complete answers to each of the following questions. Unless otherwise indicated, use complete sentences to write your answers.

1. Describe three ways by which protozoans move.
2. Explain two ways in which algae are important to the biosphere.
3. On a separate piece of paper, copy the following table. Use the information in the chapter to complete the table.

Organism	Kingdom	Sub-group	Characteristics of Sub-group	Effects on Other Organisms
Bread mold				
Seaweed				
Ameba				
Yeast				

Table 5–2

4. Why did biologists create a third kingdom of living things, the protists?

5. Why was the Kingdom Fungi also proposed?

6. How does an ameba carry on the life processes of food getting, movement, response, and reproduction?

7. How are algae similar to true plants? How are they different?

8. Explain why fungi are either parasites or saprophytes.

9. Compare and contrast asexual and sexual reproduction. List three types of asexual reproduction.

APPLYING SCIENCE

1. Protozoans in drinking water can cause illness in humans. Why do some parts of the world have this problem more than others?

2. What kinds of materials would you use to create a model of a moving ameba?

3. What is the source of food for an athlete's foot fungus?

4. The organisms in Fig. 5–18 are lichens. A lichen is a partnership between organisms from two different kingdoms, an alga and a fungus. Lichens can grow on tree barks, rocks, old fence posts, and dry, bare ground. Collect and identify some that live near you.

Fig. 5–18

BIBLIOGRAPHY

Bartusiak, Marcia. "Living in Rock and Lichen It." *Science 83*, April 1983.

Cobb, Vicki. *Lots of Rot*. Philadelphia: J. B. Lippincott, 1981.

Earle, Sylvia. "Undersea World of a Kelp Forest." *National Geographic*, September 1980.

Lewis, Lucia. *The First Book of Microbes*. New York: Franklin Watts, 1972.

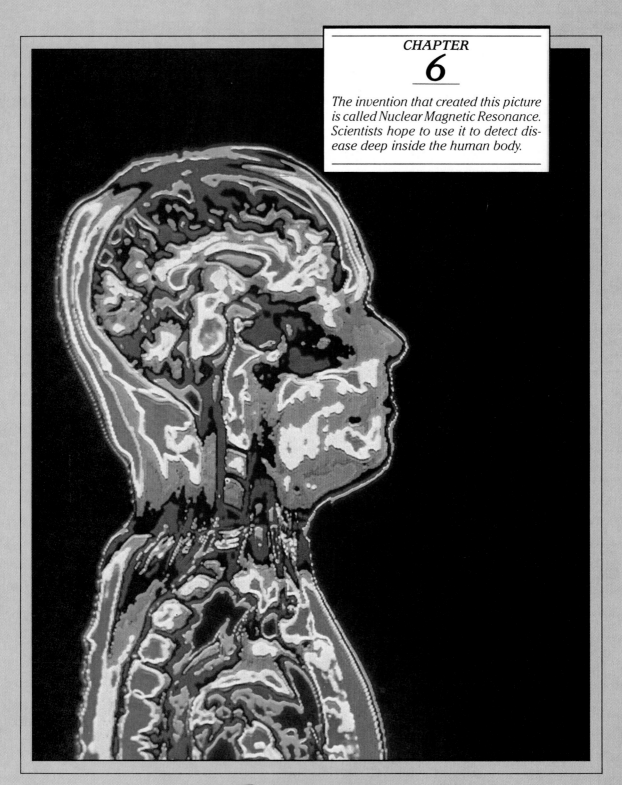

CHAPTER

6

The invention that created this picture is called Nuclear Magnetic Resonance. Scientists hope to use it to detect disease deep inside the human body.

DISEASE

1. Explain what an infectious disease is.
2. Explain how the body defends itself against invasion by infectious diseases.
3. Describe methods of preventing infectious diseases.
4. Identify some of the important causes of disease other than microorganisms.

6-1. Living Things and Disease

At the end of this section you will be able to:

- ☐ Describe three ways in which infectious diseases spread.
- ☐ Explain the germ theory.
- ☐ Name some of the types of microorganisms known to cause disease.
- ☐ Name some plant diseases and their causes.

In Ireland in the 1800's, many people lived on small farms. Potatoes were a major crop because they are easy to grow, are a good food source, and can be stored over the winter.

In both 1845 and 1846, a potato disease, called late blight, destroyed almost the whole Irish potato crop in just a few short weeks. Over one million people died of hunger. Many others left their homeland and came to America.

What is **disease?** A *disease* is a condition in which some part of a living thing is not working properly. The story about the potato blight shows that humans are not the only living things that can have a disease. Any plant or animal can become diseased.

Disease A condition in which some part of a living thing is not working properly.

INFECTIOUS DISEASE

Some diseases are caused by microorganisms. These microorganisms live in the host plant or animal and cause damage to it in some way. They may attack tissues and destroy them. Sometimes, they make poisons that kill their hosts.

Infectious disease A disease caused by a microorganism. It can spread from person to person.

This type of disease can travel from one host to another. It is called an **infectious disease.** *Infectious diseases* can be spread in three ways:

1. By direct contact. This means that touching dishes and bed linens that have been used by a sick person may cause the spread of certain diseases. Any direct contact is dangerous in these cases.

2. By water. Some diseases are spread through the water supply. Sometimes, wastes from humans and animals are not kept far enough away from the water supply. When the water is drunk, some types of disease spread. Typhoid is a disease that spreads in this way.

3. By air. Many diseases are spread on tiny droplets of water in the air. For example, suppose you have a cold and you sneeze. Millions of the microorganisms that caused your cold leave your mouth in these droplets. Anyone touching them or breathing them in can catch your cold. Diphtheria and tuberculosis are also spread in this way. See Fig. 6–1.

Carrier Someone who can spread a disease without showing symptoms of the disease.

Some people that do not show symptoms of a disease can spread a disease. Such people are called **carriers.** *Carriers* can be just as dangerous as those who are ill with the disease. Infections caused by *Staphylococcus* bacteria can be spread by carriers, for example. Animals can spread disease also. The housefly carries and spreads dozens of different kinds of disease germs. Diseases such as malaria and yellow fever are carried by some mosquitoes. Dogs, cats, wolves, cattle, squirrels, and bats can spread the rabies virus.

THE GERM THEORY

What kinds of microorganisms cause infectious diseases? One of the most common causes is bacteria. The idea that bacteria can cause disease was developed by Louis Pasteur in the late 19th century. This idea is often called the *germ theory of disease. Germ* means any microorganism that can cause disease. At first this theory, like many others, was not taken seriously. However, the idea soon gained the support of some famous scientists and physicians of the day.

Robert Koch, a German physician, showed that bacteria was the cause of a disease called anthrax in horses, cows, sheep, and humans. During his investigations, Koch developed a set

Fig. 6–1 The force of a sneeze sends germs into the air.

of rules, or steps, for proving that a specific microorganism is the cause of a specific disease. This set of rules, called Koch's Postulates, is still used today. Koch's Postulates, below, are a fine example of the scientific method in use. See Fig. 6–2.

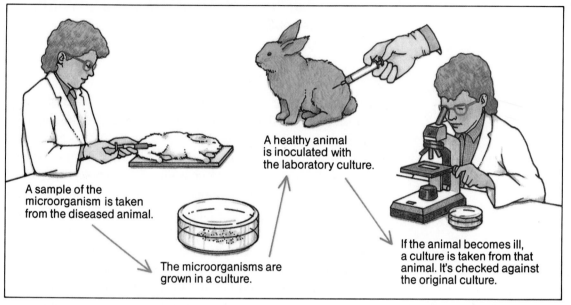

A sample of the microorganism is taken from the diseased animal.

The microorganisms are grown in a culture.

A healthy animal is inoculated with the laboratory culture.

If the animal becomes ill, a culture is taken from that animal. It's checked against the original culture.

1. Show that the microorganism believed to cause the disease is present in the organism.

2. Grow the microorganism in laboratory cultures.

3. Inject microorganisms from the laboratory culture into a healthy animal. Examine the animal for the disease.

4. If the animal gets the disease, grow the suspected organism in laboratory cultures. Check to make sure that they are the same organisms that were in the original culture.

Joseph Lister, an English surgeon, was one of the first to see the importance of Pasteur's work. Lister used chemicals that kill bacteria to make his operating rooms safe. He made sure that everything that touched the patient was very clean. His method prevented **infections** after surgery. An *infection* is an area where a microorganism is invading the tissue of a host. The chemicals that Lister used are called **disinfectants.** *Disinfectants* kill many of the germs that cause disease.

Bacteria can affect a human body in different ways. In some cases, such as tuberculosis, cells and tissues of the host are killed by the bacteria. In other diseases, damage is done by **toxins** released by the bacteria. For example, the bacteria that

Fig. 6–2 Scientists still follow Koch's procedure today.

Infection An area where a microorganism is invading the tissue of a host.

Disinfectant A chemical capable of killing many of the germs that cause disease.

Toxin A poison produced by a microorganism. It causes harm to the host.

cause tetanus live on the host where the skin is broken. These bacteria make a *toxin*, a strong poison, which is carried to the brain by the bloodstream. There, it does enough damage to the brain to cause death in humans.

Bacteria have been shown to cause a long list of human diseases. See Table 6–1. Lists just as long or longer could be made of the bacterial diseases of other animals or of plants.

VIRAL DISEASES

For a long time, scientists could not find the microorganisms that caused diseases such as smallpox and rabies. Smallpox is an easily spread and often fatal disease. It causes severe chills, headaches, fever, nausea, and pains in the back, arms, and legs. Pink-red blisters appear all over the body. They dry up and leave hollow pockmarks. The agents that cause smallpox and rabies are able to pass through very fine filters. All bacteria that scientists had identified were known to be stopped by these fine filters. Also, these microorganisms are too small to be seen with even the most powerful light microscopes. They came to be known as filterable viruses or simply viruses, from a Latin word that means *poison*. Many other human diseases, such as the flu and chicken pox, are caused by viruses. See Table 6–1.

BACTERIAL DISEASES	VIRAL DISEASES
bacterial dysentery	chicken pox
bacterial pneumonia	the common cold
boils	fever blisters
bubonic plague	influenza (flu)
gonorrhea	measles
strep throat	mumps
syphilis	polio
tetanus	rabies
tuberculosis	some types of pneumonia

Table 6–1

OTHER CAUSES OF DISEASE

Fungi can also cause diseases in humans. Some examples are athlete's foot and ringworm.

Infectious diseases can also be caused by protozoa. Amebic

dysentery, malaria, and sleeping sickness are examples. Amebic dysentery is caused by a species of ameba. Malaria is spread by the bite of a certain type of mosquito. Sleeping sickness is caused by a protozoan that is passed from one host to another by the bite of the *tsetse* fly.

Some parasitic worms also cause diseases such as hookworm and trichinosis. Diseases caused by these worms cause much human suffering.

DISEASES IN PLANTS

Bacteria, viruses, and fungi also cause diseases in plants. A disease caused by a fungus wiped out the chestnut tree in this country in the early part of this century. More recently, elm trees have been dying from a fungal disease. Thousands of elms have died. See Fig. 6–3.

A disease of tobacco was found to be caused by a virus. Much early work on viruses was done on this type, the cause of tobacco mosaic disease.

Fungi can cause very serious diseases in crops. Wheat rusts, for example, can cause large losses in wheat crops. The potato blight is caused by a funguslike protist that spreads very quickly.

Fig. 6–3 Dutch Elm disease has nearly wiped out one of North America's favorite shade trees.

SUMMARY

A disease is a condition in which some part of a living thing is not working properly. Infectious diseases are caused by microorganisms. They can be spread in several ways. Bacteria, viruses, fungi, and protozoa can cause infectious diseases in plants and animals.

QUESTIONS

Use complete sentences to write your answers.
1. What is an infectious disease?
2. Name three ways infectious diseases can be spread.
3. What is the germ theory and who first developed it?
4. Name five different types of microorganisms that cause disease, and give an example of a disease each causes.
5. Name three plant diseases and the type of organism that causes each of them.

SKILL-BUILDING ACTIVITY

USING LABORATORY EQUIPMENT

PURPOSE: To examine an oral thermometer and compare it to a laboratory thermometer.

MATERIALS:

oral fever   laboratory
  thermometer   thermometer
pencil

PROCEDURE:

A. Hold the oral thermometer by the top, opposite the bulb end. It should always be held this way.

B. To read the thermometer, hold it in your right hand. The bulb end should be at the left. Roll the thermometer between your fingers until you see a clear line. Numbers will be alongside this clear line. The long marks opposite the numbers show whole degrees. The short marks show two tenths of a degree. Look at Fig. 6–4.

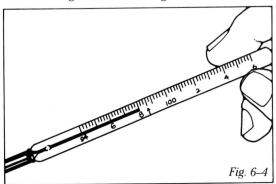

Fig. 6–4

1. What is the lowest mark on the scale? The highest? Where does the colored fluid end?

C. Look at the thermometer again. This time, find the spot where the liquid line narrows. It is between the bulb and the part of the stem that carries the numbers. There appears to be no colored liquid in this area. The liquid breaks at the narrow point. As a result, all the liquid above the break remains in place. Because of the break, you can take the thermometer out of your mouth and the temperature in the mouth still shows on the scale. That is the reason a fever thermometer must be shaken down before using it.

D. Compare the fever thermometer with the laboratory thermometer.

  2. Does the line of fluid in the laboratory thermometer have a narrow area above the bulb?

  3. Will the liquid in a laboratory thermometer remain at the highest temperature taken?

  4. Does a laboratory thermometer need to be shaken down?

E. Set the laboratory thermometer aside in its proper place. If the fever thermometer reads above 95^0F, the colored liquid must be shaken down to at least 95^0F. To do this, shake the colored liquid down with a brisk movement of your wrist. CAUTION: Make sure there is nothing near you that the thermometer could hit and cause it to break. Stay away from desks, tables, chairs, and other students.

CONCLUSION:

1. List several of the differences between a fever thermometer and a laboratory thermometer.

6-2. The Body's Defenses

At the end of this section you will be able to:

- ☐ Explain how skin and mucus defend the body against microorganisms.
- ☐ Describe how white blood cells fight disease.
- ☐ Describe the immune system and how it fights disease.

Fig. 6–5 This boy must be protected from all germs— even those from his family.

Why would a child have to spend his life inside a plastic bubble? Why would the child not be allowed to touch anyone? The reason is that some people are born with no natural defense against infectious disease. See Fig. 6–5. Most of us *are* able to fight off bacteria and viruses that could harm us. How do we do this?

SKIN AND MUCUS

The first line of defense is the skin. The skin is thick and tough. It stops microorganisms from entering the body. Where the skin is broken, germs can enter. That is why it is very important to clean cuts and scrapes. The bark of a tree protects the plant from disease organisms in much the same way. Large wounds on a tree should be closed with tree paint.

Mucus is a thick, sticky fluid that covers many surfaces inside the body and in the natural openings of the body. *Mucus stops germs from attacking tissue not covered by skin.* For example, the inside of the nose is covered by tiny hairs and a mucus lining. These hairs trap dust and germs from the air you breathe. The trapped germs are swallowed with the mucus and travel down the food tube to the stomach. In the stomach are strong acids. These acids can kill the microorganisms trapped in the mucus. Stomach acids also destroy germs brought into the body with food.

Sometimes extra mucus is made. If there is a lot of mucus in your nose, you should try to blow it out. Blowing your nose and sneezing gets rid of trapped microorganisms. That is why it is important to cover your mouth and nose when you sneeze.

What happens if you cut your skin and germs enter the cut? Then your second line of defense, the **white blood cells,** become active.

Mucus A thick, sticky fluid covering many surfaces inside the body and in its natural openings.

White blood cells Cells in the blood that protect the body from infection.

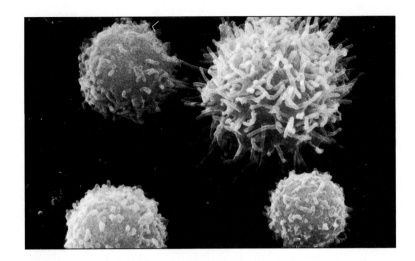

Fig. 6–6 Two types of white blood cells are visible here.

WHITE BLOOD CELLS

White blood cells are one part of our blood. See Fig. 6–6. They are made in the center part of some bones. Many of them are found in the lymph nodes and in the tonsils. Some are found between the cells in your tissues.

If microorganisms get past the skin, the first line of defense, then the white blood cells take over. It is believed that damaged tissue, such as a cut, and invading germs both release chemicals. These chemicals attract white blood cells. When this happens, the area around the cut becomes warm and appears red. The cut has become *infected.*

Amebalike white blood cells surround and destroy germs and damaged tissues. This stops the infection and cleans the area so that proper healing can take place.

THE IMMUNE SYSTEM

In the body, white blood cells have still another way of fighting disease. Some kinds of white blood cells help make special chemicals called **antibodies.** Each *antibody* fights a particular type of microorganism or foreign chemical. See Fig. 6–7.

Antibodies Chemicals made by the body to fight germs or other foreign bodies.

These antibodies become attached to the outside surface of invading organisms. The invader is destroyed by the antibody. The production of antibodies is a fast process. A few days after the invader has entered, a large amount of antibodies can usually be found in the blood.

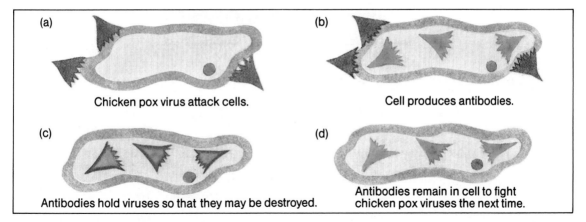

(a) Chicken pox virus attack cells.

(b) Cell produces antibodies.

(c) Antibodies hold viruses so that they may be destroyed.

(d) Antibodies remain in cell to fight chicken pox viruses the next time.

Fig. 6–7 White blood cells can make antibodies.

When a disease like chicken pox is over, many of the antibodies against it are gone. There is a group of white cells, however, that "remembers" how to make the antibody. If the virus that causes chicken pox enters the body again, these cells can make antibodies very quickly. They eliminate the virus before it can do any damage. That is why a person usually gets diseases like measles, mumps, whooping cough, scarlet fever, and chicken pox only once. This resistance to a disease is called **acquired immunity.** *Acquired immunity* to some diseases lasts a lifetime.

During a *fever,* the blood travels faster around the body. This allows the blood to carry white blood cells and antibodies to the infected cells faster. Thus a fever is another way the body defends itself against disease.

Acquired immunity   Resistance to reinfection with a disease after the body has once recovered from the disease.

SUMMARY

The body has three lines of defense against infectious disease. These are the skin and mucus membranes; the white blood cells; and the immune system. A natural immunity to a disease often follows a case of the disease.

QUESTIONS

Use complete sentences to write your answers.

1. How does each of the following protect against disease: skin, mucus, stomach acids?
2. How do white blood cells fight invading organisms?
3. What are antibodies?
4. What is acquired immunity?

INVESTIGATION

BODY TEMPERATURE

PURPOSE: To find the effect of exercise on "normal" body temperature.

MATERIALS:

alcohol	pencil
sterile cotton balls	graph paper
fever thermometer	clock

PROCEDURE:

A. Wash the thermometer with cool water and soap. Rinse well. Wipe it off with a cotton ball moistened with alcohol. The alcohol will kill any microorganisms on the thermometer.

B. To take your temperature, place the bulb end of the thermometer under your tongue. With your mouth closed, leave the thermometer under your tongue for at least three minutes.

C. After three minutes, read and record your temperature. This is *your* "normal" body temperature.

D. After you have recorded your temperature, shake down the thermometer until it reads 95°F again. Then set the thermometer down in the safe place your teacher provided.

E. Stand next to your desk and briskly jog in place for three minutes. Take your temperature again. Record this "after-exercise" temperature.

   1. How do the "normal" and the "after-exercise" temperature compare?

F. Submit your temperature readings to your teacher to post on the board. When all your classmates have submitted their temperature readings, answer the following questions.

   2. Are the "normal" temperatures of your classmates all the same?

   3. The reading of 98.6°F is usually given as "normal." How do you think this number was obtained?

G. Using all the readings of your classmates, find the class' average "normal" temperature. Also find an average "after-exercise" temperature.

   4. How close is the class "normal" to 98.6°F?

H. Plot the temperature readings of all the students on a graph. See Fig. 6–8. Be sure to label the parts of the graph. Ask your teacher to help you do this.

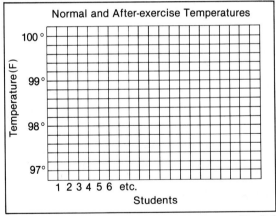

Fig. 6–8

CONCLUSIONS:

   1. What have you learned about "normal" body temperature?

   2. What is the effect of exercise on body temperature?

6-3. Preventing Disease

At the end of this section you will be able to:

- ☐ Explain what immunization is.
- ☐ Identify some public health measures.
- ☐ Explain why personal hygiene is important.
- ☐ Describe an important antibiotic.

Do you know anyone who has had smallpox, diphtheria, or scarlet fever? How about whooping cough, or polio? Your answer is probably "no." These diseases were much more common just a few years ago. What has happened to make them rare today? How are people able to prevent disease?

IMMUNIZATION

You have learned that you can become *immune* to a disease after you have had it once. You will not become ill even though the viruses or bacteria that cause the disease enter your body again. This immunity can last for years.

Scientists in the last century noticed this, too. For example, if a person had a mild case of smallpox and got well, the disease would not return. Some scientists tried infecting people on purpose, to give them a mild case of smallpox. This was dangerous because in some cases the sickness they caused was not mild.

Dr. Edward Jenner knew that people who worked around cows sometimes got a disease called cowpox. See Fig. 6–9.

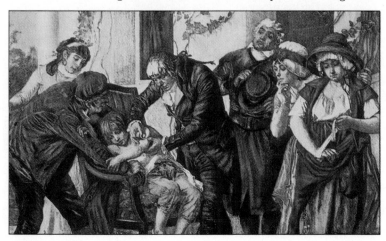

Fig. 6–9 Edward Jenner developed the first vaccine.

In humans this was not a serious disease. These people did not get smallpox. Jenner tried using material from a cowpox sore to infect a young boy. The boy got cowpox. Then Jenner used material from a smallpox sore to infect the boy again. He did not get even a mild case of smallpox.

People did not accept this method of preventing disease right away. Today it is very common. It is called **vaccination.** Killed or weakened viruses or bacteria are used to make a **vaccine.** *Vaccination* is the injection of a *vaccine* into the body to cause the body to produce antibodies. If a live germ later enters the body, the white cells "remember" how to make the antibody. The antibodies then kill the germs.

Some immunities of this type do not last for life. If a long time goes by and the body does not come in contact with the germs, it does not make antibodies any more. As a result, a "booster" shot or dose of the vaccine has to be given. This is true with tetanus, for example.

PUBLIC HEALTH

A community often takes steps to protect the health of its citizens. These are called *public health measures*. Public health officers or officials in a water department control the water supply in a city or town. These officials make sure that the pipes are in good condition. If the water is not clean, they see that it is filtered or that purifying chemicals are added to it. See Fig. 6–10 (left).

Fig. 6–10 Public health inspectors check the bacterial count of water (left) *and tag livestock that have had their vaccines* (right).

Public health officers also inspect cows, barns, dairies, and other food industries. Look at Fig. 6–10 (right). They make sure that milk and other foods in their state are free from disease-causing bacteria.

Pasteurization, a heating process, kills disease-causing bacteria in milk and some other foods. Can you guess what famous scientist developed this process? Steps are taken to stop the spread of germs in other foods, too. These steps include sanitary packaging, freezing, drying, and canning. These and other methods of treating and preserving food slow the growth of bacteria. Proper cooking of meats kills the tiny worms that may be found in the flesh of animals such as pigs. See Fig. 6–11.

Other important public health measures include controlling insects that carry certain diseases.

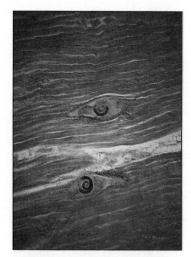

Fig. 6–11 *Heat can kill the Trichina worms that may be in pork.*

PERSONAL HYGIENE

Personal hygiene includes all the ways we keep our bodies, homes, and other belongings clean and in good shape. Your bodies, and especially your hands, touch many things each day. Washing hands with soap and water is a very important yet very easy way to help prevent the spread of disease. The soap helps the water to clean the skin. It also kills bacteria. Keeping homes clean, especially the bathroom, inhibits the growth of germs. Using soap and hot water to wash the bedding and dishes used by a sick person helps to limit the spread of illness.

Getting enough sleep and eating healthy foods are also a matter of personal hygiene. The body is better able to resist infection if it is well-rested and well-nourished. This is especially true for still-growing bodies.

ANTIBIOTICS

One of the best means of treating certain bacterial diseases is by the use of **antibiotics.** *Antibiotics* are chemicals made by living things that are able to kill some bacteria. They were found by accident in 1928 by Alexander Fleming, a Scottish scientist. He was growing bacteria in dishes. He noticed that in one of the dishes a mold was growing with the bacteria. Around the mold was a clear area in which no bacteria were

Antibiotic A chemical, produced by living things such as fungi, that is used to kill some disease-causing bacteria.

growing. See Fig. 6–12 (left). The mold, a *Penicillium*, seemed to give off a substance that killed the bacteria. The substance was called *penicillin*. Penicillin and other antibiotics have been effective in helping the body control many bacterial infections. These drugs kill a germ or stop its growth without injuring the patient. They have saved many lives when infections could not be stopped by the body's own defenses.

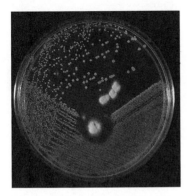

Fig. 6–12 Fleming (right) *found a mold that could kill some bacteria* (left).

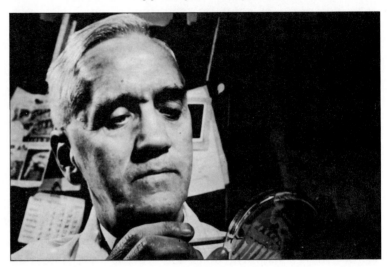

SUMMARY

Immunity to a disease can be obtained by the use of vaccines. Public health measures have helped to stop the spread of infectious diseases by making sure the water supply is pure and food is clean. Personal hygiene also helps stop the spread of disease. Since penicillin was discovered, it and other antibiotics have saved many lives.

QUESTIONS

Use complete sentences to write your answers.

1. What can a doctor do to help make you immune to certain diseases?
2. Why is it important to keep your body and your clothes clean?
3. How did Alexander Fleming find penicillin?
4. Why are diseases that are spread through the water supply not usually found in large cities in North America?

INVESTIGATION

TESTING DISINFECTANTS

PURPOSE: To compare the effectiveness of several disinfectants in preventing the growth of germs.

MATERIALS:

2 petri dishes with sterile nutrient agar	mouthwash alcohol hydrogen peroxide
5 sterile swabs	marking pencil
disinfectant solution	transparent tape toothpick

PROCEDURE:

A. Label the petri dishes with your name and class. Number them 1 and 2.

B. Remove the covers and rub your finger over the top of the nutrient agar in each dish. With the toothpick, draw two lines to separate the agar into four sections. See Fig. 6–13.

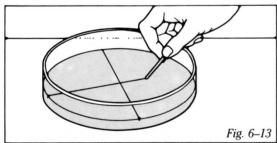

Fig. 6–13

C. Soak one cotton swab in the disinfectant solution. Remove the cover of the dish. Rub the soaked swab on one of the four sections. With a marking pencil, print the letter D on the bottom of the dish under the same section.

D. Take another swab and soak it in the mouthwash. Rub the soaked swab on a different section. Make sure that it doesn't overlap or run into the other sections. Print the letter M on the bottom of the dish under this section.

E. Take a third swab and soak it in alcohol. Rub the swab on another section. Print the letter A on the bottom of the dish.

F. Soak the fourth swab in the hydrogen peroxide. Rub it on the fourth section. Print the letter H on the bottom of the dish.

G. Replace the cover of dish 1 and tape it closed.

H. Remove the cover from dish 2 and rub a dry cotton swab over the surface of the agar. Discard the swab. Replace the cover of the dish and tape it closed.

I. Place both dishes upside down in a dark place at room temperature. Then answer the following questions:

1. What is the purpose of dish 2?

2. In what area of dish 1 do you think the fewest germs will grow? The most?

J. In two days, check the dishes. CAUTION: Do not take the covers off. Always treat bacteria cultures as disease carriers.

3. On which section of dish 1 do you find the most growth?

4. On which section do you find the least growth?

5. How does the growth in dish 1 compare to that in dish 2?

CONCLUSION:

1. Based on your results, which of the substances tested works best to stop or slow the growth of microorganisms?

6-4. Other Causes of Disease

At the end of this section you will be able to:

- ☐ Give some examples of the types of diseases that cannot be spread from one person to another.
- ☐ Explain what cancer is and list some possible causes and treatments.

Babies that have just been born can be suffering from serious diseases. Factory workers or miners with no sign of infection can have badly damaged lungs. What causes these diseases?

NONINFECTIOUS DISEASES

There are many diseases that cannot be spread from one person to another. Some are caused by lack of food. *Malnutrition*, especially lack of protein, is the most widespread and serious nutritional disease in the world. When certain vitamins are not in the diet, people can become very ill. Scurvy and rickets are examples of this kind of disease.

Other diseases, such as some cases of mental retardation, are caused by defects in chromosomes. *Phenylketonuria* (PKU) is one such disease. This is a condition in which a single important enzyme from the liver is missing. The lack of this enzyme causes one of the parts of proteins to turn into a poisonous substance. This chemical can damage a growing child's brain. See Fig. 6–14. Diabetes is also a disease that is caused by a chemical disorder in the body. It can be inherited—that is, passed from parent to child.

The air you breathe can be harmful. Look at Fig. 6–15. Factory smokestacks and cars pour a large amount of harmful

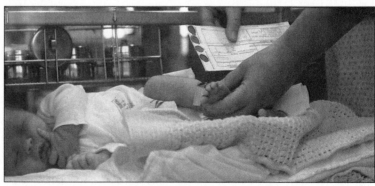

Fig. 6–14 PKU can be identified and treated shortly after birth.

Fig. 6–15 Factories and cars put dangerous wastes into the air we breathe.

chemicals and particles into the air. Breathing this air over a period of time can cause damage to the lungs. Workers such as miners breathe in coal dust that blackens their lungs. After a while, breathing becomes very difficult because their lungs are diseased.

Sometimes the immune system starts attacking the body's own cells. One theory is that a form of arthritis is caused this way. Very little is known about diseases of this type.

Some diseases seem to be related to a person's life style. Some heart diseases fall into this group. Too much *stress* may affect a person's health. It can put wear and tear on the body.

CANCERS

Cancer is caused by a loss of control of cell division and function. *Cancer* often takes the form of a tumor. A tumor is a mass of cells. Cells in a cancerous tumor divide rapidly. The cells do not do the work they are supposed to do. For example, if the cancerous tumor is in the liver, the cells do not function as liver cells. Liver cells begin to die, and the person becomes ill.

There are a variety of cancers. Plants and animals other than humans also develop cancer. The cause of cancer is not known at this time. It is known that such things as cigarette smoke can cause it to occur. Certain chemicals in our water and food are also cancer-causing agents. Another possible cause is radiation, which is a form of energy. Radiation comes from X-rays, the sun, and nuclear explosions. Certain other cancers are known to be caused by viruses.

Cancer A group of cells in the body in which there is loss of control of cell division and function.

CANCER'S SEVEN WARNING SIGNALS
Change in bowel or bladder habits **A** sore that does not heal **U**nusual bleeding or discharge **T**hickening or lump in breast or elsewhere **I**ndigestion or difficulty in swallowing **O**bvious change in wart or mole **N**agging cough or hoarseness

Table 6–2

There is no one cure for cancer yet. Usually the growth is removed by surgery. The use of limited radiation and drugs to kill the cancer cells gives good results in many cases. Some doctors think that a good outlook on life can help a patient fight any disease, even cancer. See Table 6–2.

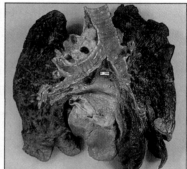

Fig. 6–16 Can you see the difference between healthy lungs (left) *and diseased lungs* (right)?

SUMMARY

Many diseases cannot be transmitted from one person to another. Some diseases are inherited. Others are caused by poor nutrition, by chemicals in water and air, or by weaknesses in the body's own immune system.

QUESTIONS

Use complete sentences to write your answers.

1. What are some of the types of diseases that cannot be transmitted from one person to another?
2. How are cancer cells different from normal cells?
3. List some possible causes and treatments of cancer.

TECHNOLOGY

HELP FOR MEDICAL DETECTION: IMAGING

Medical science has always searched for a safe way of looking inside the body from the outside. This is important for making accurate diagnoses and prescribing treatment. Doctors have been using X-rays for years, but X-rays only show a change in the size or shape of an organ. Some diseases do not affect organs in that way. Instead, they may affect the body by slowing it down or by stopping some life function. What was needed was a

technique that would give the doctors a picture or image of the way in which the organ or tissue was working. The images from the PET scanner give that kind of information.

The PET scanner is an exciting new tool that can be used to "see" and diagnose brain disorders, cancers, tumors, and heart disease. The PET scanner's computer produces images that show not only the shape of organs, as did the X-ray, but also where the activity is—that is, where energy is being used.

PET stands for Positron (**pawz**-ih-trawn) Emission Tomography (toh-**mawg**-rah-fee). A positron is a tiny, positively charged particle found in radioactive atoms. Positrons are added to nonradioactive substances, even foods such as glucose. A patient undergoing a PET scan is given some of this radioactive glucose to drink. (It emits about as much radiation as a chest X-ray and must be given in carefully controlled dosages.) The glucose provides energy for

cells. In a few minutes, it travels to parts of the body that are using energy. There, the positrons meet up with electrons found in all body tissues. When a positron meets an electron, they violently attract each other. This releases two bursts of energy called gamma rays. The PET scanner detects gamma rays coming from the patient. A computer determines where the rays are coming from. The more active areas use more glucose and produce more rays. The computer "scans" (looks over) the tissue and produces a picture in which healthy tissue is shown in a color different from unhealthy tissue. Doctors can then more accurately pinpoint the source of a person's illness.

PET scans are not, as yet, readily available as a diagnostic tool. So far, there are only a few scanners in this country, and they are being used primarily as a research tool. Two areas of research look especially promising. The PET scan is useful in tracing the flow of a special drug that can dissolve blood clots in a person's coronary arteries. Doctors can see whether restoring the blood flow is helping the heart to function. The scan may also be valuable to patients having coronary bypass operations. The PET scan can show which part of the heart is working well enough to profit from this kind of surgery.

While this technique is not commonly used, it promises important medical breakthroughs.

CHAPTER REVIEW

VOCABULARY

On a separate piece of paper, match each term with the number of the statement that best explains it. Use each term only once.

white blood cells   cancer   vaccine   antibodies
infectious disease   carrier   toxins   disease
acquired immunity   antibiotic   disinfectant   mucus
vaccination   infection

1. Cells in the body that make antibodies.
2. A disease that can be spread from person to person.
3. A resistance to a disease because of the presence in the blood of antibodies against that disease.
4. An organism that can spread a disease even if it is not ill with the disease.
5. A killed or weakened germ that, when administered, gives one immunity to a disease.
6. A medicine, produced by a living thing, that can kill bacteria.
7. Poisons made by some microorganisms.
8. Chemicals made by some white blood cells to fight disease-causing organisms.
9. A condition in which some part of a living thing is not working properly.
10. A group of body cells that are no longer carrying on their normal activities.
11. A thick fluid covering the insides of body openings.
12. An area of the body where many germs are growing.
13. A substance used to kill disease microorganisms.
14. The introduction into the body of a killed or weakened germ to cause immunity to develop.

QUESTIONS

Give brief but complete answers to each of the following questions. Unless otherwise indicated, use complete sentences to write your answers.

1. Give an example of a disease caused by each of the following types of organisms: bacterium, virus, protozoan, fungus.
2. What are some ways infectious diseases are spread?
3. Describe the body's first three lines of defense against infection.
4. How can recovering from a disease result in a person's becoming immune to further illness from that disease? What is another way a person can become immune to some diseases?

5. Why are water purification and sewage treatment plants important in safe-guarding public health?
6. Explain how microorganisms can cause disease.
7. What are some functions of white blood cells?
8. What are some causes of disease other than microorganisms?
9. What do the following diseases have in common: diabetes, arthritis, malnutrition, and PKU?
10. Name three plant diseases and the type of organisms that cause them.
11. What is cancer?
12. How can air pollution cause disease?

APPLYING SCIENCE

1. Why was the discovery of antibiotics so important?
2. How can a defect in a chromosome cause a disease?
3. If interferon could be produced in large quantities, why might it be useful as a treatment for certain cancers?
4. Write to the public health officer in your community. Ask if there are rules covering people who work in restaurants. Are there special clothes they must wear? Do they have to cover their hair? Must they wash their hands before touching food? Find out the reasons for these rules.

BIBLIOGRAPHY

Ardley, Neil. *World of Tomorrow: Health and Medicine.* New York: Franklin Watts, 1982.

Blonston, Gary; Bush, Hayden; Rensberger, Boyce. "Cancer: The New Synthesis." *Science 84,* September 1984.

Cohen, Daniel. *The Last Hundred Years: Medicine.* New York: M. Evans and Co., Inc., 1981.

Knight, David C. *Viruses, Life's Smallest Enemies.* New York: William Morrow and Co., 1981.

Langone, John. "Monoclonals: The Super Antibodies." *Discover,* June 1983.

Merlino, Kim Solworth. "Fighting Germs." *3-2-1 Contact,* December 1982/January 1983.

Nourse, Alan E., M.D. *A First Book: Your Immune System.* New York: Franklin Watts, 1982.

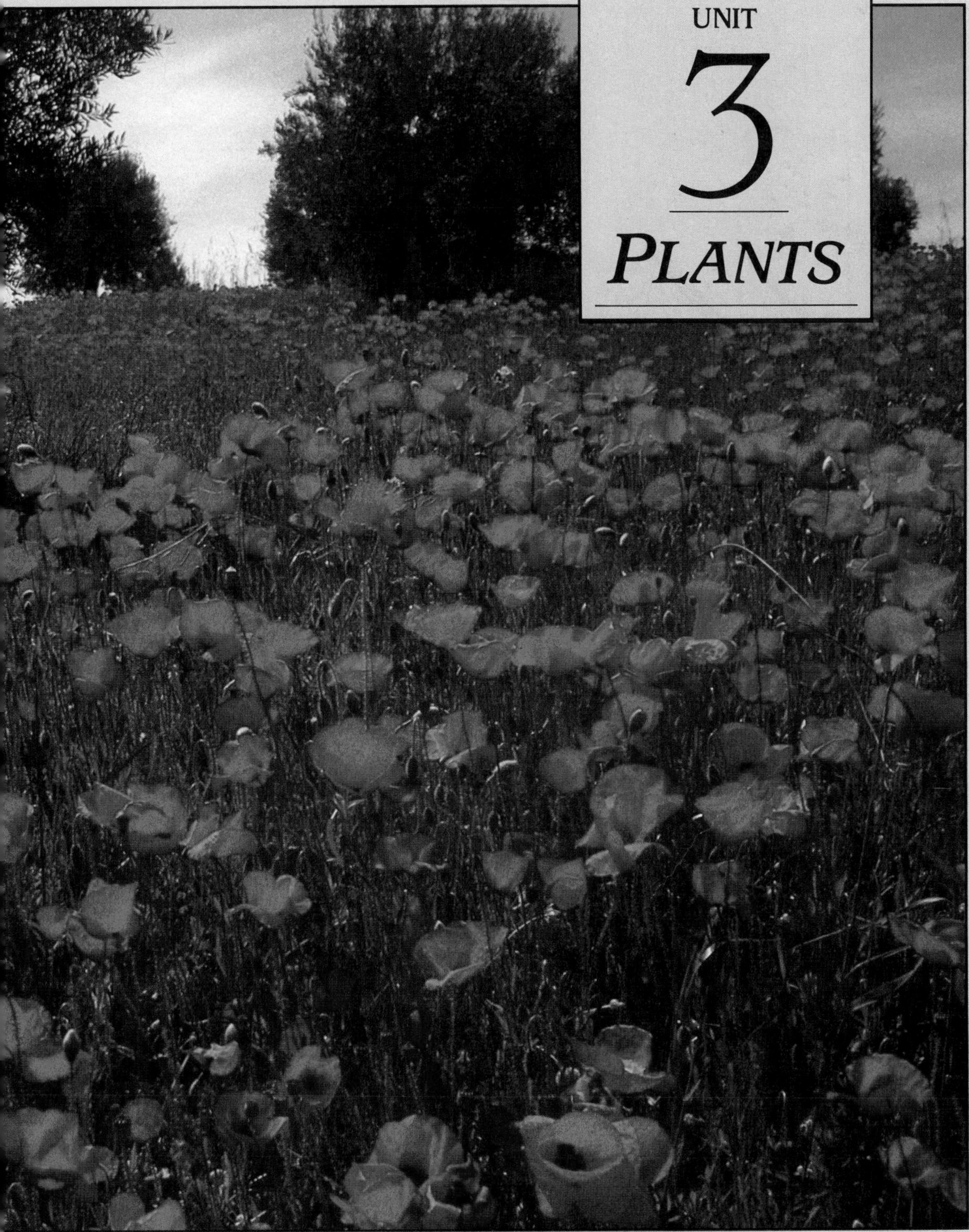

UNIT 3

PLANTS

CHAPTER
7

In spring, the frond of a fern uncurls to reveal its lacelike form. Most ferns of today are small, dainty woodland plants. But millions of years ago, they dominated the earth's land areas.

CLASSIFYING PLANTS

CHAPTER GOALS

1. Identify characteristics common to all plants.
2. Distinguish between vascular and nonvascular plants.
3. Identify representatives of the subgroups within the Plant Kingdom.
4. Discuss how plants have adapted to different environments.
5. Compare the life cycles of some representative plants.

7-1. The Plant Kingdom

At the end of this section you will be able to:

- ☐ List the traits shared by all plants.
- ☐ Tell how land plants are adapted to their environment.
- ☐ Contrast vascular and nonvascular plants.

What do baseball bats, cotton pajamas, and some medicines have in common? They are all made from plants or plant products. Plants have always been very important to people. They are a source of food for all living things, even other plants. We use plants for clothing, tools, dyes, medicines, shelter, and many other purposes.

CHARACTERISTICS

The Plant Kingdom includes organisms that contain chlorophyll. Chlorophyll allows plants to use the sun's energy to make their own food. All plant cells have hard cell walls. In addition, most plants are either rooted to the ground or attached to something. Plants cannot move from place to place as most animals do.

Plants generally share the same needs. They must have light, water, carbon dioxide, oxygen, and minerals. Plants use these substances to carry on their life processes.

The ancestors of most plants were probably tiny green algae. Algae are members of Kingdom Protista. Although they are not plants, algae do have chlorophyll. They, too, can make their own food. Algae live in a water environment.

ADAPTING TO LAND

On land, light and air are above the ground; water and minerals are below the ground. As a result, land plants must have different structures above and below the ground. The structures below the ground anchor the plant. They also absorb water and minerals. These structures are the *roots.* See Fig. 7–1. The structures above the ground capture sunlight and air to make the food for the entire plant. These are the *leaves.* Leaves and roots are organs of a plant. *Stems, seeds,* and *flowers* are also organs. In some plants, stems produce leaves and carry materials throughout the plant. Flowers are used for reproduction in flowering plants.

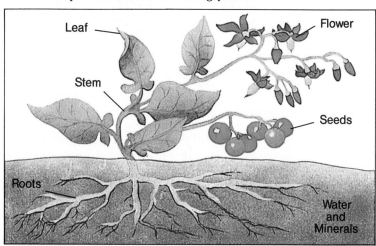

Fig. 7–1 The structures of a typical land plant.

Fig. 7–2 Land plants need vascular tissues in order to grow large.

Plants have many specialized tissues just as animals do. Among these tissues are *growth tissues,* which make the plant larger and form the other tissues. Plants have *covering tissues* to protect the plant from injury and water loss. The bark of a tree is an example. There are also *food-making tissues* and *storage tissues.* The organs of a plant usually contain *strengthening tissues* to support it.

VASCULAR AND NONVASCULAR PLANTS

Some plants have **vascular** tissue. This tissue is made up of cells organized to transport water, minerals, and food. Some vascular tissue also holds stems and leaves up in the air. Such plants are called *vascular* plants. Because vascular tissue can provide support and move materials from one part of the plant to another, some of these vascular plants can grow quite large.

Plants with vascular tissue do not have to live in a moist environment. They can live in a variety of environments because they are able to obtain, transport, and sometimes store water. The presence of vascular tissues also determines how a plant is classified. Vascular plants belong to the phylum *Tracheophyta.*

Plants that do not have this vascular tissue are called **nonvascular** plants. *Nonvascular* plants belong to the phylum *Bryophyta.* Because they cannot transport water, bryophytes are small and must live in moist areas. They do not have true roots, stems, or leaves, as do vascular plants. Fig. 7–3 shows more detailed subdivisions of the Plant Kingdom.

Vascular Having specialized tissue for transporting water, minerals, and food.

Nonvascular Lacking specialized tissue for transporting water, minerals, and food.

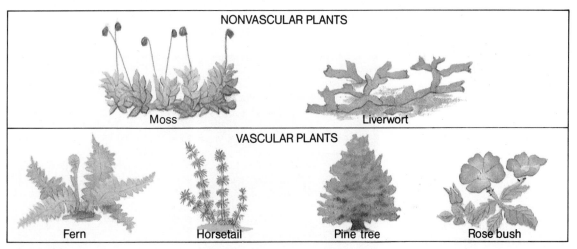

Fig. 7–3 Plant classification is based on the presence or absence of certain structures.

SUMMARY

All plants have the same basic characteristics and needs. They have special organs and tissues to help meet these needs. Some plants have special vascular tissue to help meet the needs of transport and support.

QUESTIONS

Use complete sentences to write your answers.

1. How are all plants the same?
2. How do land plants obtain water?
3. What provides support for some land plants?
4. How are vascular and nonvascular plants different?

7-2. Mosses and Liverworts

At the end of this section you will be able to:

- ☐ Identify the characteristics of mosses and liverworts.
- ☐ Describe their habitat and explain how they are adapted to their environment.
- ☐ Describe the life cycle of a moss.

A long time ago, people believed that the shape of a plant's leaf indicated the type of medicine it could make. As a result, people steeped the *hair cap moss* like a tea, and used it as a hair tonic. Mosses and their close relatives, the liverworts, are nonvascular plants, or bryophytes. Bryophytes were probably among the first organisms to live on land. See Fig. 7–4.

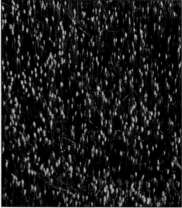

Fig. 7–4 Hair cap mosses (left) *and a liverwort* (right) *called Marcantia.*

CHARACTERISTICS

As you learned in the last section, bryophytes do not have true roots, stems, or leaves. They do have rootlike structures called *rhizoids* (**riz**-oidz) for absorbing water. Rhizoids take in water by osmosis. Unlike most roots, rhizoids are small and do not reach very far.

Bryophytes lack vascular tissue for transporting materials throughout their structures. They must move water and minerals from cell to cell by diffusion. This occurs very slowly. For this reason, bryophytes are very small. Only in a small plant could water reach all the cells by diffusion. Bryophytes must live in places where there is a lot of moisture. Moisture may come from rain, dew, or ground water. Bryophytes usually

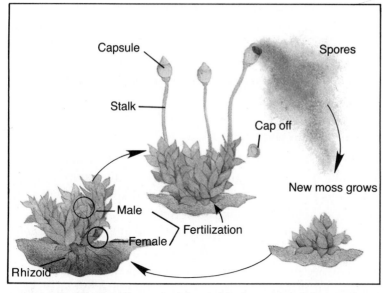

Capsule

Spores

Stalk

Cap off

New moss grows

Male

Fertilization

Female

Rhizoid

Fig. 7–5 The life cycle of a moss plant has several stages.

grow in the shade. They would quickly dry out and die in direct sunlight.

A *moss* plant consists of a slender, stemlike stalk, usually less than 5 cm tall. See Fig. 7–5. This stalk is surrounded by tiny leaflike structures. Each is only one or two cells thick. Rhizoids project out into the soil from the base of the stalk.

Liverworts have a thin, tough, flat, leaflike body. Their shape sometimes resembles the human liver. This is how they got their name.

LIFE CYCLE

Mosses and liverworts carry out both asexual and sexual reproduction during their lives. The stages in the reproduction of an organism are called its *life cycle*. See Fig. 7–5.

Sexual reproduction requires two special reproductive cells. The reproductive cell from the male parent is called the **sperm.** The reproductive cell from the female parent is called the **egg.**

Sperm and *egg* cells develop at the top of the moss plant. In some species, both the sperm and the eggs are on the same plant. In others, they form on separate plants. The sperm swim through a coating of dew to the egg cell. Without water, the sperm could not reach the egg. One sperm joins with one egg. The process is called **fertilization.** This completes the sexual stage of reproduction.

Sperm A reproductive cell from a male parent.

Egg A reproductive cell from a female parent.

Fertilization The joining of a sperm cell and an egg cell.

Spore A special reproductive cell of certain organisms that grows into a new organism when conditions are right.

During the asexual stage, a tall stalk develops from the *fertilized* egg. At the top of the stalk is a capsule in which hundreds of **spores** develop. When the *spores* are ripe, the capsule breaks. The spores are scattered by the wind. If the spores land in a moist place, a new moss plant will develop. The life cycle of liverworts is very much like that of mosses.

ADAPTATIONS

The mosses found on rocks, tree trunks, and in shady places are really many plants grouped together. This grouping helps the plants to hold each other up. The tiny spaces between the plants can retain water like a sponge.

During hot, dry weather, the mosses' leaflike structures twist up and close. See Fig. 7–6. By doing this, they don't lose as much water by evaporation. Although mosses may look dry and lifeless during this time, they are still living. When it rains, the leaflike structures unfold. The plant appears to come back to life.

Thick growths of mosses and liverworts help form and protect the soil beneath. They grow in cracks and help break down the rocks. They also slow down the water runoff after a heavy rain and prevent erosion. Their remains mix with the broken rocks, helping to make soil.

SUMMARY

Bryophytes were probably the first organisms to live on land. Mosses and liverworts are nonvascular plants in the phylum Bryophyta. They are small and must live in moist, shady areas. In the course of their lives, bryophytes carry out both sexual and asexual reproduction. Bryophytes require water for sexual reproduction.

QUESTIONS

Use complete sentences to write your answers.

1. What are three parts of a moss or liverwort plant?
2. In what environments do mosses and liverworts live?
3. In a moss, how does the sperm get to the egg to fertilize it?
4. During which part of the life cycle of a moss, sexual or asexual, do spores develop?

Fig. 7–6 Mosses can survive brief periods of dry weather.

INVESTIGATION

OBSERVING MOSSES

PURPOSE: To collect mosses and to observe a moss plant closely.

MATERIALS:

live moss plants	cup of water
hand lens or	transparent cups
dissecting	pebbles
microscope	sand
toothpicks	soil
dried moss plants	water

PROCEDURE:

A. With your teacher's permission, collect samples of several mosses. At the end of this investigation you will use them to create a terrarium display. Use your toothpick to separate a single moss plant from the clump. See Fig. 7–7.

B. Sketch and label the parts of the moss. Keep in mind that the stalks with capsules are asexual structures.

  1. How are the leaflike structures arranged on the stalk?

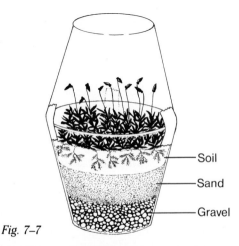

Fig. 7–7

  — Soil
  — Sand
  — Gravel

2. Can you find any capsules from which the cap or lid has fallen? See if you can shake any dust out of that capsule.

3. What do you think this dust is?

C. Search among the moss for some stalks that have yellowish, starlike cups at their tips.

  4. What are these cups? Look down the stalks and see if you can find last year's cup.

  5. How many cups can you see on this stalk?

D. With the hand lens, observe the dry moss.

  6. How is it different from wet moss?

E. Place the dry moss in a cup of water.

  7. What happens to the water?

F. Examine some of this wet moss with the hand lens.

  8. How does it look now?

G. Construct a terrarium that will allow the mosses to live. Put gravel in one of the cups to a depth of 3 cm. Add 2 cm of sand over it. Spread 2 cm of soil over the sand. See Fig. 7–7.

H. Add water to the cup to about the height of the sand. Place the mosses on the soil. Tape another cup in place to create a lid.

CONCLUSIONS:

  1. What environmental conditions do mosses prefer?

  2. Why is it useful for the moss to be able to change form from when it's dry to when it's wet?

¡COMPUTE!

SCIENCE INPUT

Classification systems usually go from the most general to the least general. For example, the plant world is usually divided into two very large general groups, the GYMNOSPERMS and the ANGIOSPERMS. The GYMNOSPERM group is characterized by plants that usually have needlelike leaves and cones which produce unprotected seeds. (This group, however, does in some instances have plants that produce primitive flowers and fleshy fruits.) ANGIOSPERMS are those plants whose leaves are broad and have one or two types of vein configuration. Both of these large groups are subdivided into units that further identify plants having slightly different characteristics. Each of these is again subdivided into seed groups called MONOCOTYLEDONS and DICOTYLEDONS. Differences between the monocots and dicots include the number of embryonic leaves in the seed, fruit type produced, the flower-petal multiples, and root types.

COMPUTER INPUT

In this exercise, a computer version of a traditional pencil-and-paper game is used to help you learn plant classification. Instead of entering an entire word, you will instead choose a letter of the word you believe to be the answer. The program has been designed to "hint" at the correct answer. It will also tell you how many wrong letters you chose, giving you a good idea of whether you really know the answer or if you have an incorrect word in mind. These hints and comments about your responses are called "feedback." Feedback is information about your answers. It helps you learn more efficiently. Your teacher's comments and grades on your assignments are another example of feedback. They give you information you can use to judge the quality of your work. The computer's feedback helps in this way, too. It acts as a private tutor and can give you immediate feedback.

WHAT TO DO

Make a data chart that lists brief definitions for the new words and concepts in Chapter 7. A sample chart appears below. Make your definitions short and clear.

Enter Program Plant Classification, using the information from your data chart to complete statements 310 to 319. Save the program on disc or tape, and then run it. The program will give you the definition and then ask you to enter, letter by letter, the word or concept being defined. If the letter you choose is in the word, it will be printed on the screen in its proper place. If the letter is not in the word, you will be asked to input another letter. At the end, the computer will tell you how many wrong guesses you made.

Sample Data Chart

Term	Definition
Seed	Complete embryo plant
Gymnosperms	Vascular plants with unprotected seeds
Angiosperms	Vascular plants with protected seeds
etc.	

DIVISIONS AND CHARACTERISTICS: PROGRAM PLANT CLASSIFICATION

GLOSSARY

HARDWARE   the machinery itself, including the CPU (Central Processing Unit) keyboard and monitor (screen).

SOFTWARE   the programs used in the computer. Usually, these are available on a floppy disc or cassette tape.

PERIPHERALS   computer accessories. A peripheral might be a printer, a joystick, the TV screen, or the keyboard. Everything except the actual electronic circuits are peripherals.

PROGRAM

```
100   REM PLANT CLASSIFICATION
110   FOR X = 1 TO 20: PRINT : NEXT
120   READ W$,H$ : IF W$ = "0" THEN $
      END
130   T$ = " ":Q = 0
140   FOR X = 1 TO LEN (W$):T$ = T$ +
      "-":NEXT
150   PRINT : PRINT : PRINT : PRINT H$"-
      "T$: PRINT : INPUT "CHOOSE A
      LETTER --> ";L$
160   G$ = T$:T$ = " ":Q = Q + 1:Q1 =
      0
170   FOR X = 1 TO LEN (W$)
180   IF MID$ (W$,X,1) = L$ AND Q1 = 0
      THEN Q = Q - 1:Q1 = 1
190   IF MID$ (W$,X,1) = L$ THEN T$ =
      T$ + L$: GOTO 210
200   T$ = T$ + MID$ (G$,X,1)
210   NEXT
220   FOR X = 1 TO LEN (W$)
230   IF MID$ (T$,X,1) = "-" THEN GOTO
      150
240   NEXT
250   PRINT W$;"- ";Q;" WRONG
      LETTER(S)": PRINT : PRINT
260   FOR X = 1 TO 15: PRINT : NEXT :
      GOTO 120
300   REM DATA STATEMENTS
310   DATA    SEED, COMPLETE
      EMBRYO PLANT
320   DATA 0,0
```

PROGRAM NOTES

If you change the nature of the data statements, this program could be used in a variety of ways. You could choose words and concepts from other chapters, or even from other subjects. You could, for example, change the word to the name of an historical figure and change the definition to an event or achievement associated with that person.

BITS OF INFORMATION

Feedback is an important function of computer learning systems. It is a term first applied to both computers and humans by Norbert Wiener, a pioneer in computer electronics. He explained feedback as a process of communication involving a response to an original statement or input that helps the person or machine make an adjustment if necessary. Wiener's best-known book is called *Cybernetics,* which is a term used to define the science of control and communication.

7-3. Club Mosses, Horsetails, and Ferns

At the end of this section you will be able to:

- ☐ List the characteristics of club mosses, horsetails and ferns.
- ☐ Describe their habitat and how they are adapted to their environment.
- ☐ Describe the life cycle of a fern.

At one time in the earth's past, horsetails and ferns were the most abundant type of land plants. Great forests of these plants grew and died. Their remains were buried deep beneath the mud and sand of ancient swamps. Gradually, these remains were changed into coal. We burn this coal as a source of energy today.

Horsetails, club mosses, and ferns are some of the oldest kinds of vascular plants. Vascular plants have true roots, stems, and leaves. They belong in the phylum *Tracheophyta*.

CLUB MOSSES AND HORSETAILS

The first club mosses lived on earth almost 385 million years ago. They were some of the most numerous plants on land at that time. Some were very large trees, forming the earth's first forests.

The horsetails, or scouring rush plants, appeared after the club mosses. They too were some of the major plants at that time. Later they declined in number. The horsetails living today are small. See Fig. 7–8. Their stems are hollow and have

Fig. 7–8 Club mosses (left) *and horsetails* (right) *have small leaves and are usually less than a meter tall.*

"joints." At each joint is a whorl of scalelike leaves. The ancient horsetails had a similar structure, but they were like large trees. These eventually died out.

Both club mosses and horsetails can reproduce asexually by growing new shoots. These new shoots can come from a root or from a running stem that trails on top of the ground. The life cycles of club mosses and horsetails are similar to the life cycle of ferns, which is discussed next.

FERNS

The ferns appeared on earth at about the same time as the horsetails. Along with horsetails, ferns were once among the dominant land plants. Ferns have declined somewhat since then. Yet, there are many modern species of ferns living today. See Fig. 7–9.

Fig. 7–9 A small woodland fern (left) *and a large tree fern* (right) *from Mexico.*

Ferns are vascular plants. Special transport tissue carries water and food throughout the plant. Plants that have transport tissue can grow very large. Some ferns in tropical areas are actually tree-sized.

Most ferns in North America are about one meter or less in height. Each dainty, lacy structure that you can see is actually only a leaf of the plant. It is called a *frond.* Both the stem and the roots of most ferns grow underground. Several fronds grow from the underground stem. From the stem, clusters of roots work their way down into the soil.

Ferns, like mosses, also have two reproductive stages to their life cycle. One part is sexual and the other part is asexual. Look at Fig. 7–10 on page 168. The sexual stage is hard to observe because the plant is so tiny at this time. Sperm and

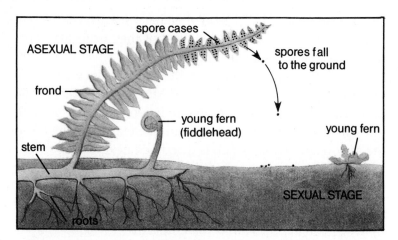

Fig. 7–10 The life cycle of a fern includes a sexual and an asexual stage.

eggs develop on the underside of the flat-lying plant. The sperm must swim through a layer of moisture to fertilize the eggs.

A large plant develops from the fertilized egg. This is the asexual stage. Later, small, dark dots appear on the underside of the fronds. These are the cases in which the spores develop. There may be hundreds of cases. When the spores are ripe, the cases break. The spores drift away on the wind.

Sometimes during hot, dry weather, there is not enough water for the sexual stage to occur. When this happens, the asexual stage continues to live. Sometimes this stage can live for many years until there is enough moisture for the sexual stage to occur.

SUMMARY

Club mosses, horsetails, and ferns are some of the oldest vascular plants on earth. They have true roots, stems, and leaves. These plants have both asexual and sexual stages to their life cycle. They need moist environments for the sexual stage of their life cycle.

QUESTIONS

Use complete sentences to write your answers.

1. How are club mosses, horsetails, and ferns different from mosses and liverworts?

2. How does moisture play a part in the reproduction of club mosses, horsetails, and ferns?

3. What happens to the life cycle of a fern during a dry spell?

OBSERVING FERNS

PURPOSE: To observe a fern plant closely.

MATERIALS:

bracken fern	slide
hand lens	cover slip
microscope	eyedropper
prepared slide of	water
sexual stage of	scalpel
fern	

PROCEDURE:

A. Examine the bracken fern plant. Draw and label its frond, stem, and roots. Take a closer look at a single frond. Use the hand lens.

  1. Can you see whether the edges of the little "leaflets" on the frond are folded under?

B. If you said yes to 1, lift up one of these edges. Fig. 7–11 is a photograph of the underside of a bracken fern frond.

  2. What grows beneath these folded edges?

C. Carefully unfold the edge of a little leaflet. Empty its contents onto a slide. Add a drop of water and a cover slip. Examine the slide under low power of the microscope. Draw what you see.

  3. To what stage of the life cycle do these belong?

D. Examine the prepared slide of the sexual stage under the microscope. At this stage in the life cycle of a fern, the plant is so small that you can almost see through it. Draw a picture of what you observe and label it "sexual stage plant."

E. Using the bracken frond again, cut across the stemlike structure of the frond with the scalpel.

  4. Do you see a figure in it?

  5. What does it look like?

CONCLUSIONS:

  1. What parts of the fern plant did you observe?

  2. Is the fern a vascular plant? What evidence did you observe?

  3. What parts of the fern life cycle did you observe?

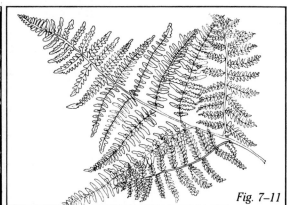

Fig. 7–11

7-4. Plants with Naked Seeds

At the end of this section you will be able to:

- ☐ Explain how seed production helps land plants survive.
- ☐ Describe the characteristics of gymnosperms and name some examples.
- ☐ Explain the special adaptations of conifers to dry environments.

Counting the rings in a tree is one way to find out how old it is. Scientists use a coring drill to look at a tree's rings. This does not kill the tree. Scientists who count the rings of bristlecone pine trees in Nevada have found them to be the oldest living organisms on earth. Some of these trees are almost 5,000 years old! See Fig. 7–12. Bristlecone pines are members of the group of plants called **gymnosperms.**

Gymnosperms Plants with seeds that are not protected by another covering in addition to the seed coat.

Fig. 7–12 Bristlecone pines grow high up in the mountains on the California–Nevada border.

PLANTS WITH SEEDS

Gymnosperms have many traits that allow them to live in drier environments than the plants we have discussed so far. One reason for this is the development of **seeds.** Think of a *seed* as a tiny plant with a food supply and wrapped up in a protective coat. This is a very good way for a plant to reproduce. Seeds can remain inactive for a long period of time. The seeds can start to grow only when conditions are right. See Fig. 7–13.

Seed A complete, tiny, young plant, surrounded by a stored food supply and protected by a seed coat.

The word gymnosperm means "naked seed." This is because the seed has no covering, unlike the pit and fleshy part of a peach. Gymnosperms are seed-bearing plants that do not produce flowers. Their seeds are not protected by another covering in addition to the seed coat.

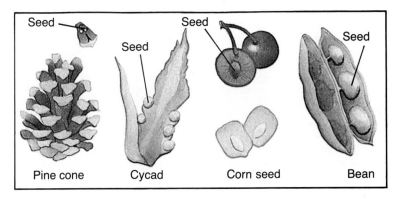

Seed

Seed

Seed

Seed

Pine cone   Cycad   Corn seed   Bean

Fig. 7–13 Pines and cycads produce "naked seeds," while cherries, corn, and beans have a protective covering.

GROUPS OF GYMNOSPERMS

There are several groups of gymnosperms. One class is the *cycads*. There were more cycads living during the time the dinosaurs lived than now. In fact, there is evidence that some dinosaurs ate cycad leaves and seeds. Cycads have palmlike leaves and large cones. Cycads grow mainly in tropical areas. One type grows in Florida. See Fig. 7–14.

Ginkgos are another interesting class of gymnosperm. There is only one living species left of this group, which was once quite large. The ginkgo, or maidenhair tree, is often planted in cities. This is because it is resistant to air pollution. See Fig. 7–14.

Fig. 7–14 Cycads (left) are extinct in their natural habitat. Ginkgos (right) are originally from China.

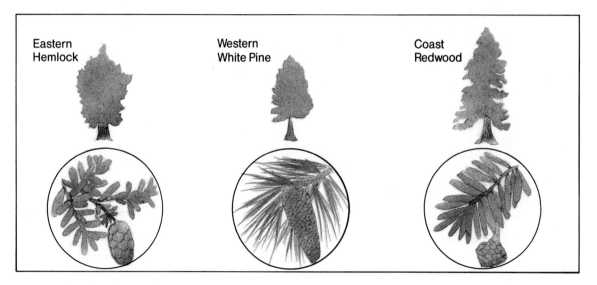

Eastern Hemlock

Western White Pine

Coast Redwood

Fig. 7–15 Conifers often grow in dense forests in cold, northern climates. Conifers produce separate male and female cones. The female cones are pictured here.

The most well-known group of gymnosperms is the *conifers*. They produce special reproductive structures called cones. Conifers are often called evergreens because they do not shed all their leaves in winter. However, not all evergreens are conifers. The most familiar conifers are spruces, firs, cedars, pines, and redwoods. See Fig. 7–15. The biggest, tallest, and oldest trees on earth are all conifers.

ADAPTATIONS

Conifers are well suited to living in dry climates. Their leaves are needlelike. See Fig. 7–15. This leaf shape reduces water loss from the tree. The leaf's surface area is so small that not much water can evaporate from it. The leaf has a thick, protective covering to help keep in water. However, needlelike leaves also reduce the amount of food the tree can make. To make up for this, evergreens have to make food all year round. They do not drop all their leaves at the same time. Needles are always being shed and replaced.

Cones, the reproductive organs of conifers, are either male or female. The male cones are small and produce **pollen** grains. *Pollen* contains the sperms cells. Pollen is another adaptation to a drier climate. Pollen allows fertilization to take place without the presence of water. In conifers, sperm cells do not have to swim to join the egg cells. The pollen is carried by the wind to the larger female cones. The female cones contain the egg cells. When fertilization takes place, seeds

Pollen Structure(s) containing the sperm cell(s) of a vascular, seed-bearing plant.

Fig. 7–16 In North America, 75 percent of our lumber is obtained from conifers.

develop at the base of the scales of the cone. The seeds have thin, papery, winglike structures. When the seeds are ripe, the scales of the cone open. Then the wind carries the winged seeds through the air. If the environmental conditions are right, when the seeds fall, they will grow into new plants.

Conifers are the most important lumber-producing trees. Products such as wood, paper, pulp, turpentine, and other chemicals, are produced from conifers. See Fig. 7–16.

SUMMARY

Seed-producing plants are the most successful plants on earth today. One reason is that the seeds can remain inactive for a long time. One group of the seed-producing plants are the gymnosperms. Of the four classes of gymnosperms, conifers are the most common. Their needlelike leaves and life cycle allow them to live in dry environments.

QUESTIONS

Use complete sentences to write your answers.

1. Why is the seed an important adaptation for land plants?
2. How are gymnosperms suited to dry climates?
3. Name three groups of gymnosperms and give an example of each.
4. Where are the male and female organs of a conifer found?

7-5. Plants with Covered Seeds

At the end of this section you will be able to:

☐ Describe the characteristics of angiosperms.

☐ Explain why angiosperms can live in a wide variety of environments.

☐ Compare the structures of monocot and dicot plants.

If asked to make a list of fruits, most people would probably include apples, oranges, pears, bananas, peaches, and plums. Would you think of tomatoes, eggplants, cucumbers, peanuts, and grains? These are fruits, too.

CHARACTERISTICS

Angiosperms Vascular plants with seeds enclosed in a protective organ.

A *fruit* provides a protective covering for a seed. Plants that have their seeds enclosed in a protective organ belong to the class of seed plants called **angiosperms.** *Angiosperm* means "covered seed." Angiosperms are also called flowering plants because they produce flowers. Flowers are the reproductive organs of these plants. They may be big, colorful, and sweet-smelling, like those used in flower arrangements. Flowers may also be small and not very noticeable, like those of some trees and grasses. See Fig. 7–17.

Fig. 7–17 There is great variety in the flowers, seeds, and fruits produced by angiosperms.

Angiosperms are the most abundant class of plants on the earth today. There are more different kinds of angiosperms than all the other plant groups combined. Scientists have studied the fossil record to find out how old plants are. An-

Black Cherry   Red Oak   Wood Lily

giosperms became widespread at about the time that the dinosaurs were dying out.

Angiosperms can now be found in almost every part of the globe. They grow in deserts, on mountaintops, in tropical rain forests, in salt marshes, in lakes and streams, and even in polar regions.

ADAPTATIONS

Scientists are not really sure why angiosperms are so successful. They do have a very efficient vascular system. This would allow their leaves to be bigger. With bigger leaves, a plant can make more food and therefore grow faster. Also, bigger leaves allow a young angiosperm to grow with less sunlight than that needed by young gymnosperms.

Flowers are the reproductive organs of these plants. Male parts produce pollen grains, which contain the sperm cells. As with gymnosperms, water is not needed for fertilization. Flowers for which the wind carries the pollen are usually small and not easily noticed. Some animals, especially insects, may carry pollen from one part of a flower to the other or from flower to flower. Flowers for which animals carry the pollen are usually large and colorful. They sometimes have an odor that attracts animals. They may also have a sweet nectar that animals can drink. See Fig. 7–18.

Some angiosperms also depend upon animals to scatter, or disperse, their seeds. The burs of a burdock stick to the fur of an animal. See Fig. 7–18. They may be carried great distances before they fall off. Many fruits are eaten, seeds and all, by

Fig. 7–18 How do insects and mammals help disperse seeds?

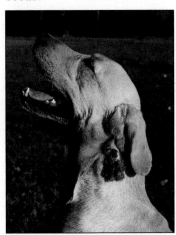

birds and other animals. The seeds pass out of their bodies undigested. You, too, may have helped seed dispersal. Did you ever throw an apple core into a field or spit out watermelon seeds? If the conditions where the seeds landed were right, they will have grown into new plants. Wind and water may also scatter the seeds of many angiosperms.

MONOCOTS AND DICOTS

Monocot A plant that produces seeds with one part.

Dicot A plant that produces seeds with two halves.

Angiosperms may be either **monocots** (**mon**-oh-kawtz) or **dicots** (**die**-kawtz). The *monocots* include grasses, corn, wheat, daffodils, lilies, and palms. Oaks, maples, willows, roses, beans, tomatoes, and dandelions are *dicots.* These terms refer to the structure of the seed. A *cotyledon* (kawt-ul-**eed**-un) is a part of the seed that stores food for the young plant. Monocots have one of these storage areas; dicots have two. A peanut is a dicot seed. There are two halves to the seed. However, a corn kernel has only one part. It is a monocot.

There are other characteristics that separate monocots and dicots. Monocot leaves have veins that run parallel to the entire length of the leaf. Look at the veins in the leaf of corn in Fig. 7–19. Monocot flowers usually have petals in multiples of three (3, 6, 9, etc.).

Dicot leaves have branching veins. They sometimes resemble a fine hairnet. The leaves of a maple are typical. See Fig. 7–19. The dicot flowers commonly have either four or five petals or multiples of four or five. Geraniums are an example with five petals to the flower.

Fig. 7–19 What characteristics separate monocots from dicots?

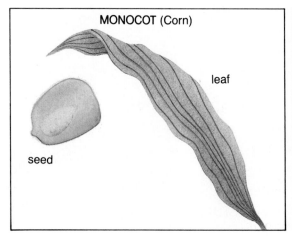

MONOCOT (Corn)

leaf

seed

DICOT (Red maple)

seeds

leaf

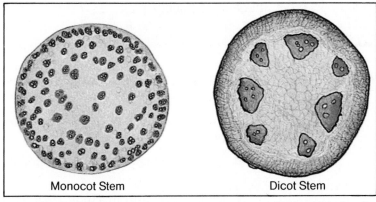

Monocot Stem

Dicot Stem

Fig. 7–20 Monocots and dicots also differ in the arrangement of the cells that transport water and food.

Another difference between monocots and dicots is in the arrangement of the vascular tissues in the stem. In a monocot, such as corn, these tissues appear in bunches scattered throughout the stem. However, in a dicot stem, such as that of a tree, the vascular tissues are arranged in rings, one inside of the other. See Fig. 7–20.

IMPORTANCE TO PEOPLE

Maple, oak, cherry, and birch trees are important sources of lumber. Some of these are called hardwoods because the wood is very strong. More importantly, angiosperms produce all our food crops such as grains, vegetables, and fruits, and also food for the animals that we eat.

SUMMARY

Plants that produce fruit to cover their seeds belong to a group called angiosperms. Angiosperms are the most abundant type of plants on the earth today. Angiosperms are divided into two groups, monocots and dicots.

QUESTIONS

Use complete sentences to write your answers.

1. Which three characteristics separate angiosperms from all other plants?
2. Why can angiosperms live in such a wide range and variety of environments?
3. What are some ways pollen is transferred from flower to flower?
4. How are monocot plants different from dicot plants?

COMPARING AND CONTRASTING

PURPOSE: To observe and identify differences between monocot and dicot plants and seeds.

MATERIALS:

hand lens	corn seed (soaked)
grass plant	scalpel or single-
dandelion plant	edged razor blade
bean seed (soaked)	bean plant

PROCEDURE:

A. Separate one leaf from each plant. Make a sketch of each leaf. Be sure to include the vein pattern of the leaf.

1. Which plant has veins that run parallel down the length of the leaf?

2. Which plant has veins that form a network pattern?

3. Which plant is a monocot? Which plant is a dicot?

B. Examine the corn and the bean seeds given to you.

4. Which seed can be separated into two parts? What are these parts called?

5. What is the function of the two parts of the seed?

C. Locate the tiny plant inside the seed and sketch it.

6. Which seed cannot be separated into parts?

7. Which seed is from a monocot plant? Which is from a dicot?

D. Examine the bean plants to answer the following questions.

8. What type of veins do the bean leaves have?

9. Is this bean plant a monocot or a dicot?

10. Does this answer match the answer you gave to question 3?

CONCLUSIONS:

1. Write a brief paragraph contrasting the monocot and dicot plants that you observed.

2. Would a dandelion seed have one or two parts?

3. On what information do you base your prediction?

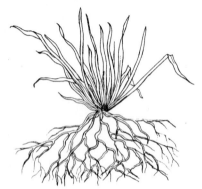

Fig. 7–21   Grass Plant   Dandelion Plant   Bean Plant

CAREERS IN SCIENCE

BOTANIST

Botanists study all aspects of plants. They examine the cells that make up plants and investigate the processes that plants carry out in order to live, such as photosynthesis and reproduction. Other botanists study the factors that kill plants, like insects and disease, and try to find ways to solve these problems. A knowledge of botany is applicable to many different fields—from ecology to agriculture and other businesses. Possible work sites include botanical gardens, parks, commercial and research greenhouses, and landscaping companies.

A Bachelor of Science degree is usually necessary for an entry-level position in botany. Higher-level positions, available at universities, in government, and in international agencies such as the United Nations, usually require a Ph.D.

For further information, contact: American Institute of Biological Sciences, 1401 Wilson Boulevard, Arlington, VA 22209.

AGRICULTURAL EXTENSION SERVICE WORKERS

Agricultural extension service workers provide advice to farmers on the best techniques in agriculture. They might describe a new method of planting crops that would reduce soil erosion, or they could provide the latest information on growing fruit trees, on creating a farm pond, or on beekeeping.

Agricultural extension workers work in the city and suburbs as well as the country. They are, for example, called in as consultants to school agricultural programs, or they may develop a career as lawn and shrub experts.

Extension workers have a bachelor's degree with a major in agriculture or home economics. Jobs are available through state agricultural colleges at which extension services are located. Many people who teach at these colleges have advanced degrees.

For further information, write: Federal Extension Service, U.S. Department of Agriculture, Washington, DC 20250.

CHAPTER REVIEW

VOCABULARY

Unscramble the terms. On a separate piece of paper, match each unscrambled term with the number of the definition that best explains it.

ESPMR	COOOMNT	CDTIO	OMNGRASIPSE
ESDE	GEG	ZTIFOETINLARI	UARSLVAC
OYMSGENSPRM	LOPNEL	ONACUVNSLAR	PSOER

1. A reproductive cell from a male.
2. A complete, tiny, young plant, surrounded by a stored food supply and protected by a seed coat.
3. Vascular plants with seeds that are not protected by another covering in addition to the seed coat.
4. A plant that produces seeds with one part.
5. A reproductive cell from a female parent.
6. Structure(s) containing the sperm cell(s) of a vascular seed-bearing plant.
7. A plant that produces seeds with two halves.
8. The joining of a sperm cell and an egg cell.
9. Vascular plants with seeds enclosed in a protective organ.
10. Having specialized tissue for transporting water, minerals and food.
11. Plants lacking a special tissue to transport materials.
12. A reproductive cell that will grow when conditions are right.

QUESTIONS

Give complete but brief answers to each of the following questions. Unless otherwise indicated, use complete sentences to write your answers.

1. How are vascular and nonvascular plants different?
2. Is the peanut a monocot or dicot seed? Why?
3. List the needs of all plants.
4. Why can gymnosperms live in drier areas than ferns can?
5. What are some special structures that adapt land plants to their environment? What is the function of these structures?
6. How does a moss survive hot dry weather?
7. Place in order the following parts of the life cycle of a moss: asexual stage, fertilized egg, stalk with capsule, capsule breaks, spores develop, sexual stage.
8. How are mosses and liverworts different from other types of plants?

9. What plant has a whorl of scalelike leaves around each "joint"?
10. What kinds of environmental conditions do club mosses, horsetails, and ferns prefer?
11. Why is the sexual stage in a fern's life cycle so hard to observe?
12. Into what does a fertilized egg develop?
13. What characteristic(s) of gymnosperms and angiosperms is (are) the same?
14. How does a conifer's needles adapt it to its environment?
15. How is pollen transferred in flowering plants?
16. Why are angiosperm seeds so widely dispersed?

APPLYING SCIENCE

1. What properties of mosses make them a good packing material for transporting living things?
2. Research some uses of angiosperms and arrange your information on a chart. Headings may include Food, Clothing, Lumber, Fuel, Landscaping, Beverages, and Drugs and Medicines.
3. Find out why the passage of the Morrill Act of 1862 was important to agriculture.
4. Make a map of an area around your home or school. On the map identify the plants, using their common names, and noting whether the plants are gymnosperms or angiosperms. If they are angiosperms, identify whether they are monocots or dicots. Does the type of vegetation on your map indicate differences in environmental conditions? For example, is there more water in one place than another? Does one place get more shade or more sunlight?

BIBLIOGRAPHY

Bellamy, David. *The World of Plants.* New York: Grolier, 1976.

Cole, Janna. *Plants in Winter.* New York: Crowell, 1973.

Earle, Olive L. *The Strangler Fig and Other Strange Plants.* New York: Morrow, 1967.

Eickmeier, William G. "Desert Resurrection." *Natural History.* January 1984.

Hogner, Dorothy C. *Endangered Plants.* New York: Crowell, 1977.

Mack, Richard N. "Invaders at Home on the Range." *Natural History.* February 1984.

Stone, Doris M. *The Lives of Plants.* New York: Scribner's, 1983.

Wenner, Michael. *Man's Useful Plants.* New York: Macmillan, 1976.

CHAPTER

8

Plants are living organisms. They grow and reproduce. They make their own food by capturing energy from the sun. They have a complex internal network for moving food and water.

PLANT LIFE PROCESSES

CHAPTER GOALS

1. Describe reproduction in seed-producing plants.
2. Explain how growth and growth responses take place in vascular plants.
3. Describe the structure and functions of the transport system in vascular plants.
4. Describe the process of photosynthesis in plants and explain its importance to other living things.

8-1. Plant Reproduction

At the end of this section you will be able to:

- ☐ Compare the reproductive organs of seed-producing plants.
- ☐ Describe sexual reproduction in seed-producing plants.
- ☐ Identify ways pollination takes place.
- ☐ Describe vegetative propagation.

Would you like to pick a flower that weighs 11 kg? Impossible? Deep in the jungles of an island called Sumatra grows the rafflesia (rah-**flee**-zee-uh). This plant has the largest known flower in the world. It may measure as much as a meter wide. Its dark-red petals are almost 2.5 cm thick. However, you really wouldn't want to pick this flower because it smells so bad. The odor attracts flies that help to transfer its pollen.

SEED-PRODUCING ORGANS

Flowers are the reproductive organs of angiosperm plants. Cones are the reproductive structures of many gymnosperm plants. Both gymnosperms and angiosperms produce seeds by sexual reproduction.

In sexual reproduction, two special cells, an egg and a sperm, join together. This joining is called fertilization. In seed-producing plants, sperm cells are contained in pollen. Pollen looks like yellow dust. It is made by the male reproductive

Fig. 8–1 On a Scots Pine, the male cones are above the female cone.

Pistil The female reproductive part of a flower.

Stamen The male reproductive part of a flower.

structures. The female reproductive structures produce egg cells.

Most conifers produce two types of cones, male and female. Look at Fig. 8–1. Male cones are small and produce the pollen. The larger cones are the female organs. They are composed of many woody scales. The eggs are found on the surface of these scales.

Angiosperms produce flowers. Most flowers have the same general structure. See Fig 8–2. The *sepals* protect the reproductive organs of the flower before it opens. Some flowers have brightly colored *petals* and produce a sugary substance called nectar. Both attract birds and insects, which help in *pollination.*

In the center of the flower is the **pistil.** The *pistil* is the female reproductive organ. It is shaped somewhat like a light bulb. The slender neck is called the *style.* It supports the sticky *stigma.* The large, round part at the base of the style is called the *ovary.* Inside the ovary are one or more *ovules.* The ovules later become seeds. Some flowers have hundreds of ovules.

The pistil is surrounded by male reproductive organs called **stamens.** There are two separate parts to a *stamen,* the *filament,* and the *anther.* The filament is a threadlike stalk that supports the knoblike anther. The anther produces pollen grains.

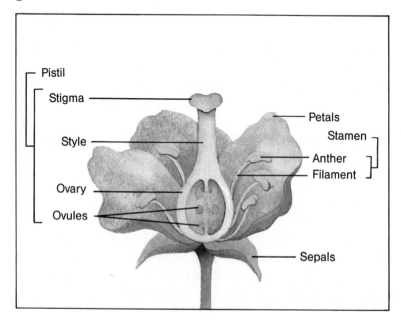

Fig. 8–2 The structure of a typical flower.

POLLINATION

For fertilization to occur, pollen must be transferred from the male structures to the female structures. This process is called **pollination.** Wind, water, insects, birds, and other animals help to transfer the pollen.

Many bees visit flowers for their sweet-tasting nectar. Bees use the nectar to make honey. Bees also eat pollen. When bees gather pollen, some of the sticky pollen grains attach to the bees' hairy bodies. When a bee goes to the next flower, some of the pollen is brushed off. The top of the pistil is sticky, too. If the pollen is brushed off onto the pistil, pollination takes place. See Fig. 8–3.

Birds, especially hummingbirds, also help pollinate flowers. Some flowers are even visited by bats. These flowers are large and usually have a strange smell that only bats like.

Many flowers do not have to depend upon insects, birds, or animals for pollination to take place. Some pollen is so light that it can be carried by the wind. Cone-bearing trees are wind-pollinated. Wind-pollinated plants must produce large amounts of pollen. This helps to increase the chances that some pollen will reach the pistil. See Fig. 8–3.

Some plants are cross-pollinated. Cross-pollination occurs any time pollen lands on a stigma that has come from another

Pollination The process in which pollen is transferred from the male organ to the female organ.

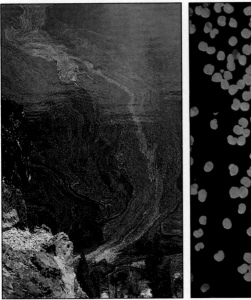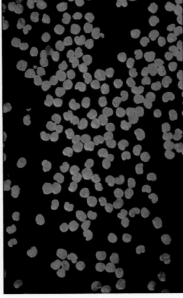

Fig. 8–3 Wind-carried pine tree pollen settles on the surface of a lake (left). *Pine pollen grains magnified 150 times* (right).

plant—usually the same kind of plant. When this happens, the new plant will have characteristics of both parents. It will not be identical to either parent.

Other flowers are self-pollinated. This means that pollen from the anther lands on the stigma of the same plant. Garden peas are an example of a self-pollinating flower. New plants that grow from self-pollinating flowers have the same characteristics as the parent plant.

SEEDS AND FRUIT

Once a pollen grain lands on the stigma, it begins to grow a tube. See Fig. 8–4. The pollen tube grows through the style until it reaches an ovule in the ovary. A sperm from the pollen grain moves down the tube into the ovule. Within the ovule, the sperm joins with the egg. After the egg is fertilized, a new cell is formed. The cell divides many times to form a tiny

Fig. 8–4 The development of a seed enclosed within a fruit begins with pollination.

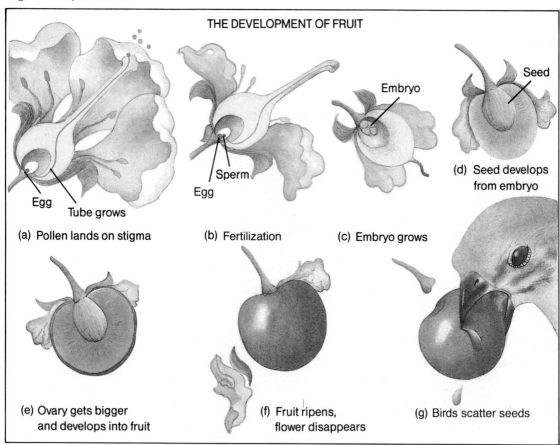

THE DEVELOPMENT OF FRUIT

(a) Pollen lands on stigma — Egg, Tube grows

(b) Fertilization — Sperm, Egg

(c) Embryo grows — Embryo

(d) Seed develops from embryo — Seed

(e) Ovary gets bigger and develops into fruit

(f) Fruit ripens, flower disappears

(g) Birds scatter seeds

young plant called an **embryo.** A food supply and a seed coat for the *embryo* are also formed.

While the seed is developing, the ovary gets bigger and develops into a **fruit.** A *fruit* is the ripened ovary of a flower. It contains the seeds. Apples, oranges, pears, and watermelons are fruits. Nuts, grains, tomatoes, and cucumbers are also fruits.

Fruits help to protect and nourish the seeds. In addition, fruits help scatter the seeds. The winged fruits of the maple tree are carried by the wind. The burs of a burdock stick to the fur of animals. The burs may be carried great distances before they fall off. Some fruits such as pods burst open when they are ripe. The seeds are cast away from the plant. Many fruits are eaten, seeds and all, by birds and other animals. The seeds pass right through the animals' digestive systems. They may be dropped far from where they were eaten.

Seeds begin to grow when the environmental conditions are favorable. When seeds are scattered far and wide, there is a better chance that some will find good conditions.

VEGETATIVE PROPAGATION

Some plants can reproduce by asexual reproduction. Cuttings of roots, stems, or leaves can grow into new plants. This is called **vegetative propagation.**

Bulbs are one example of vegetative propagation. See Fig. 8–5 on page 188. A bulb is a short, underground stem. The bulbs divide into several pieces that can be separated and planted. Onions, tulips, daffodils, and lilies grow from bulbs.

Vegetative propagation also takes place in the white potato plant. The potato is an underground stem. The "eyes" of the potato are the tiny buds of the stem. Farmers cut up the potatoes and plant pieces that have eyes. A new plant quickly sprouts from each of these eyes.

The strawberry plant and some grasses can reproduce by runners. A runner is a stem that "runs," or grows, along the ground. When one of its buds touches the ground, it roots and produces a new plant. New roots and leaves grow from the bud.

Vegetative propagation is often used to grow plants faster and more reliably than with seeds. Many seedless varieties of

Embryo A developing organism in its earliest stages of development.

Fruit A ripened ovary that contains the seeds.

Vegetative propagation When new plants are produced from a part of a plant other than a seed.

Fig. 8–5 *Vegetative propagation by cuttings, the eyes of potatoes, bulbs, and runners.*

flowering plants have been developed in this way. These plants produce seedless fruits, such as oranges and grapes. They can be reproduced only by vegetative propagation.

SUMMARY

Seed-producing plants reproduce sexually. Cones and flowers are their reproductive organs. The wind and some animals help to pollinate seed-producing plants. Seed-producing plants can also reproduce asexually by vegetative propagation.

QUESTIONS

Use complete sentences to write your answers.

1. Which parts of a conifer have the same functions as the anther and pistil of a flower?
2. How is pollination different from fertilization?
3. How does pollen from the anther reach the eggs in the ovary?
4. How do people help spread seeds?
5. What is vegetative propagation?

INVESTIGATION

EXAMINING A FLOWER

PURPOSE: To identify the parts of a flower.

MATERIALS:

flower	1 glass slide
hand lens	1 cover slip
scalpel	microscope
millimeter ruler	alcohol

PROCEDURE:

A. Examine a flower with a hand lens. Locate the sepals, petals, stamens, and pistil. Draw the flower and label these parts.

   1. What color are the sepals?

   2. How are the sepals different from the petals?

B. Carefully remove the sepals and petals from the flower. Closely examine the pistil. Look at the top of the pistil. This is the stigma.

   3. What is an advantage of the stigma's being sticky?

   4. Is any pollen on the stigma?

   5. If there were, what would you expect to see inside the style?

C. Slice the pistil in half lengthwise along the style. Use the scalpel, as shown in Fig. 8–6.

   6. What are inside the ovary?

   7. What do these contain?

D. Measure the length of the style.

   8. What is the length of the style?

   9. How long must a pollen tube grow to reach one of the ovules?

   10. What is the function of the pollen tube?

E. (CAUTION: Anyone who has pollen allergies should wear a protective mask for the remainder of this activity.) Remove a stamen. Tap the anther over a slide until you can see pollen on the slide. In some flowers, the anther must be broken open to release the pollen. Make a wet mount of the pollen grains by adding some alcohol. Add a cover slip. Examine them with the microscope under low power, then switch to high power.

   11. How would you describe the size of pollen grains?

   12. What shape are they?

   13. What color are they?

CONCLUSIONS:

   1. List the parts of the flower that you identified.

   2. In what ways can flowers vary from plant to plant?

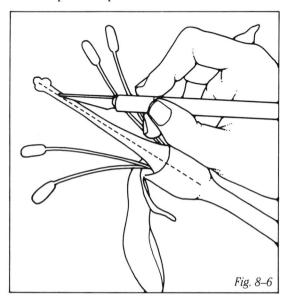

Fig. 8–6

Fig. 8–7 *The mangrove tree is a rapid grower.*

8-2. *Plant Growth*

At the end of this section you will be able to:
- ☐ Describe the sequence of events in seed growth.
- ☐ Identify the growth areas of a plant.
- ☐ Describe some growth responses in plants.

Some plants grow so quickly they almost appear to do so before your eyes. One example is the mangrove tree. See Fig. 8–7. These trees grow at the edge of the ocean in Florida. When a new mangrove tree begins to grow, it grows about 2.5 cm every hour.

SEED GERMINATION

When a seed begins to grow, or *germinate,* the first thing it does is take in water. This causes the seed to swell. The embryo inside starts to grow. Usually, the root tip emerges out of the seed first. See Fig. 8–8. Seeds can store very little water. This is why it is important for the water-absorbing part of the plant, the roots, to grow first.

Next, the young stem and leaves emerge from the seed. These first leaves are called seed leaves or **cotyledons** (kaw-tih-**lee**-dunz). *Cotyledons* are different from the other leaves that will form later. In peas and beans, the cotyledons are very thick. They store a lot of food. The cotyledons of squash and radishes are thinner and look more like leaves.

Cotyledon A seed leaf.

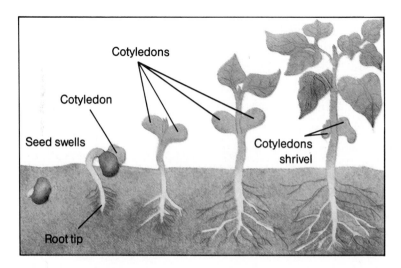

Fig. 8–8 *The germination of a bean seed.*

After the cotyledons emerge, the first true leaves begin to get larger. In some seeds, these leaves are already present inside the embryo. The leaves turn green when sunlight hits them. Then they get even larger. The seedling continues to use the food from the cotyledons until these leaves can provide enough food. Eventually, the cotyledons shrivel.

GROWTH

Plants grow when their cells get longer or by plant division. The tiny embryo inside a seed produces a seedling by the stretching of its cells. However, for further growth to occur, new cells must be added by cell division.

The young plant continues to grow in length by cell division at the stem and root tips. The cells formed at the stem tip grow into new stems and leaves. The cells formed at the root tip result in new roots. As the plant matures, the stem cells get longer. These two processes of cell division and then elongation result in continued plant growth. For some plants, such as **herbs,** this is the main type of growth.

Herbs do not live for many years. Their stems are soft. Grasses, garden vegetables, and some flowering plants are examples of herbs. See Fig. 8–9.

Herb A plant whose stem does not have a lot of supportive tissue.

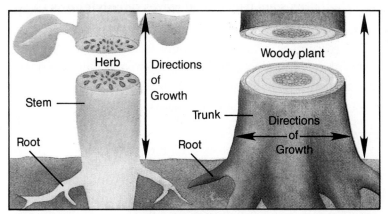

Fig. 8–9 Herbs grow at the stem and root tips. In addition, woody plants grow wider in their stems.

Trees, shrubs, and vines have a lot of support and transport tissue in their stems. These plants are called **woody plants.** *Woody plants* live and grow for many years. In addition to root and stem growth, woody plants also have a layer of dividing cells in their stems. New transport cells are made each year by this dividing cell layer in the trunk and stems. This allows the trunk of a woody plant to grow wider.

Woody plant A plant whose stem has a lot of supportive and transport tissue arranged in rings.

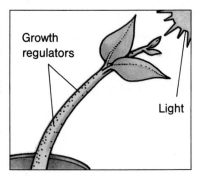

Fig. 8–10 Faster growth on one side of the stem causes the plant to bend. A bending spring is a model for this.

Stimulus Any change in the environment that causes a response in an organism. The plural of stimulus is stimuli.

Tropism The growth of a plant in response to a stimulus.

GROWTH RESPONSES

Have you ever noticed that leaves seem to "turn" toward the sunlight? Why do roots always grow down into the ground? Chemicals called growth regulators control these responses.

Stimuli (**stim**-yoo-lie) such as light and gravity affect the production of growth regulators. These *stimuli* cause growth responses called **tropisms** (**troe**-piz-ums). Growth toward the stimulus is a positive *tropism*. Plants growing by a window always lean toward the light. See Fig. 8–10 (left). Growth away from the stimulus is a negative tropism.

Light causes growth regulators to move to the shaded side of the stem. As a consequence, this side grows faster than the side facing the light. See Fig. 8–10 (right). The difference in the speed of growth causes the stem to bend.

The responses of a plant to light is called phototropism. Plants also respond to gravity. This reponse is called geotropism. Some parts of the plant, such as the roots, show positive geotropism. Other parts show negative geotropism.

SUMMARY

Plants grow by cell division. The new cells can grow by getting longer. Seedlings grow first by elongation. Woody plants have a special layer of dividing cells that allows growth in width. Growth regulators influence the growth of plants.

QUESTIONS

Use complete sentences to write your answers.

1. Where on a plant does growth mainly take place?
2. List, in order, the events that occur in seed growth.
3. What is a tropism? Give an example of a tropism.

SKILL-BUILDING ACTIVITY

MEASURING PLANT GROWTH

PURPOSE: To measure the growth of a plant at specific locations and to determine the areas in which growth takes place.

MATERIALS:

a bean plant with at least an 8-cm stem

millimeter ruler

ink

fine point pen or toothpick

germinating bean seeds with roots 1 cm long

straight pins

paper towel

jar with lid

a piece of cardboard

PROCEDURE:

A. Place the ruler next to the stem of the bean plant. Starting at the tip of the stem, make an ink mark every 5 mm along the stem. Be sure that you mark only the stem and not part of the youngest leaf.

B. Wait one or two days. Then complete step C. While you are waiting, begin step D.

C. Measure the spaces you marked the day or so before. Record your results.

 1. Did any of the spaces get larger?

 2. Where does growth appear to occur on the stem?

D. Cut the piece of cardboard so that it will fit standing up inside the jar. It should be able to stand without falling over. Set this cardboard aside to use later.

E. Lay a germinated seed on the table. Starting at the tip of the root, make an ink mark every 2 mm. Work quickly so the root doesn't dry out.

F. Stick a pin through the center of the seed.

Then pin the seed to the cardboard so that the root points downward. Place the cardboard in the jar.

G. Add enough water to the bottom of the jar until the root tip is 2.5 cm above the water. See Fig. 8–11.

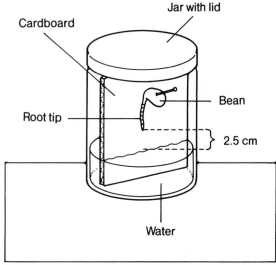

Fig. 8–11

H. Screw on the jar lid and keep the jar in a shady place. Wait one day.

I. Next day: Remove the seed from the jar. Measure the space you marked on the roots the day before. Record your results.

 3. Did any of the spaces increase in size?

 4. Where does growth appear to occur in the roots?

CONCLUSIONS:

 1. What processes were you able to observe in this activity?

 2. What statement can you make about which parts of plants grow in length?

Fig. 8–12 Cactuses are adapted to take in water quickly and store it for a long time.

8-3. Transport in Plants

At the end of this section you will be able to:

- ☐ List the ways plants use water.
- ☐ Describe the parts and functions of a plant's transport system.
- ☐ Explain one theory for the transport of sap.

In the desert grow the oldest plants that are not trees. See Fig. 8–12. A saguaro cactus plant alive today may be 200 years old. There is something else unusual about these plants. They can survive for three years without rain. But when the big rains come, the saguaro can take up almost a metric ton of water from the earth. Some saguaros become so filled that they nearly burst.

WATER FOR PLANTS

Most plants must regularly get water from the soil. Water is necessary for all plant life functions. Without it, the plant cannot live.

Water helps plants keep cool. As the sun shines, and as the plant carries on its internal life processes, it can become very warm. The plant can get so warm, in fact, that its cells can be killed. Water evaporates from the surface of a plant. This causes the plant to cool. The cooling helps to keep the plant's temperature down. You experience this same phenomenon when you get chilled after bathing or swimming.

In addition to water, plants must also get certain minerals, carbon dioxide, and oxygen. Most plants have special structures, the roots, for taking in and transporting these materials.

TRANSPORT SYSTEM

The root is part of a plant's transport system. Numerous tiny **root hairs** take in water and minerals. *Root hairs* make the root appear furry. Water and dissolved substances diffuse through the layers of root cells to the main transport tubes.

Roots continue to grow as long as a plant lives. Since the soil contains pieces of rock and other hard objects, this growth is not always easy. However, roots are stronger than they look.

Root hairs Tiny, tubelike structures growing out of a root that allow the plant to take in large amounts of water and minerals.

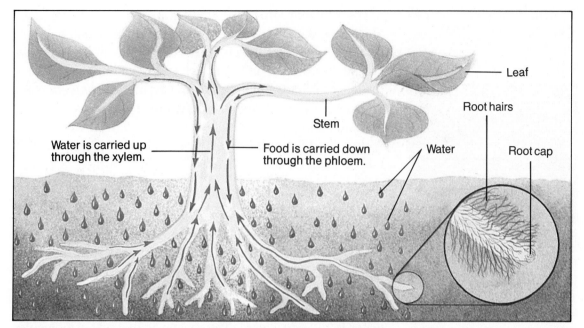

Leaf

Root hairs

Water is carried up
through the xylem.

Stem

Food is carried down
through the phloem.

Water

Root cap

Fig. 8–13 The vascular system
of a typical plant.

They push through the soil with great force. They may even grow into cracks in rocks and eventually split the rocks apart. To protect the root as it grows, there is **root cap** at the tip of every root. New *root cap* cells are always being made to replace the cells that wear out.

How do the water and minerals taken in by the roots get up to the leaves, where food is made? How does the food made in the leaves get down to the roots? In some plants, these substances travel through a transport, or vascular, system. The transport system is a series of tubes. See Fig. 8–13. These tubes connect the roots, through the stem, to the smallest branches and leaves. The veins you see in a leaf, or the "strings" in a stalk of celery, are tubes in this vascular system. You may recall, from Chapter 7, that some members of the Plant Kingdom do not have vascular tissue.

The transport system of a plant has two main parts. One part is the **xylem** (**zie**-lem) tubes, or water-carrying tubes. *Xylem* tubes are hollow and have thick walls. They help support the stem. The xylem tubes transport a watery mixture called *sap*. Sap contains water, sugars, and minerals. The xylem tubes carry sap upward in the plant. Fig. 8–14 on page 196 shows a cross-section of a plant stem. The xylem tubes look like large, open circles. They are stained light-brown.

Root cap Cells at the tip of a root that are loosely packed and protect the root from being damaged as it grows.

Xylem Tubes that transport water and minerals upward in a plant.

Phloem Tubes that transport dissolved food materials.

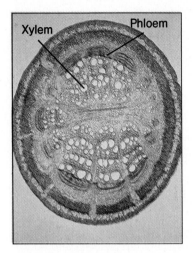

Fig. 8–14 A cross-section of a stem shows xylem and phloem tissues.

The **phloem** (**floe**-em) tubes transport food in the plant. They are not hollow and have very thin walls. In Fig. 8–14, the phloem is stained a dark-blue color. Phloem cells have a smaller diameter than xylem cells. Most of the food made in the leaves is transported downward through the *phloem* to the stem and roots. There, the food is either used or stored. The phloem tubes mainly carry food downward. However, in early spring, food is transported upward through the phloem. This food is used to start new growth.

TRANSPORT THEORY

Scientists have done many experiments in trying to find out how water moves up the xylem. The current theory is that the leaves are the cause of this movement. Thus, scientists believe that the xylem is filled with water from the roots to the leaves. As the plant loses water through its leaves, this column of water is pulled upward. The reason is that water molecules hold tightly to each other. Each water molecule pulls on the next water molecule. The effect is like that of a string of magnets. As the column of water moves upward, it pulls more water from the soil into the roots. Some additional water enters the roots by osmosis.

SUMMARY

Plants need food and water to carry out their life processes. Water is taken in by the roots. In vascular plants, food and water are moved by the transport system. Scientists have developed a theory to explain how water moves up inside a plant.

QUESTIONS

Use complete sentences to write your answers.
1. What are the functions of water in a plant?
2. What is the function of xylem tubes? Phloem tubes?
3. What is the function of the roots of a plant?
4. Explain how water is thought to move upward through the xylem.

INVESTIGATION

CAN TRANSPORT IN PLANTS BE SEEN?

PURPOSE: To observe the effects of transport in plants.

MATERIALS:

glass beaker or jar	celery stalk with
water	leaves
red food coloring	paper towel

PROCEDURE:

A. Fill a glass one-quarter full of water. Add enough drops of food coloring to the water to turn it bright red.

B. Trim the bottom end of the celery stalk. Place it with the trimmed end down in the glass with the food coloring. See Fig. 8–15. Leave the celery stalk in the glass overnight.

    1. What do you predict will happen?

C. The next day, carefully examine the celery stalk and its leaves.

    2. How do the celery stalk and leaves look?

    3. Was your prediction, from step B, correct?

D. Remove the celery stalk from the water and place it on a paper towel. Cut off part of the stalk that was underwater and examine it.

    4. Describe what you see.

E. Cut off another section of the stalk.

    5. Describe what you see.

    6. What do you see that proves water moves upward in plants?

    7. What is the name of the tissue through which the colored water traveled?

CONCLUSIONS:

1. When you pull off the "strings" of a celery stalk, what actually are you removing?

2. What is their function?

3. Design an experiment to find out how fast the celery transports water. What procedure would you follow?

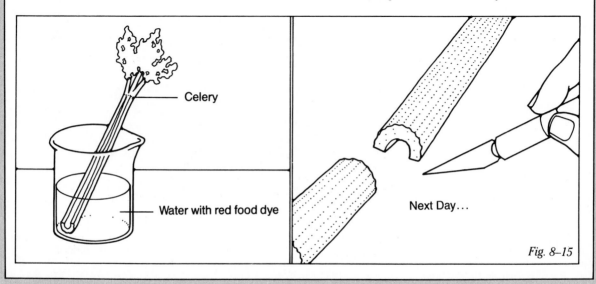

Celery

Water with red food dye

Next Day...

Fig. 8–15

8-4. Food-Making by Plants

At the end of this section you will be able to:

☐ Identify the structures and functions of a typical leaf.

☐ Describe the steps in the food-making process.

☐ State how the plant uses the food it makes.

Leaves are the solar collectors of a plant. They are also "kitchens" or "food factories" where plants use the collected sunlight to make food.

How would you describe leaves to someone who has never seen them? Would you describe their shapes or sizes? Their colors? You might begin with the jobs leaves do. Most leaves have similar purposes and structures.

LEAF STRUCTURE

If you examine broad leaves you will notice that they have a stalk and a flattened blade. The blade provides a large surface area for collecting sunlight. If you examine a broad leaf more closely, you will notice a complex network of "lines" in the blade. These lines are veins that branch out from the stalk. The veins transport materials to and from the leaf.

Within the thin leaf are many special cells and tissues that work together. See Fig. 8–16. The outermost layers of a leaf are only one cell thick. These layers are called the *epidermis* (ep-ih-**der**-mis). The epidermis of the leaf is like the skin on an animal. It protects the inside of the leaf from injury and

Fig. 8–16 A cross-section reveals the structures inside a typical leaf.

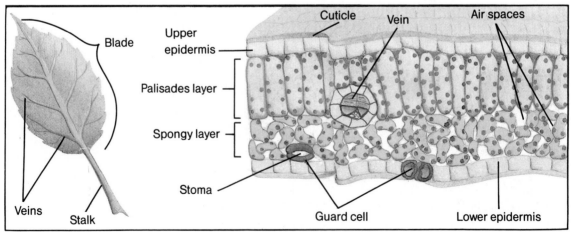

Blade · Veins · Stalk · Upper epidermis · Palisades layer · Spongy layer · Stoma · Cuticle · Vein · Air spaces · Guard cell · Lower epidermis

foreign organisms. It also prevents the loss of too much water. In many plants, a waxy layer called the *cuticle* covers the epidermis. The cuticle also prevents the loss of water.

The *palisade* (pal-ih-**sade**) *layer* is just below the upper epidermis. The cells of this layer are long and narrow. Below the palisade layer are the irregularly shaped cells of the *spongy layer.* The palisade layer and spongy layer contain chloroplasts. The chloroplasts contain chlorophyll. Most of the food-making of the leaf takes place in these two layers.

The cells of the spongy layer are loosely packed together. There are spaces between these cells. These openings are called **stomata** (stoe-**mah**-tuh). In most plants, the *stomata* are found on the lower surface of the leaf. Carbon dioxide enters the leaf and oxygen and water leave through the stomata. To prevent too much water loss, the stomata are able to close. Two bean-shaped **guard cells** form the border of each stoma. The *guard cells* open and close the stoma. See Fig. 8–17.

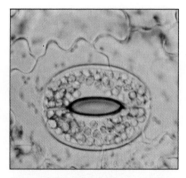

Fig. 8–17 *Unlike other epidermis cells, guard cells contain chlorophyll.*

Stoma Tiny opening in the epidermis. The plural of stoma is stomata.

Guard cells Two bean-shaped cells in the epidermis that form a stoma.

FOOD MAKING

The guard cells and other cells of plants contain green, oval structures called chloroplasts. Chloroplasts contain the pigment chlorophyll. Chlorophyll has the ability to capture the energy in sunlight to make food. This food-making process is called **photosynthesis** (foe-toe-**sin**-thuh-sis).

Plants need sunlight to make chlorophyll. When a plant is placed in the dark, it no longer makes chlorophyll. After several days, its leaves begin to lose their green color. The plant will soon die if it doesn't receive sunlight.

So far, we know that chlorophyll and sunlight are needed for plants to make food. If you have ever taken care of plants, you know that they also need water to survive.

Sunlight, chlorophyll, water, and carbon dioxide are the "ingredients" necessary for *photosynthesis.* See Fig. 8–18 on page 200. Green plants put these substances together in a special way.

Chlorophyll uses energy from sunlight to split water molecules into hydrogen and oxygen. The oxygen is given off into the environment. Most of the oxygen needed by animals comes from plant photosynthesis. Chlorophyll also captures energy

Photosynthesis Process by which green plants use light energy to make simple sugar and release oxygen.

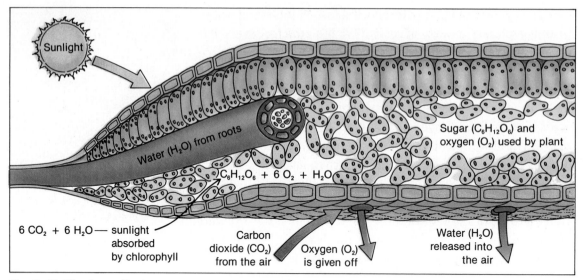

Fig. 8–18 The ingredients and the final products of photosynthesis.

from the sun and stores it. This stored energy is used to combine the hydrogen split from the water molecule with carbon dioxide. After several additional steps, simple sugar and water are the final products. The entire process of photosynthesis can be summarized as an equation:

**Green plants + light energy, carbon dioxide, and water →
simple sugar, oxygen, and new water**

Water needed for photosynthesis enters the plant through its roots. It moves to the leaves through the veins. It passes into the air spaces in a leaf as water vapor. The water then diffuses into the food-making cells of the leaf. Carbon dioxide enters the leaf from the air through the stomata. It also diffuses into the food-making cells of the leaf.

The main product of photosynthesis is simple sugar. The sugar may be used by the leaf where it was made. If not, it is sent to other parts of the plant through the veins. Several different materials can be made with this sugar. In some green plants, many sugar molecules are linked together to form starches. These starches serve as stored energy for the plant. The sugar can also be changed into other substances, such as fats. The simple sugar can be combined with other substances to form proteins.

Oxygen and water are also given off by photosynthesis. The plant does not need them. These wastes diffuse into the atmosphere through the stomata.

Plants release energy by the process of respiration. Respiration goes on in plants all the time. During the day, a plant makes food by photosynthesis. It needs energy to make the food as well as carry on its other life processes. Some of the newly made food will be changed immediately into energy by respiration. Usually, more food is made than is used by the plant. The food is stored. During the night, respiration can still go on even though there is no sunlight for photosynthesis to take place.

OTHER ORGANISMS AND PHOTOSYNTHESIS

Green plants are not the only organisms that carry on photosynthesis. In fact, only about 10 percent of all photosynthesis is carried on by green plants. Algae, which are mainly found in the ocean, carry on the other 90 percent of all photosynthesis. Many of these algae are microscopic.

Unlike plants and algae, animals are not able to make their own food. They all rely on plants for food. For example, many people like to eat hamburgers, but the cow from which the beef came fed on plants. See Fig. 8–19.

Animals also depend upon plants for the oxygen they breathe. Animals need oxygen to release the energy from the food that they eat. This is called respiration.

Fig. 8–19 How do animals depend on photosynthesis?

SUMMARY

Leaves are specially adapted to make food by the process of photosynthesis. The energy in the food may be immediately released by the process of respiration. Some food may be stored for future use by the plant.

QUESTIONS

Use complete sentences to write your answers.

1. How does the leaf (a) prevent too much water loss, (b) give off oxygen and water vapor?
2. What is the function of chlorophyll?
3. List, in order, the steps of photosynthesis.
4. What does the plant do with the food it makes?

INVESTIGATION

A PRODUCT OF PHOTOSYNTHESIS

PURPOSE: To show the presence of sugar in the shoots of onions.

MATERIALS:

goggles	sugar solution or
sprouted onion	molasses
bulbs grown in	water bath
light	water
2 test tubes	test tube holder
Benedict's solution	test tube rack
	graduated cylinder

PROCEDURE:

A. (CAUTION: Wear goggles when doing this activity.) You will need to work with a partner for this activity. Copy Table 1 onto a separate piece of paper.

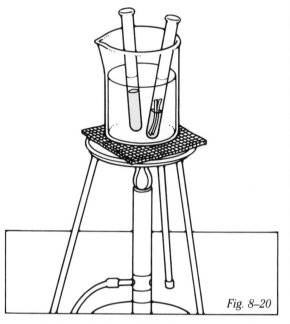
Fig. 8–20

	Sugar Solution	Green Shoots
Sugar Test		

Table 8–1

B. Add 5 mL of sugar solution to one test tube. Then add 10 mL Benedict's solution to this test tube. Set this in the test tube rack.

C. Cut two 3-cm sections of the green shoots of sprouted onion bulbs. Add them to the second test tube.

D. Add 10 mL Benedict's solution to this test tube.

E. Heat both test tubes in a water bath until the contents begin to boil. See Fig. 8–20.

F. If the test was positive for sugar, the Benedict's solution will change color from blue to green and finally to reddish orange. The reddish-orange color indicates the presence of simple sugars. Record your observations in the data table. Place a (+) in the box if the test for sugar was positive. If the test was negative, place a (−) in the box.

1. Why did you test a sugar solution for sugar?

2. Did the green onion shoots contain sugar?

CONCLUSIONS:

1. Why did the onion bulbs have to be sprouted in the light for this activity?

2. If the green shoots contain sugar, from what process can you infer the sugar came from? Describe this process.

SUPERPLANTS: THE 21ST CENTURY DIET?

With the help of genetic engineering, your future shopping list may include such exotic items as the "super-tomato" and the "mea-tato." Genetic engineering is a new way of improving on the qualities that nature has given plants. Through this process scientists can produce plants with higher protein content, or with shapes more convenient for harvesting, or with a higher degree of resistance to pests and disease. All these charac-

teristics are determined by the plant's genes. The genes for a desired trait can actually be placed inside a plant. The process of genetic engineering usually involves snipping the desired trait off one organism's genes and splicing it onto another organism's genes. That trait may be color, size, texture, protein content, and so on. Genes act like a set of instructions for every living thing. For example, a corn plant with large, yellow kernels has genes for that trait. Different genes would produce corn with small, tender, white kernels. Much genetic engineering is focused on making plants more resistant to disease, such as the projects aimed at producing rust-resistant wheat. Success in such efforts would reduce crop failures and help feed the world's hungry millions.

A big problem with changing crop plants genetically is that they have thousands of genes, and it is hard to know which genes control which traits. Bacteria are easier to work with because they have fewer genes. Nitrogen-fixing bacteria live in the roots of peas, beans, alfalfa, peanuts, and clover, and they function as a natural fertilizer to these crops. These bacteria take nitrogen gas out of the air and change it into a form that plants can use. Scientists are already using genetic engineering to speed up the fertilizing activity of the bacteria. The next advance would be to splice the nitrogen-fixing gene from this bacteria onto the genetic structures of the crop plants themselves.

Genetic engineering has stirred up controversy. For instance, researchers figured out how to prevent frost from damaging potato plants. At about 30°F, certain bacteria on potato leaves behave like "ice seeds." Ice crystals grow around them and cause cell damage. When scientists removed certain genes from this bacteria, it no longer formed ice. If such altered bacteria are placed in a potato field, the potato plants won't freeze until the temperature is 23°F.

But what if the altered bacteria spread out of the field? What if they protect other plants from frost—plants like weeds? These questions and others have been raised by those concerned about possibly harmful effects of genetic engineering. Should plants be altered to suit human needs? Could genetically engineered plants drive natural plants out of the environment? Will the benefits of genetic engineering outweigh the risks?

CHAPTER REVIEW

VOCABULARY

On a separate piece of paper, write TRUE next to the number of each statement that is true. Next to the number of each false statement, write FALSE, then make the statement true by writing the correct term in place of the underlined incorrect term.

1. Xylem tubes transport dissolved food materials.
2. Phloem tubes transport water and minerals upward.
3. Cells at the root tip that protect the root from being damaged make up the root cap.
4. Pollination is the process by which pollen is transferred from the male organ to the female organ.
5. Vegetative propagation is a method by which new plants are produced from seeds.
6. A cotyledon is a ripened ovary that contains the seeds.
7. A fruit is a seed leaf.
8. Any change in the environment that causes a response in an organism is a stimulus.
9. The growth of a plant in response to a stimulus is a tropism.
10. An embryo is a developing organism in its earliest stages of development.
11. Root hairs are tiny, tubelike structures growing out of a root.

QUESTIONS

Give brief but complete answers to each of the following questions. Unless otherwise indicated, use complete sentences to write your answers.

1. What are two examples of vegetative propagation? What advantage to a plant does this process have?
2. Why is a leaf called an organ of photosynthesis?
3. On a separate piece of paper, copy each of the terms listed below. Next to each term, write a brief phase to explain its function in a plant's leaf.

vein	cuticle	stoma	spongy layer
upper epidermis	lower epidermis	guard cells	chloroplasts

4. Summarize the process of photosynthesis as an equation.
5. Why do you occasionally have to turn a plant that is growing by a window? Explain what causes this tropism.

6. Why must pollination occur before fertilization can take place?
7. Why do some fruit growers also keep bees in their orchards?
8. What is the relationship of a fruit to a seed?
9. What are the male and female reproductive parts of a flowering plant? Of a cone-bearing plant?
10. What special process goes on in stem tips and root tips?
11. Why must most plants frequently be able to get water? What plant structures help a plant to get water?
12. Why do plants lose water through their leaves? How does this help transport water up through the stem?

APPLYING SCIENCE
1. Why are celery "strings" so hard to chew? What part of the plant are they?
2. Why might flowers and plants be removed from a hospital room at night?
3. Maple syrup is made by tapping the xylem of a sugar maple tree in the spring. How can this be if the phloem transports food?
4. Using your knowledge of transport in plants, devise and carry out an activity that will turn a white carnation blue or green. This is what florists do to get these colors that do not grow naturally.
5. To see how strong seed growth is, in a jar place a few bean seeds and cover them with soil. Add some water. Place a piece of thin cardboard cut to fit so it sits directly on top of the soil. Then place a few pebbles on top of the cardboard. Mark the position of the cardboard on the outside of the jar with a grease pencil or crayon. Note the position of the cardboard as the seeds grow.

BIBLIOGRAPHY
Branley, Franklyn M. *Roots Are Food Finders.* New York: Crowell, 1975.

Gutnik, Martin J. *How Plants Make Food.* Chicago: Children's Press, 1976.

Holmes, Anita. *The 100 Year Old Cactus.* New York: Four Winds, 1983.

Mabey, Richard. *Oak & Company.* New York: Greenwillow, 1983.

Nussbaum, Hedda. *Plants Do Amazing Things.* New York: Random House, 1977.

Pringle, Laurence. *Being A Plant.* New York: Harper & Row, 1983.

Rahn, Joan E. *Seeing What Plants Do.* New York: Atheneum, 1972.

Wilson, Ron. *How Plants Grow.* New York: Larousse, 1980.

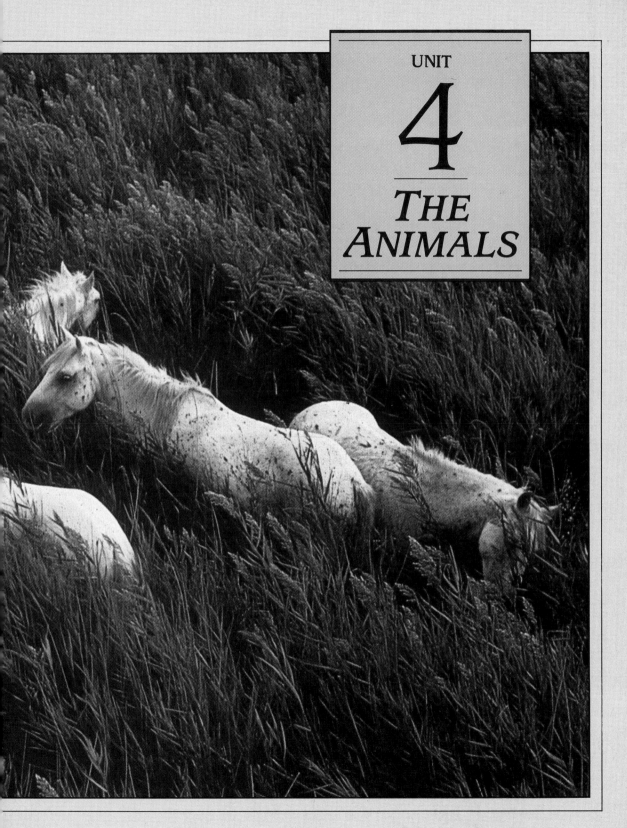

What do you see in this picture? Is it a plant or animal? Is it an abstract sculpture? Actually, these are many tiny coral animals, which live in colonies at the bottom of the sea.

SIMPLE INVERTEBRATES

CHAPTER GOALS

1. Contrast vertebrate and invertebrate animals.
2. Describe the characteristics of sponges, coelenterates, and several kinds of worms.
3. Explain how organisms in each group carry on the life processes.
4. Compare and contrast the organ systems present in each group.

9-1. What Is an Animal?

At the end of this section you will be able to:

☐ Describe ways in which animals are different from other living things.
☐ Explain the difference between *vertebrate* and *invertebrate* animals.
☐ Describe how sponges carry on the life processes.

You are a member of the Animal Kingdom. How can you tell? You are not a Moneran, because you are made of many cells and most of your cells have a nucleus. You are not a Protist, because you are many-celled and have very complex organ systems. You are not a plant, because you don't have cell walls and you cannot make your own food. You are not a fungus, because you digest your food inside your body in a complex digestive system. You are an animal.

CHARACTERISTICS OF ANIMALS

There are only a few differences between animals and other living things. But they are important ones. An animal is a many-celled organism that cannot make its own food. Most animals take in their food and digest it inside their bodies. Their cells do not have cell walls. Their cells form tissues and organs more complex than those of other living things. The nervous system especially is more advanced. Finally, most animals move about freely.

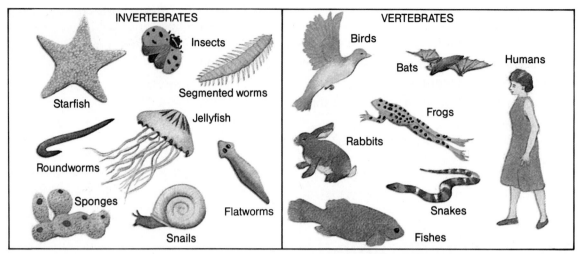

Fig. 9–1 The Animal Kingdom can be divided into invertebrates and vertebrates.

There are over 1.2 million different species of animals known today. New species are found each year. Because of the great numbers of animals, classification is not easy. Scientists have grouped animals into about twenty-six phyla. Of these we will study the nine largest phyla.

Animals are usually divided into two groups. Animals without backbones are called **invertebrates** (in-**vur**-teh-braytz). All the animal phyla except one are in this group. *Invertebrates* include roundworms, flatworms, jellyfish, sponges, insects, and marine animals such as lobsters, and mussels. Look at Fig. 9–1. This chapter will cover five phyla of invertebrates.

Animals with backbones are called **vertebrates** (**vur**-teh-braytz). All *vertebrates* are placed in the same phylum. Examples of vertebrates are fishes, frogs, snakes, birds, and humans. There are a small number of animals in this phylum that do not have a true backbone. These include acorn worms, sea squirts, and lancelets.

Invertebrates Animals without backbones.

Vertebrates Animals that have backbones.

THE SPONGES

Sponges belong to the phylum *Porifera* (puh-**rif**-eh-rah). Most *sponges* are found in salt water but a few types live in fresh water. They are usually attached to one place for most of their lives. These animals have body cells that do special jobs, but the cells do not form tissues. Sponges are the least complex invertebrates.

Sponges have many different shapes. Look at the one in Fig. 9–2. Their body walls are two layers thick. In between these

Sponge A simple invertebrate animal that has one body opening.

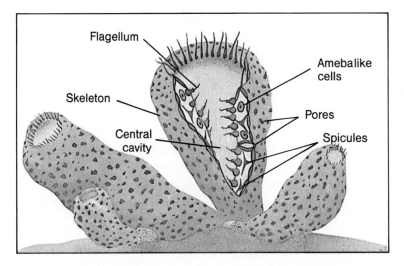

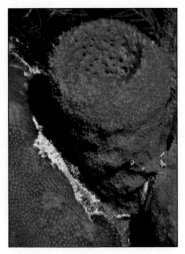

Fig. 9–2 Sponges have a simple body plan.

two layers is a jellylike substance. In the center of the body is a space called a *central cavity*. The walls of most sponges are held up by a network of *fibers* or *spicules*. These act as a *skeleton* and give the sponge its shape. The name *Porifera* comes from the number of tiny openings in the wall called *pores*. These pores lead to a system of canals through the body wall.

People collect and dry some types of sponges that have fibers as their skeletons. The cells of these sponges die and decay, leaving the fibers. The fibers of these sponges and the spaces between them can hold a good deal of water. The sponges are used for bathing and cleaning.

Each individual cell of a sponge can react to stimuli. However, there are no special tissues to allow the whole animal to react. Oxygen is taken directly into each cell from the water around it.

HOW SPONGES GET FOOD

Adult sponges live anchored to underwater surfaces. They cannot move around to obtain their food. Instead, they draw their food to them. The central cavity and pores of a sponge are lined with special cells. Look at Fig. 9–2. Each of these cells has a flagellum. The movement of all the flagella draws water and food into the sponge through its pores. The cells take in pieces of food from the water. The food is then digested by these special cells. Digested food and wastes leave the cells. Amebalike cells carry the digested food to other cells in

Fig. 9–3 Harvested sponges hung to dry on a boat in Florida.

Regeneration The ability of some living things to grow new parts.

the sponge. The wastes enter the water around the cells. The water travels through a canal system and enters the central cavity. From there, the water and waste leave the sponge through a special large pore.

REPRODUCTION IN SPONGES

Sponges can reproduce sexually by producing eggs and sperm. They can also form new sponges asexually by *budding.* Look at Fig. 9–2 again. The buds break off, drift away, and attach themselves in a new location. Sometimes the buds do not break off. Instead, they stay on the parent sponge and form a large group of sponges.

When sponges reproduce sexually, an egg and sperm join. The young sponge, or larva, spends some time in the open water before settling to the bottom. There it attaches and develops into an adult sponge.

Living things that do not have very complex organ systems often show **regeneration.** If a part is cut off, they *regenerate,* or grow, a new one. A sponge is able to do this. In fact, if a sponge is cut into pieces, the pieces will usually grow into whole new sponges. This is done in places where sponges are picked and sold as bath sponges. See Fig. 9–3.

SUMMARY

There are over one million types of animals on earth. Scientists have grouped them into about twenty-six phyla. All but one of these phyla contain invertebrates, or animals without backbones. One phylum is made up of vertebrates, or animals with backbones. Sponges are the least complex of the invertebrates. Their cells do special jobs but do not form tissues or organs.

QUESTIONS

Use complete sentences to write your answers.

1. How do animals differ from organisms in the other four kingdoms?
2. What is meant by the terms vertebrate and invertebrate?
3. Why are sponges sometimes called "simple" invertebrates?
4. What role do pores play in food-getting in a sponge? Why are the cells with flagella important?

9-2. Coelenterates

At the end of this section you will be able to:

☐ Describe the general structure of the coelenterates.

☐ Compare and contrast the structure of coelenterates and sponges.

☐ Describe the ways coelenterates carry out the life processes.

Have you ever seen pictures of people skin diving? Or have you ever visited a large aquarium? People usually look for the big, free-swimming fish. However, if you look carefully, there is a lot more to see. Starfish, crabs, and strange plants are part of the scene. Either along the bottom or on the rocky shores, you will see animals that look like plants. See Fig. 9–4. In fact, many of these animals were thought to be plants as recently as 200 years ago. These animals include the corals and the jellyfish.

Fig. 9–4 A sea anemone is an animal that spends much of its life anchored to one spot.

HOLLOW-BODIED ANIMALS

The phylum *Coelenterata* (sih-**len**-ter-**aht**-ah) includes animals that are somewhat more complex than the sponges. The phylum name means *hollow sac.* This refers to the body structure of these animals. Each one forms a hollow sac with only one opening. See Fig. 9–5 on page 214. In contrast, sponges

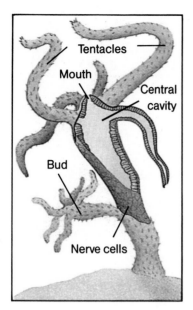

Fig. 9–5 *The hydra has a hollow body with one opening.*

have many openings. In coelenterates, the walls of the sac are made of two layers of cells, with a jellylike material between them. The single opening to the hollow central cavity serves as a mouth. Surrounding it is a ring of *tentacles* (**ten**-tah-kuls). These are like long, thin arms. They are lined with special *stinging cells.* These cells can shoot out a sharp, pointed thread. The point injects a poison that stuns or kills small animals. Other stinging cells shoot out threads that wrap around food particles. In some of these animals, the threads are sticky and can trap small pieces of food. Look at Fig. 9–6.

The cells of coelenterates are more specialized than those of sponges. Coelenterate cells form tissues that do certain jobs. There are covering tissues on the outside of the body. The lower ends of the cells in both layers act like muscle fibers. They move the body and the tentacles. There is a network of nerve cells that responds to stimuli. The nerve cells are located in the jellylike material between cell layers. They do not form a true nervous system, however.

BODY FORMS OF COELENTERATES

Coelenterates have two different body forms. One is a tube-like form that is attached to a rock on the ocean bottom, or even a plant. At one end is a disc that attaches the animal. At the other end is the mouth, surrounded by tentacles. The *hydra* has this form. It is known as the polyp form. The second form is a free-swimming, bell-shaped structure known as a medusa.

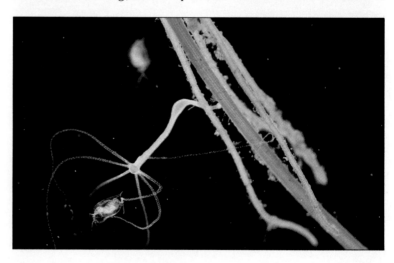

Fig. 9–6 *A hydra captures a Daphnia with its stinging tentacles.*

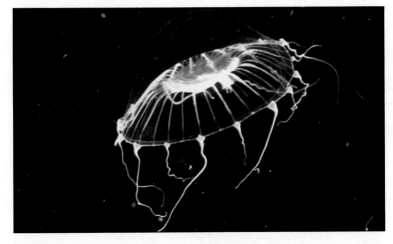

Fig. 9–7 In its medusa form, the jellyfish can swim freely.

Jellyfish have this form. See Fig. 9–7. Some types of coelenterates show both body forms during their life cycles.

When viewed from above, many coelenterates look like a wheel with the tentacles coming from a central point. The bodies of coelenterates such as the hydra or jellyfish have two different ends. However, if a coelenterate is split in half lengthwise, the two sides are exactly the same. Thus, coelenterates like the hydra and jellyfish can respond to food or danger from any side or direction.

HOW COELENTERATES GET FOOD

Coelenterates use their tentacles to help them get food. The stinging threads from the special cells kill or trap tiny living things.

Once the food has been trapped, the tentacles bend toward the mouth. The food is pushed through the mouth into the central cavity. Some cells in the inner layer give off enzymes. The enzymes start breaking down the food. Amebalike cells transport the digested food to other body cells. Wastes leave the central cavity through the mouth opening.

REPRODUCTION OF COELENTERATES

Coelenterates may reproduce asexually by budding. The drawing of the hydra in Fig. 9–5 shows such a bud forming on the animal's side.

Coelenterates also reproduce sexually. There are male coelenterates, which produce sperm, as well as egg-producing females. The life cycle of the *Aurelia* jellyfish shows both types

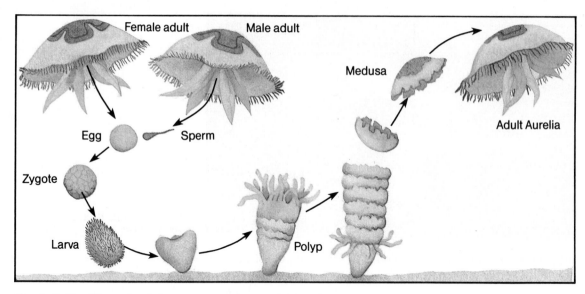

Fig. 9–8 *The life cycle of the* Aurelia *jellyfish.*

of reproduction. It also has both body forms. Can you find them in Fig. 9–8?

During one stage of its life, *Aurelia* is in the *polyp* form and looks like a hydra. As this stage grows, it forms buds. The buds are shaped like a stack of saucers. These "saucers" later break away and drift off to become the bell-shaped, free-swimming form. This form, called the *medusa,* reproduces sexually. The young, or *larvae,* swim free. They attach to a surface and start the cycle again by forming polyps.

SUMMARY

Coelenterates have more specialized cells than sponges. Their bodies are simple sacs with walls made of two layers of cells. The body has one opening surrounded by tentacles. These tentacles have special stinging cells. Coelenterates form tissues but no organs.

QUESTIONS

Use complete sentences to write your answers.

1. In what ways are the bodies of sponges and coelenterates similar? How are they different?
2. Compare the food-getting methods of a sponge and a hydra.
3. List and explain the functions of three types of tissues found in coelenterates.

INVESTIGATION

OBSERVING THE HYDRA

PURPOSE: To observe how a coelenterate carries out certain life functions.

MATERIALS:

shallow dish or well slide
hand lens or dissecting microscope
2 medicine droppers

hydra culture
Daphnia culture
dilute acetic acid (5% solution)
light source
cardboard square (10 cm x 10 cm)

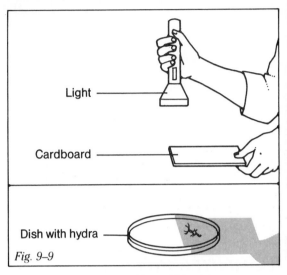

Light ——

Cardboard ——

Dish with hydra ——

Fig. 9–9

PROCEDURE:

A. Use a medicine dropper to transfer a hydra from the culture to a shallow dish of water or a well slide.

B. Sketch the hydra on a separate piece of paper. Label the following parts: tentacles, tubelike body, mouth.

1. Does the hydra attach to the glass or does it swim freely about?

2. How many tentacles does your hydra have? How does this number compare to the hydras of other students?

3. Are there any buds growing from the side of your hydra?

C. Touch the tentacles and then the body with the point of a pin.

4. Describe what happens when the hydra is touched.

5. Name the life process of which this is an example.

D. Shine a light on the dish. Hold a piece of cardboard as shown in Fig. 9–9. The hydra should be on the dark side of the dish.

6. How does the hydra respond?

E. Now shine the light on the side of the dish containing the hydra.

7. How does the hydra respond?

F. Place a drop of the *Daphnia* culture in the dish with the hydra. Observe how the hydra feeds.

8. What happens when a *Daphnia* comes in contact with the tentacles?

9. How does food get into the hydra's mouth?

G. Return the hydra to the culture and prepare a fresh specimen. To observe the stinging cells, use the second dropper. Add a drop of weak acetic acid to the water.

10. Describe what happens when the hydra senses the acid.

CONCLUSION:

1. Which life processes were shown by the hydra you observed?

SCIENCE INPUT

Simpler multicellular invertebrates, such as those from the Porifera, Coelenterata, and worm Phyla, live in similar environments (moist earth, water), have similar functioning body tissues, simple or no organ systems, and similar life cycles. In Chapter 9, you have learned about a number of multicellular invertebrates. In this exercise, you will use your knowledge of these organisms to write definitions for them suitable for Program X-word.

COMPUTER INPUT

The computer can be programmed in a "game format"—that is, in the form of a game. In this case, the game is a crossword puzzle. Your entry of the correct response will fill in one part of the puzzle.

A crossword puzzle is an arrangement of interlocking vertical and horizontal words. How is it possible to print out such an arrangement on the monitor or TV screen? The screen is not like a blank piece of paper. It is more accurate to think of it as a piece of graph paper, with positions marked off electronically rather than in ink. A program can indicate in exactly which square a word or mark should occur.

In Program X-word you will be shown a definition and will be asked to enter the name of the invertebrate that the definition identifies. The definition will appear on the bottom of your screen from left to right in the usual reading pattern. Your answer will appear in its proper position in the puzzle on top of the screen.

If you have answered the first fourteen items correctly, then the last two will have already been printed on the screen. That is the reason

why data statements 380 and 390 read "in vertical column." It is a clue to the word.

WHAT TO DO

Below is a list of poriferans, coelenterates, and worms. On a separate piece of paper, write a brief definition for each. For example: anemone—a flowerlike coelenterate; or Nematoda—a phylum of roundworms. Or you might want to give a clue to a word instead of a definition; for example, fish—jelly_ _ _ _; or, sponge—something often used in cleaning.

When entering your program, be sure your data statements are in the format required by the program, as shown in lines 290, 360, and 370. After entering Program X-word, save it on disc or tape, and then run it. The crossword puzzle will not go on to the next item until you have entered the correct answer.

List of Terms for Crossword Puzzle

anemone	Porifera	Annelida
Ascaris	sponges	larva
buds	tentacles	coelenterates
coral	parasite	Nematoda
filter	sea	fish
hydra	*Trichina*	

GLOSSARY

VTAB   instruction to the computer to locate the letters, numbers, or symbols in a vertical column.

HTAB   instruction to the computer to locate the letters, numbers, or symbols in a horizontal row.

PROGRAM

```
100    REM X-WORD
110    FOR CS = 1 TO 24: PRINT : NEXT
       CS
120    DIM IN$(16),CH$(16),VI(16),HI(16),
       R$(16)
130    PRINT : PRINT : PRINT : PRINT
       " USE THE LISTING OF TERMS ON
       YOUR": PRINT "WORKSHEET AS
       'INPUT' FOR COMPLETING THE":
       PRINT "CROSSWORD PUZZLE."
140    FOR TD = 1 TO 3000: NEXT TD:
       FOR CS = 1 TO 24: PRINT : NEXT
       CS
150    FOR X = 1 TO 16: READ
       CH$(X),VI(X),HI(X),IN$(X): NEXT X:K
       = 0
160    FOR N = 1 TO 16
170    VTAB 19: CALL - 958: VTAB 19:
       PRINT N"        "CH$(N)
180    VTAB VI(N): HTAB HI(N): INPUT
       " ";R$(N)
190    IF R$(N) < > IN$(N) THEN K = K +
       1: GOTO 180
200    IF R$(N) = IN$(N) THEN K = K + 1:
       NEXT N
210    VTAB 19: CALL - 958: IF K = 16
       THEN PRINT ,"GREAT!",
215    VTAB 19: CALL - 958: IF K > 16 OR
       K = 20 THEN PRINT ,"COULD BE
       BETTER!",
220    VTAB 19: CALL - 958: IF K > 20
       THEN PRINT ,"TRY AGAIN!",
230    PRINT : PRINT " YOU IDENTIFIED 16
       INVERTEBRATE": PRINT
       " ORGANISM IDEAS IN "K"
       ATTEMPTS!": END
240    DATA        ,2,3,LARVA
250    DATA        ,3,7,SEA
260    DATA        ,3,11,TRICHINA
270    DATA        ,4,7,COELENTERATES
280    DATA        ,5,3,HYDRA
290    DATA        FLOWER-LIKE COELEN-
       TERATE ,5,10,ANEMONE
300    DATA        ,6,5,PARASITE
310    DATA        JELLY_ _ _ _,7,6,FISH
320    DATA        ,7,11,ANNELIDA
330    DATA        ,8,4,BUDS
340    DATA        ,8,11,CORAL
350    DATA        ,9,9,FILTER
360    DATA        A PHYLUM OF ROUND-
       WORMS,10,10,NEMATODA
370    DATA        SOMETHING OFTEN USED
       FOR CLEANING,11,5,
       SPONGES
380    DATA        IN VERTICAL COLUMN-
       ,21,2,ASCARIS
390    DATA        VERTICAL COLUMN-
       ,22,2,TENTACLES
400    END
```

BITS OF INFORMATION

To understand computer graphics—the "painting" of pictures or images on the TV screen or monitor—it is helpful to keep the idea of a crossword puzzle in your mind. A computer screen is made up of vertical and horizontal picture elements, called pixels for short. A pixel is a point or very tiny box on the screen. To make a computer image, the program must indicate how each pixel will be colored. The more pixels that are filled, the more realistic the image appears to be. The fewer pixels that are filled, the more the elements of the image will have a squarish appearance.

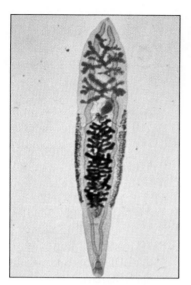

Fig. 9–10 You can see the internal organs of this flatworm.

9-3. Flatworms, Roundworms, and Segmented Worms

At the end of this section you will be able to:

☐ Compare the body plans of the worms and the less complex invertebrates.
☐ Describe the characteristics of the three phyla of worms.
☐ Explain how the three types of worms carry on certain life processes.

The flatworms, roundworms, and segmented worms show a number of differences from the invertebrates you have just studied. Sponge cells show some division of labor. Coelenterates show some division of labor and the grouping of cells into tissues. The other phyla you will study have many tissues, and they form organs and organ systems. Their body plans are also very different from those of the less complex phyla.

Most animals, including the worms, have a "head" and a "tail." They also have an upper side and a lower side. This body plan allows animals that move about a lot to have most of their sense organs in the head region. In this way, they can sense danger and try to avoid it. They can also move toward food. How is this body plan different from that of the sponges and coelenterates?

THE FLATWORMS

Flatworms get their name because of the flattened shape of their bodies. The flatworms are placed in the phylum *Platyhelminthes* (plat-eh-hel-**min**-thees). Most of them live in the bodies of other animals as parasites. Blood flukes, liver flukes, and tapeworms are types of flatworms. See Fig. 9–10. Each spends part of its life cycle in an animal host. Humans become infected by eating meat or fish that contains these parasites.

Planaria is a type of flatworm that is not a parasite. Members of this group are found in both fresh and salt water and on the land. The planarian has a pair of light-sensitive *eyespots* on its upper surface. See Fig. 9–11. The appearance of the eyespots give the worm a "cross-eyed" look. Beneath the eye-

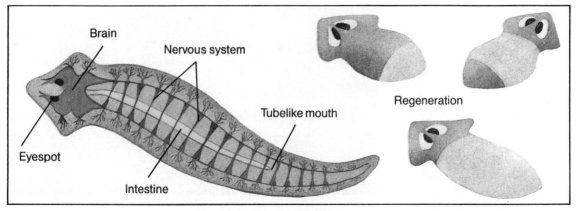

Brain

Nervous system

Eyespot

Intestine

Tubelike mouth

Regeneration

spots is a mass of nerves that forms a simple *"brain."* Throughout the body are nerves forming a *nervous system.*

The planarian has a *tubelike mouth* sticking out of its lower surface. It is the only body opening. The tube sucks pieces of food into the **intestine,** or the food tube. This is a special organ for digestion. Digested food passes out of the food tube and moves to other cells. Oxygen is taken in from the outside by diffusion and moves to each cell. There is no blood to carry food or oxygen. Solid wastes leave through the tubelike mouth. Other wastes leave the planarian's body through special pores. There are well-developed muscles that help the worm move.

Planarians regenerate easily. Fig. 9–11 shows what happens when one of these worms is cut. Planarians produce eggs and sperm to reproduce sexually.

THE ROUNDWORMS

Worms with smooth, round bodies are placed in the phylum *Nematoda* (nem-uh-**tode**-uh). They are commonly called *roundworms.* Roundworms are found everywhere. They are very common in soil and can be found in both fresh and salt water. Look at Fig. 9–12 on page 222. Many are parasites of plants and animals. Puppies, kittens, and young children are often infected with roundworms.

The roundworm's body has two openings. At one end is the mouth where food enters. Food is digested and absorbed as it passes through the *intestine.* Undigested wastes continue until they pass out through another opening called the **anus.** This plan is much more efficient than a food tube with only one opening. It is the beginning of a true *digestive system.*

Fig. 9–11 The planaria is able to regenerate parts of its body.

Intestine The hollow tube in which digested food is absorbed.

Anus The opening of the digestive system through which solid wastes are passed out of the body.

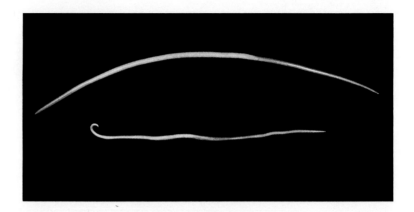

Fig. 9–12 Roundworms have two openings to their digestive systems.

THE SEGMENTED WORMS

Worms in the phylum *Annelida* (an-neh-**lid**-ah) have bodies that are divided into *segments*. These worms live in the oceans, fresh water, and in the soil. The earthworm is a common example of a segmented worm. An adult segmented worm has about 100 segments. This type of worm has several well-developed organ systems. Look at Fig. 9–13.

There is a simple brain located in the third segment. From it, a nerve cord runs down the underside of the worm. In each segment, there is a tiny nerve center that connects to the muscles. Signals from the nerves control muscle action. Earthworms do not have eyes or ears, but they can react to sounds

Fig. 9–13 Earthworms have several well-developed body systems.

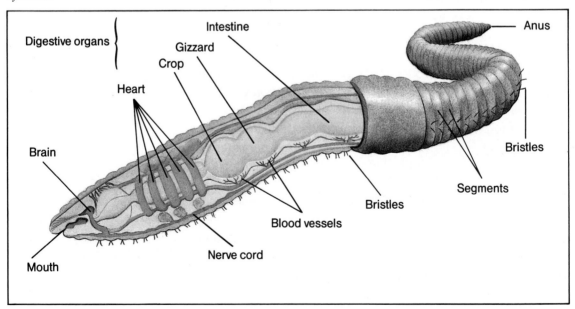

and to light. Earthworms can also sense vibrations in the earth around them.

Earthworms get food by burrowing through the ground and taking in soil. Their complete *digestive system* is made up of several organs. The body has two openings, a mouth and an anus. As soil passes through the worm, pieces of food are digested. Wastes and soil particles move out through the anus. The earthworm's wastes enrich the soil. In addition, air and water enter the soil as the earthworm burrows through it. This is very helpful to growing plants.

The thin, moist skin of the earthworm allows oxygen to enter the tissues. It also lets waste carbon dioxide leave. Thus the skin acts as a breathing system.

Bringing oxygen to every cell in the earthworm's body cannot be taken care of by this method only. The worm's body is too thick, and cells inside the body would not get enough oxygen. Most animals, including the earthworm, have a system of tubes filled with fluid called blood. These tubes act as a pump, or "heart," pushing the blood along by contracting and relaxing. Blood carries oxygen through this *circulatory system* to all parts of the body.

Fig. 9–14 This Bristle worm is a marine segmented worm.

SUMMARY

Most of the other animals besides the sponges and coelenterates have a body form with a head and tail and upper and lower surfaces. The flatworms have one body opening, muscles, and a nervous system. Roundworms and segmented worms have two body openings. Segmented worms have muscles, a nervous system, and a system for circulating blood.

QUESTIONS

Use complete sentences to write your answers.

1. In what way are the bodies of animals simple or complex?
2. What is a major difference in the body structure of flatworms and roundworms?
3. How does the flatworm bring food and oxygen to all of its body cells?
4. How do oxygen and digested food reach the cells on the inside of an earthworm's body?

SKILL-BUILDING ACTIVITY

OBSERVING AND RECORDING DATA

PURPOSE: To observe the behavior of a typical flatworm in response to stimuli.

MATERIALS:

planaria culture	acetic acid (5%
medicine dropper	solution)
petri dish (black)	food (chopped liver
hand lens	or hard-boiled
dissecting pin	egg)

PROCEDURE:

A. On a separate piece of paper, copy Table 9–1 below. Record the response of the flatworm to each of the stimuli tested in the activity.

Stimuli	Response
Vibration	
Touch	
Light	
Food	
Acid	

Table 9–1

B. Partly fill the petri dish with water. Transfer one or two flatworms to the dish.

C. On a separate piece of paper, sketch the flatworm. Label the head, tail, eyespots, and feeding tube.

1. Are the eyespots located on the upper or lower surface of the worm?

2. On which surface are the mouth and feeding tube located?

3. Is the feeding tube extended?

4. Are you looking at the upper or lower surface of the worm?

5. Describe what happens when the petri dish is tapped.

6. Describe what happens when the worm is touched lightly with a pin.

D. Place the cover on the petri dish. Turn the cover so that the flatworm is under the clear portion. Observe what happens.

7. How does the flatworm react to light?

E. Turn the cover so that the flatworm is again in the light. Repeat twice more.

8. Does the flatworm respond the same way?

9. How does the flatworm know when it is in the light?

F. Remove the cover. Place a small piece of food on the opposite side of the dish from the flatworm. Replace the cover so the food is beneath the clear area. Wait.

10. Describe what happens.

11. How did the flatworm detect the food?

12. Use the hand lens to observe how the flatworm feeds. How does it take in food?

13. Describe what happens when a drop of weak acetic acid is placed near the flatworm.

CONCLUSIONS:

1. What do these tests indicate about the nervous system of the planarian?

2. What do they indicate about the muscular system of the planarian?

CAREERS IN SCIENCE

ANIMAL ECOLOGIST

Animal ecology is the study of the effects of the environment on animals. It is a field of work that combines several areas of scientific knowledge. For instance, an animal ecologist may study the effect of a severe drought on an animal population. Another might study the special cycles some animals pass through. Lemmings, for instance, increase in number over a three- or four-year period. Then, their population "crashes" suddenly, and their numbers decline rapidly. This decline affects other animals, such as owls, that depend on lemmings for food.

Anyone planning a career as an animal ecologist will need an advanced degree. Course work combines the fields of zoology and environmental science. Jobs are available at universities, in government research programs, and with environmental research groups.

For further information, contact: American Institute of Biological Sciences, 1401 Wilson Boulevard, Arlington, VA 22209.

ANIMAL TRAINER

Animal trainers work in a variety of settings. Some train wild animals that appear in circus acts. Some conduct obedience classes for household pets, while still others train Seeing Eye dogs to guide people who are blind. Marine animal trainers train dolphins to do tricks in shows at parks, zoos, and aquariums. Trainers also work with animals that appear in movies or television commercials.

There is no formal method of becoming an animal trainer. An interest in animals and a great deal of patience is essential. Experience at an animal shelter, veterinary hospital, or animal grooming establishment is helpful. Courses in animal physiology and experimental psychology would be helpful for understanding the behavior, limitations, and special needs of those animals that can be trained.

For further information, write: National Congress of Animal Trainers and Breeders, Rte. 1, Box 32H, Grayslake, IL 60030.

VOCABULARY

On a separate piece of paper, match each term with the number of the statement that best explains it. Use each term only once.

regeneration   intestine   anus
vertebrates   invertebrates   sponge

1. The hollow tube inside an animal in which digested food is absorbed.
2. Animals that have backbones.
3. A simple invertebrate animal that has only one body opening.
4. The ability of some organisms to grow new parts.
5. Animals without backbones.
6. The opening of the digestive system from which solid wastes are excreted.

QUESTIONS

Give brief but complete answers to each of the following questions. Unless otherwise indicated, use complete sentences to write your answers.

1. What traits separate animals from other living things?
2. How are vertebrates different from invertebrates? Give two examples of each.
3. Arrange these types of animal phyla in order from the least complex to the most complex: segmented worms, coelenterates, roundworms, sponges, flatworms.
4. Look at the pictures of the invertebrate animals in Fig. 9–15. What is the name of each?

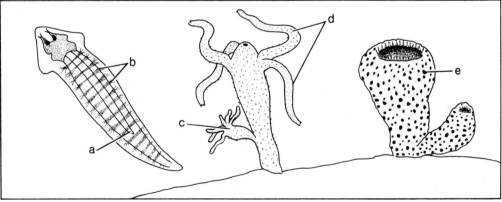

Fig. 9–15

5. Match each lettered structure in the picture to one of these terms: tentacles, bud, pore, mouth, nervous system.
6. Define each of these terms, then name one kind of animal in which it is found: pore, stinging cell, eyespot.
7. Do sponges or earthworms have organs and system?
8. Compare the nervous systems of the sponges and the planarians.
9. Compare the digestive systems of flatworms and roundworms.
10. Why do animals like the earthworm need a circulatory system?
11. What organ systems do earthworms have?

APPLYING SCIENCE

1. *Trichina* is a roundworm that is a parasite of pigs. It spends part of its life cycle in the muscle tissue of pigs. Why must pork be well-cooked when used as food for humans?
2. The Portuguese man-of-war is a large coelenterate. Why should swimmers leave the water when these floating jellyfish are seen nearby?
3. Why is a digestive system with an opening at either end more efficient that a system with only one opening?
4. After a heavy rain, earthworms are often found on the ground surface in large numbers. Explain why.
5. Scientists and doctors are interested in the way "simple" invertebrates can grow new parts. Why might they be interested?

BIBLIOGRAPHY

Elson, Lawrence M. *The Zoology Coloring Book.* New York: Barnes and Noble, 1982.

Gerthe, Henry. "A Splendor of Sponges." *Discover,* September 1983.

Jackobson, Morris. *Wonder of Sponges.* New York: Dodd, Mead, 1976.

Levine, J. "Deep Sea Corals." *Science Digest,* July 1983.

"Living Lights," *National Geographic World,* March 1981.

Shostak, Stanley. *The Hydra.* Coward-McCann & Geoghegan, 1977.

A Flamingo Tongue snail moves slowly over branches of coral in the sea around the West Indies. The snail is a member of Phylum Mollusca.

COMPLEX INVERTEBRATES

CHAPTER GOALS

1. List major characteristics of complex invertebrates.
2. Compare and contrast how mollusks, echinoderms, and arthropods carry out the life processes.
3. Trace the development of body systems through the invertebrate phyla.
4. Compare the advantages and disadvantages of having an exoskeleton.
5. Identify reasons for the success of the insects.

10-1. Mollusks

At the end of this section you will be able to:

- ☐ List some common traits of *mollusks*.
- ☐ Describe each of the three classes of *mollusks*.
- ☐ Compare the methods of feeding and movement found in *mollusks*.

The shellfish of Phylum *Mollusca* (moe-**lus**-kuh) have long been a food source for many animals. Humans will eat almost any type of mollusk: scallops, clams, snails, oysters, octopus, or squid. Which of these do you eat?

WHAT IS A MOLLUSK?

Mollusks (**mol**-usks) are invertebrate animals. Most live in the sea although some live in fresh water or on land. Most mollusks are easily recognized. They have a hard shell that surrounds and protects their soft bodies. See Fig. 10–1 on page 230. This shell is formed by a layer of tissue called the **mantle.** The *mantle* takes chemicals from the water and uses them to form the three layers of the shell.

Mollusks have some of the same features as the segmented worms. They all have a muscular digestive system with two openings and well-developed nervous and circulatory systems. Mollusks have a respiratory organ called a **gill** for obtaining oxygen. The *gill* may also be involved in food-getting.

Mollusk Invertebrate animal, usually having a hard shell that surrounds and protects the body.

Mantle The outer layer of tissue that produces the shell on a mollusk.

Gill Structure in some water-dwelling animals that absorbs oxygen from the water.

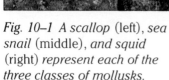

Fig. 10–1 A scallop (left), *sea snail* (middle), *and squid* (right) *represent each of the three classes of mollusks.*

The mollusk phylum is divided into three classes. This classification is partly based on the thick, muscular "foot" that is used for movement. The structure of their shells is also used in classifying mollusks. See Fig. 10–1. The three groups of mollusks we will discuss are:

1. *Hatchet-footed,* two-shelled mollusks. Clams, oysters, mussels, and scallops are members of this group.

2. *Stomach-footed,* one-shelled mollusks. Land and water snails, slugs, and conches belong to this group.

3. *Head-footed* mollusks. Most members of this group have an internal shell. Octopuses and squid are examples of head-footed mollusks.

HATCHET-FOOTED MOLLUSKS

Hatchet-footed mollusks have two-part shells. The name of the group comes from the shape of the muscular foot. When the foot is fully extended, it looks like the blade of a hatchet. See Fig. 10–2. Each part of the shell is called a valve, so this group is called the *bivalves* (*bi* means two). Clams, scallops, and mussels are bivalves. The two valves are hinged and held together by a muscle. The animals can open the shell to allow the muscular foot to be stretched outward. The foot is then pushed into the sand to form an anchor. The muscle contracts, pulling the mollusk along.

In order to feed, bivalves push a special tube out between the valves. Water is drawn into the mollusk through this tube. Next, the water flows over the gill. Any food particles in the

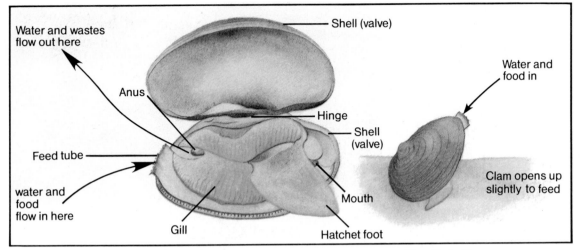

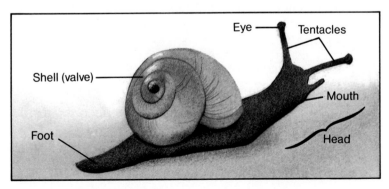

Fig. 10–2 The structures of a clam, a bivalve. The shells are open, and some of the gills have been removed.

water are trapped in a sticky substance that covers the gill. This food is then moved into the mouth. Food wastes from the anus are released into the water.

STOMACH-FOOTED MOLLUSKS

Stomach-footed mollusks have a one-part shell. For this reason, these mollusks are called *univalves* (*uni* means one). Pond snails are an example of this group. Have you ever seen a snail moving along the wall of an aquarium? The foot gives off a layer of slime over which the snail slides. The muscular foot contracts in wavelike ripples to move the animal along. These animals seem to crawl on their stomachs; thus the name of the group.

Univalves have a well-defined head area. See Fig. 10–3. The head has tentacles, eyes, and a mouth. Some univalves "graze" on algae or plants. Others can bore into the shells of bivalves, such as clams, and eat the soft interior.

Fig. 10–3 The snail is a stomach-footed mollusk. It has a single shell.

Fig. 10–4 The scientific name for a head-footed mollusk, such as this octopus, is cephalopod.

HEAD-FOOTED MOLLUSKS

The "foot" of these mollusks is divided into eight or more tentacles. The tentacles extend from the head and surround the mouth. They can be used for moving, thus the name head-footed. The tentacles also have suckers for grasping and holding food. See Fig. 10–4.

The squid in Fig. 10–1 has a shell inside its body. The octopus doesn't have a shell at all. Another member of the group, the chambered nautilus, looks like an octopus stuffed into a snail's shell. Squids and octopuses can move quickly by squirting a jet of water from their bodies. To escape from danger, these animals can release a cloudy fluid that looks like ink. This helps to hide them from pursuing predators. Some can also change color to hide from their enemies.

SUMMARY

Mollusks have a hard, protective shell covering their soft bodies. The structure of this shell and the way the mollusk moves help classify them into three groups.

QUESTIONS

Use complete sentences to write your answers.

1. Describe two different ways of classifying mollusks.
2. What are three groups of mollusks? To which groups do clams, squids, and garden snails belong?
3. How does a mollusk obtain oxygen?
4. How do bivalves obtain food? How do univalves feed?

INVESTIGATION

OBSERVING SNAILS

PURPOSE: To observe the behavior of a typical univalve mollusk.

MATERIALS:

glass jar with lid	aquarium water
pond snails	magnifier
empty snail shells	marking pen or pencil

PROCEDURE:

A. Fill the glass jar two-thirds full of aquarium water. Add two or three snails. Place the cap on the jar.

B. Place the jar on its side so that the snails come to rest on the sides of the jar. Prop the jar on both sides with books so the jar will not roll. See Fig. 10–5. Wait until the snails stretch out of their shells and begin to move around.

C. Use the magnifier to examine and sketch one snail. Label the following parts: shell, head, tentacles, "foot."

   1. From their position and motion, what do you judge the job of the snail's tentacles to be?

D. Use the marker to mark a "start" line on the jar. Trace the snail's path for 5 minutes. Measure the distance it crawls.

   2. How far did it crawl in 5 minutes?

   3. What is your "snail's pace" per hour? (Hint: Multiply your answer by 12.)

E. Observe the snail's motion from beneath. Look at its mouthparts.

   4. Does the snail appear to be eating?

   5. Describe the way the foot moves.

F. Remove the snail from the container.

   6. How does the snail react when removed from the jar?

   7. How would you describe the pattern of the shell?

   8. If the shell were "unrolled," what shape would it be?

G. Feel the trail the snail followed.

   9. Describe the feel of any material left by the snail.

H. Compare the shell of a small snail with that of a larger snail.

   10. What differences do you find in the pattern of the shells?

   11. How does a snail shell grow?

I. Examine an empty snail shell.

   12. Compare the inner and outer surfaces of the shell.

   13. Why is the inner surface smooth?

CONCLUSIONS:

   1. Why are snails called univalves?

   2. Why are snails called stomach-footed?

   3. Why do people add snails to an aquarium?

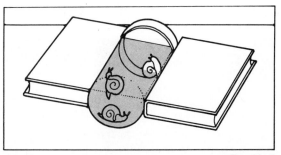

Fig. 10–5

10-2. Echinoderms

At the end of this section you will be able to:

- ▣ List some traits of *echinoderms*.
- ▣ Explain how starfish move and eat.
- ▣ Describe the body systems of *echinoderms*.

The sea urchin is an animal that looks like a pincushion. See Fig. 10–6. This is an animal called a *sea urchin*. Its shape comes from a hard, almost ball-shaped *internal skeleton*. Long, needlelike *spines* stick out all over the sides and top of the sea urchin.

ANIMALS WITH SPINY SKINS

Echinoderm Spiny-skinned invertebrate.

The spiny-skinned animals belong to Phylum *Echinodermata*. The spines of an **echinoderm** (eh-**kie**-no-derm) are part of its internal skeleton. This skeleton is made of plates that fit together so the animal can bend and move. It is covered by a rough, skinlike covering.

This phylum includes about 6,000 different types of starfish, sea urchins, sand dollars, sea cucumbers, and sea lilies. All these animals live in the ocean.

A sand dollar is like a flattened sea urchin with very short spines. Its surface feels almost like stiff velvet. A star-shaped pattern is clearly visible on the sand dollar in Fig. 10–6. In all adult *echinoderms,* the body is divided into five parts around a central disc. This pattern is called *radial symmetry.*

Some types of echinoderms look quite different. *Sea cucumbers* have been described as "large pickles" lying on their sides. See Fig. 10–7. These animals have a mouth at one end surrounded by tentacles. Their bodies are soft and squishy. Many of them feed on debris on the ocean floor. Some *sea*

Fig. 10–6 A sea urchin (left), *sand dollar* (middle), *and sea star* (right) *are echinoderms.*

  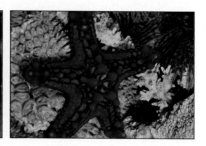

Fig. 10–7 Sea cucumbers (left) *and sea lilies* (right) *are also echinoderms.*

lilies look more like plants than animals. They are attached to the ocean bottom by a large stalk. The body on top of the stalk has long, feathery "arms" that trap food particles.

THE FISH THAT IS NOT

Starfish are among the best known echinoderms. Most starfish have a central disc with five arms, or rays, spreading out like spokes on a wheel. Some types, however, have as many as twenty arms. The undersides of these arms are covered with hollow tubes called **tube feet.** These tubes are connected to an internal network of canals that runs through the starfish. The movement of water through this network allows the *tube feet* to act like tiny suction cups. The tube feet are used for movement and food getting. The starfish can wrap its arms around a clam, using its suction feet to hold on. The tube feet pull the clam open.

Then, a strange thing happens. The starfish's stomach stretches out of its body through its mouth and enters the clam. It actually digests the soft body of the clam while it is inside the clamshell. The digested material is then sucked into the starfish.

A starfish can eat up to a dozen clams a day. For this reason, the crew on fishing boats try to kill any starfish they find. They used to chop up the starfish and throw the pieces back into the ocean. Unfortunately, any pieces that had part of the central disc grew into new starfish. This is how it was discovered that starfish can regenerate. Starfish have to be dried to be killed. You may have seen dried starfish skeletons for sale.

Tube feet Structures on the bottom surface of echinoderms, used for grasping.

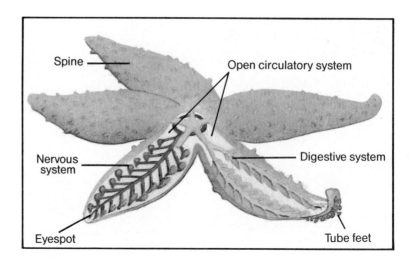

Fig. 10–8 The structure of a starfish.

Labels in figure: Spine, Open circulatory system, Nervous system, Digestive system, Eyespot, Tube feet

BODY SYSTEMS

Echinoderms, such as the starfish, have a well-developed and specialized digestive system. But they do not have specialized circulatory, respiratory, or excretory systems. Circulation of material is done by the fluid that fills the space between the body wall and the digestive system. See Fig. 10–8. This is called an *open circulatory system.* Oxygen and carbon dioxide are exchanged by diffusion through the upper body surface. Wastes are also excreted this way.

Starfish are either males or females. Fertilization is external. Sperm and eggs are released into the water where they may join. The starfish has a very simple nervous system. There is no brain. A ring of nerves is found around the mouth. A nerve cord runs the length of each arm. At the end of each arm is a light-sensitive *eyespot.*

SUMMARY

Echinoderms are spiny-skinned animals that live in the ocean. They may have a network of tube feet, used for movement and feeding. Many of their body systems are quite simple.

QUESTIONS

Use complete sentences to write your answers.

1. How are the soft parts of echinoderms protected?
2. How does a starfish obtain food?
3. Which echinoderm body system is the most complex?

10-3. Arthropods

At the end of this section you will be able to:

- Describe the characteristics of *arthropods*.
- Discuss the advantages and disadvantages of an *exoskeleton*.
- Explain how *arthropods* are classified.

The suit of armor in Fig. 10–9 was made for a boy of about twelve. The metal covered him from head to toe. The suit had joints so that he could bend his arms and legs and move. Unfortunately, he soon outgrew it. The animals in Phylum *Arthropoda* (ahr-throe-**pode**-ah) also have "suits of armor." They, too, soon outgrow them.

Fig. 10–9 A suit of armor is similar to an exoskeleton.

CHARACTERISTICS

Phylum Arthropoda contains over one million species of animals. This is by far the largest phylum of animals. **Arthropods** (**ahr**-throe-podz) are found in every part of the biosphere. They live in oceans, in fresh water, and on land. Arthropods are also the only invertebrate animals that can fly.

Arthropods share many common characteristics. Their bodies are made of several segments. They have several pairs of jointed legs. All arthropods have a hard external covering

Arthropod An invertebrate animal that has jointed legs and an exoskeleton.

Fig. 10–10 In order to grow, a crab—like all arthropods— must molt, or cast off its exoskeleton.

Exoskeleton The hard outer covering that is typical of arthropods.

Chitin The semihard compound that forms an exoskeleton.

called an **exoskeleton** (eks-oh-**skel**-eh-tun). This protective covering is composed of a semihard material called **chitin** (**kite**-ihn). The material that makes up your fingernails is similar to *chitin*. The *exoskeleton* acts like a suit of armor to protect the soft body organs. In addition, the exoskeleton prevents the animal from losing too much water and drying out.

There are, however, disadvantages to having an exoskeleton. Once it is formed, the exoskeleton cannot grow larger. In order for an arthropod to grow, it must shed its exoskeleton and grow a new one. See Fig. 10–10 on page 237. During this time, an arthropod may be easily killed by its enemies. Arthropods cannot grow very large. The weight of a large exoskeleton would be too heavy for the animal to move.

CLASSES OF ARTHROPODS

Just as there are many shared traits among arthropods, there are also many differences. These differences are used to separate arthropods into five different classes: (1) *Crustaceans*

THE ARTHROPODS				
Class	Body Divisions	Appendages	Respiration	Examples
Crustaceans	2 body sections	5 pairs of legs	gills	lobster, crab, crayfish
Arachnids	2 body sections	4 pairs of legs	breathing tubes	spiders, mites, ticks
Insects	3 body sections	3 pairs of legs	breathing tubes	grasshoppers, ants, bees, butterflies, moths
Centipedes	head and many body sections	1 pair of legs on most segments	breathing tubes	centipedes
Millipedes	head and many body sections	2 pairs of legs on each segment	breathing tubes	millipedes

Table 10–1

(krus-**tay**-shunz); (2) *Arachnids* (ah-**rak**-nidz); (3) *Insects;* (4) *Centipedes* (**sen**-tih-peedz): the name means "hundred legs"; and (5) *Millipedes* (**mil**-ih-peedz): the name means "thousand legs."

Counting the legs is the easiest way to tell to which class the arthropod belongs. But there are also many other differences between the classes. See Table 10–1.

One difference among arthropods is the number of visible body sections. Crustaceans, such as the lobster, and arachnids, such as the spider, have two visible body sections. Insects, such as ants, have three. Centipedes and millipedes have many body sections.

Besides legs, most arthropods have other structures attached to their bodies. Most of these additional structures are jointed and also able to bend and move. All arthropods have jointed mouthparts for feeding. Centipedes, millipedes, crustaceans, and insects may have *antennae.* Spiders have poison fangs, and they also have other structures for sensing. Most ants have one or two pairs of wings. Insects will be discussed further in Section 10–4.

Arthropods also differ in how they obtain oxygen. Most crustaceans live in water. They have gills beneath their exoskeleton. The gills absorb oxygen from water as it flows over them. Carbon dioxide is released into the water. The other classes are mostly land dwellers. They have a network of small tubes that run throughout the body. See Fig. 10–11. These tubes make up the respiratory system. Oxygen diffuses into the tissues while carbon dioxide diffuses out.

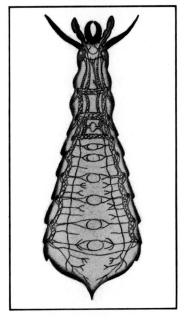

Fig. 10–11 Land-dwelling arthropods have a well-developed network of tubes for exchange of gases.

SUMMARY

All arthropods have exoskeletons and similar body parts. They are classified by these body parts. The exoskeleton of arthropods is protective, yet it also has disadvantages.

QUESTIONS

Use complete sentences to write your answers.

1. What three characteristics are common to all arthropods?
2. What characteristic is used to classify arthropods?
3. What are two disadvantages of an exoskeleton?

SKILL-BUILDING ACTIVITY

CLASSIFYING SOME INVERTEBRATE ANIMALS

PURPOSE: To identify examples of invertebrates using a classification key.

MATERIALS:
none

PROCEDURE:

A. Field biologists often use classification keys to find the name of an organism they have collected. Go through the key in this procedure one organism at a time. Start with question 1 and follow the instructions. Each question can be answered by either yes or no. Your answer will lead you to the next question. At the end of a series of questions, you will have the name of the organism and its phylum. Keep a list of the animals and their phyla as you identify them.

1. Does the animal have an exoskeleton with jointed legs? If yes, it is an arthropod. Go to 1a. If no, go to 2.
   a. Does the animal have many body segments with a pair of legs on each? If yes, it is a centipede. If no, go to 1b.
   b. Does the animal have at least five pairs of legs? If yes, it is a crustacean. This is a blue crab.

2. Does the animal have spines covering its skin? If yes, it is an echinoderm, the sea urchin. If no, go to 3.

3. Does the animal have a hard shell protecting a soft body? If yes, it is a mollusk, a snail. If no, go to 4.

4. Does the animal have a flat, ribbon-like body? If yes, it is a flatworm, Phylum Platyhelminthes. This is a tapeworm. If no, go to 5.

5. Is this animal wormlike with a smooth, tapered body? If yes, it is a roundworm, Phylum Nematoda. If no, go to 6.

6. Does the animal have a wormlike body with many sections? If yes, it is a segmented worm, Phylum Annelida. This is a bristleworm. If no, go to 7.

7. Does the animal have tentacles around its mouth opening? If yes, it belongs to Phylum Coelenterata. This is a jellyfish. If no, go to 8.

8. Does the animal have pores all over its body? If yes, it belongs to Phylum Porifera. This is a sponge.

CONCLUSIONS:

1. For each of the animals in Fig. 10–12, list the name, phylum, and identifying characteristics.

2. What characteristics are the basis for the classification of invertebrates?

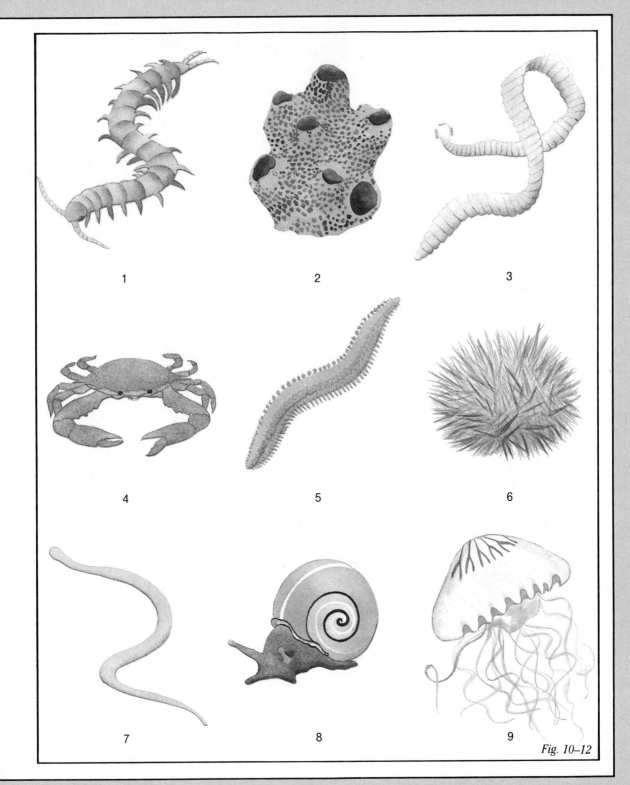

1

2

3

4

5

6

7

8

9

Fig. 10–12

SCIENCE INPUT

Since the number of insect species is greater than all other kinds of animals put together, some species of insects are present almost everywhere—and, as you learned, most insects go through metamorphosis. Metamorphosis is a series of changes in the structure of an animal as it grows. We are probably most familiar with the metamorphoses of tadpoles into frogs and caterpillars into butterflies. The changes the organism undergoes can be a very essential part of its ability to adapt to a changing environment. Understanding these adaptations is an important part of any research on insects.

COMPUTER INPUT

People sometimes make analogies (comparisons) between life processes and computer processes. We can do this because both are systems—that is, each is an organization of parts that work together and has a beginning and end. Just as there is a code, or pattern, to the development of every living thing, there must be a code, or pattern, to every computer system so that everything necessary to its existence is included. A computer flow chart, designed by the programmer before writing the program, creates the pattern and flow of logic to the computer program. Through the use of a flow chart, we can follow the purposes and steps of the program. It is particularly useful when writing a long program. With a flow chart, it is easier to keep the overall pattern in mind even as you write the many steps and subroutines in between.

WHAT TO DO

Copy the blank flow chart onto a separate piece of paper. The meaning of the flow chart symbols is given in the glossary. You are going to create a flow chart for insect metamorphosis. The entries for the flow chart are listed out of order. Rearrange the items and fill in the flow chart with the appropriate steps. Several of the boxes are filled in as an example.

FLOW CHART

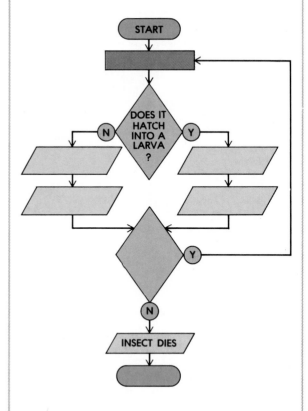

FLOW CHART ENTRIES

STOP
DOES IT HATCH INTO A LARVA?
INSECT DIES
INSECT LAYS EGGS
EGG HATCHES INTO NYMPH
PUPA BECOMES ADULT INSECT
DOES ADULT REPRODUCE?
LARVA ENTERS PUPA STAGE
NYMPH CHANGES INTO ADULT
START

GLOSSARY

 start or stop.

 shows the direction the program will follow.

 a single step in the program.

 input or output.

 a decision point: A question must be answered or two things are compared—"Y" means yes and "N" means no.

PROGRAM NOTES

While the flow chart and its symbols are associated with computer programming, the concept is one with which you might be more familiar than you think. Do you ever play board games? You start at "Go" and then progress according to a plan designed into the game. If you land on a certain square, you may have to GOTO another location. If you have fulfilled certain conditions, THEN you may continue and even repeat your turn. A challenging way to understand more about flow charts is to use its ideas and symbols to create your own adventure game.

BITS OF INFORMATION

Computers are now so much a part of our daily lives that we almost take them for granted. Some supermarket cashiers work computerized checkouts, many people do their banking through computerized banking machines or even with their personal computers, and your schedule of classes was probably drawn up by a computer. The change has been comparatively sudden, however. ENIAC, built in 1946, was 2 stories high and weighed 30 tons. It was considered a great marvel but certainly not something possible for everyone or even for their local schools. Today, one of the innovations in personal computers is the notebook computer, which may weigh as little as 4 kilos and can fit on your lap. Not everyone has a computer in her or his home, but they are increasingly available in many public places, such as your library and school.

10-4. The Insects

At the end of this section you will be able to:

- Explain why insects outnumber all other animal groups.
- Describe the structure of a typical insect.
- Contrast *complete* and *incomplete metamorphosis*.

Fig. 10–13 The social organization of a bee hive centers around the queen, at the center of the photo.

Life in a beehive is a very complex but very orderly existence. One of the bees is called the queen. See Fig. 10–13. Her only job is to produce new bees. She may lay as many as one million eggs a year. Most of the bees in a hive are female workers. They do all the work except reproduction. Some workers gather food. Others protect the hive. Some build the honeycomb. Still others feed and care for the queen and the *larvae*. What other animals can you name that live in complex societies?

IT'S A BUGGY WORLD

The number of insect species is greater than all other kinds of animals put together. So far, more than 675,000 species have been identified. This large number of insect species has been grouped into 20 to 25 orders. Representatives from a few of the more common orders are shown in Fig. 10–14.

Fig. 10–14 Representatives of some of the most common insect orders.

There are a number of reasons why insects have been so successful:

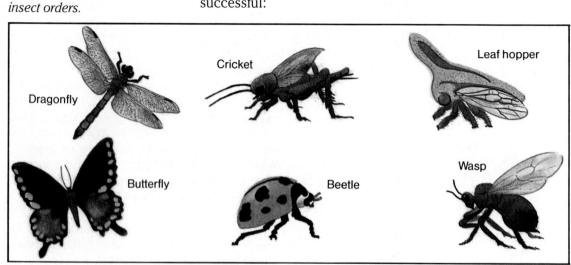

Dragonfly

Cricket

Leaf hopper

Butterfly

Beetle

Wasp

1. One reason is their small size. They can occupy parts of the environment where they have little competition from larger animals.

2. Protective coloring and body shapes help many insects hide from their enemies.

3. Insects eat a great variety of foods. Thus, they are fairly certain always to find food.

4. Many types of insects, such as ants, bees, and termites, live in groups. This social manner of living helps ensure the safety and survival of each member.

5. Probably the most important reason for the success of insects is their high rate of reproduction. One female insect may lay thousands of eggs at a time. Not all the eggs will become adults. However, these large numbers make the chances for survival of the species very good.

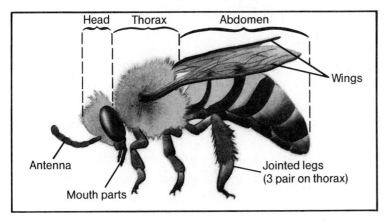

Fig. 10–15 The structures of a typical insect.

CHARACTERISTICS

Insects differ from other arthropods in several ways. One way is that only insects have three distinct body sections. They are called the *head, thorax,* and *abdomen.* See Fig. 10–15. Your body also has these sections, but the thorax and abdomen are joined to form the trunk of your body.

Only an adult insect has three pairs of jointed legs attached to its thorax. If the insect has wings, they are also attached here. The abdomen may have up to 11 segments. There are no legs attached to the abdomen.

The head holds one pair of antennae and the mouthparts. The mouthparts are also jointed. They are shaped for the type of food the insect eats. Grasshoppers, crickets, cockroaches,

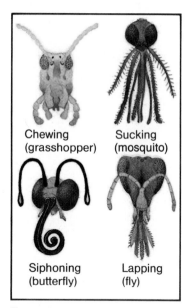

Fig. 10–16 Variation among mouth structures allows insects to obtain food from many different sources.

Metamorphosis A series of major changes in the structure of an animal as it grows.

and termites have mouthparts designed for chewing. Many lice and all of the insects called true bugs have sucking mouthparts. Butterflies and moths feed on plant nectar. Their mouthparts are shaped like tubes for drawing up the nectar. Flies have lapping mouthparts. See Fig. 10–16.

Insects have the most complex body systems of all arthropods. Their nervous systems include specialized sense organs. They can see, taste, hear, smell, and feel. The digestive system has organs for breaking down and absorbing food. The circulatory system is open. There is no system of blood vessels as there is in more complex animals. Instead, the entire body cavity is filled with a bloodlike fluid. The movements of the heart and other body muscles move this fluid around. The blood does not carry oxygen to the cells. As in all arthropods, a system of tubes runs throughout the body. These tubes carry oxygen to the cells and remove carbon dioxide.

METAMORPHOSIS

Most insects go through a series of changes as they grow from the egg to the adult. These changes are called **metamorphosis** (meh-tah-**morf**-oh-sis). There are two types of *metamorphosis. Incomplete metamorphosis* consists of three stages: an egg, a *nymph* **(nimf),** and an adult. See Fig. 10–17. The eggs are formed by sexual reproduction. The nymph hatches from the egg. The nymph may look very much like the adult except that it has no wings. As the nymph grows, it must

Fig. 10–17 The stinkbug is an insect that goes through incomplete metamorphosis.

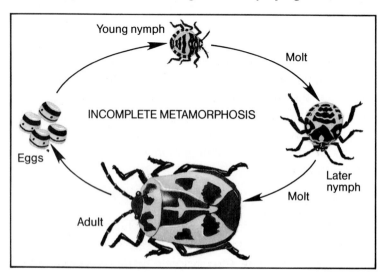

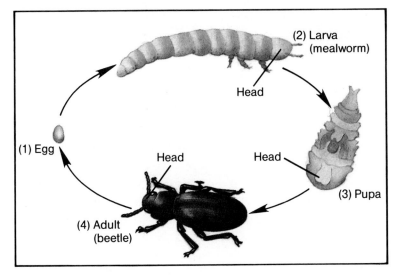

(1) Egg

(2) Larva
(mealworm)

Head

Head

Head

(3) Pupa

(4) Adult
(beetle)

Fig. 10–18 Complete meta-morphosis has four stages. This is the life cycle of a mealworm.

shed its exoskeleton several times before becoming a full-sized adult. Grasshoppers, cockroaches, and earwigs are insects that go through incomplete metamorphosis.

Complete metamorphosis consists of four stages. See Fig. 10–18. From the eggs hatch wormlike *larvae* (**lahr**-vee). The larvae of butterflies and moths are called caterpillars. Fly larvae are called maggots. Mealworms are the larvae of grain beetles. The larvae eat almost constantly and grow quickly. After a period of time, they enter a stage called the *pupa* (**pew**-pah). A hard, protective case forms. Inside the case, the larva changes into its final adult form. Finally, after a period of time, the fully developed adult insect emerges from the case.

SUMMARY

Insects are the most varied and numerous group of living things. Their size, structure, reproductive patterns, and behavior have all contributed to their success at survival. Insects go through either of two types of metamorphosis as they develop from egg to adult.

QUESTIONS

Use complete sentences to write your answers.

1. List five possible reasons for the success of insects.
2. What body structures do all insects have?
3. How do incomplete and complete metamorphosis differ?

INVESTIGATION

OBSERVING MEALWORMS

PURPOSE: To observe the structure and reproductive pattern of a typical insect.

MATERIALS:

magnifier	glass plate
mealworm culture	tweezers
mirror	

PROCEDURE:

A. Obtain one mealworm from the culture.

B. Use the magnifier to examine the mealworm carefully. Sketch what you see. Label the following parts: *head, segments, antennae, legs, exoskeleton, mouthparts.*

   1. Describe the exoskeleton.

   2. How many segments are there? Is this true of other students' mealworms?

   3. How many pairs of legs are there? How many pairs of antennae?

   4. Which segments hold the legs? Is this typical of other mealworms?

C. Observe the mealworm as it moves, both from the side and from beneath. To view its underside, place the mirror flat on the desk. Place the mealworm on the glass plate. Hold the glass plate over the mirror. By looking in the mirror, you can observe the movement. See Fig. 10–19.

   5. Describe the motion of the legs.

   6. Does the mealworm move in a straight line?

   7. What happens when the mealworm meets an obstacle?

D. Examine the mealworm culture. Try to locate the following: eggs, old exoskele-

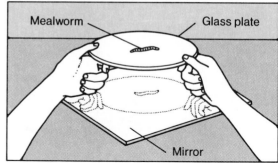

Fig. 10–19

ton, pupa, adult beetle. Sketch those you find.

   8. Does the mealworm undergo complete or incomplete metamorphosis?

E. Design other experiments to test the behavior of the mealworm. Some suggestions are: Do mealworms prefer certain colors? Do they prefer moist or dry areas? Do they prefer light or dark? Obtain your teacher's approval of your plans before you do any testing.

F. Write up each experiment as follows: Purpose: This may be stated as a question—Do mealworms prefer light or dark? Materials: List everything needed to test your idea. Procedure: Do your testing step by step so someone else could repeat what you did. Conclusions: What were your results? Did you answer your question?

CONCLUSIONS:

   1. Why is the name "mealworm" not very suitable for this animal?

   2. Why are mealworms classified as arthropods?

   3. Why are mealworms classified as insects?

CAREERS IN SCIENCE

ENTOMOLOGIST

More than 75 percent of all the animal species alive today are insects, and entomologists are the biologists who study them. Entomologists describe, identify, and classify insects. They examine the physiological and biochemical reactions occurring within insect cells or molecules. Such research has led to useful discoveries in a number of areas, including human medicine.

Entomologists work as teachers and as researchers in colleges, universities, and industry. State and local government agencies hire entomologists to do research and to conduct biological surveys.

Most employers require a minimum of a bachelor's degree in biology or entomology for entry-level jobs. More responsible, better-paying jobs usually require an M.A. or Ph.D.

For further information, contact: Teen International Entomology Group, Department of Entomology, Michigan State University, East Lansing, MI 48824.

MOLLUSK FISHER

The Phylum Mollusca includes many edible and commercially valuable creatures, from oysters to squid. The most familiar mollusks are the bivalves, among which are clams, abalones, and oysters. These stationary creatures can be hauled in from small craft by hand, using long-handled rakes or dredging nets, or from large boats by mechanical means such as hydraulic dredges. Mussels can literally be picked by hand during low tide, but most commercial harvesters use nets to haul them in. Bay and sea scallops, which are mobile, and so more difficult to catch, are usually brought in by large boats called draggers.

Mollusk fishing is difficult, tiring work requiring great skill and ability. The people who do it work in extremes of temperature and weather.

For further information, contact: Atlantic States Marine Fisheries Commission, Suite 703, 1717 Massachusetts Avenue NW, Washington, DC 20036.

CHAPTER REVIEW

VOCABULARY

On a separate piece of paper, match each term with the number of the phrase that best explains it. Use each term only once.

mantle	larva	pupa	chitin
thorax	metamorphosis	tube feet	nymph
exoskeleton	gills		

1. The middle section of an insect's body.
2. Structures on echinoderms used for grasping and moving.
3. A series of major changes in the structure of an insect as it grows.
4. The outer layer of tissue on a mollusk that produces the shell.
5. A wormlike stage of development in the life history of some insects.
6. Structures in some water-dwelling animals that absorb oxygen from the water.
7. The semihard compound that forms the exoskeleton.
8. An immature stage of development in incomplete metamorphosis.
9. The hard outer covering typical of all arthropods.
10. The hard, protective case surrounding the insect during a stage of complete metamorphosis.

QUESTIONS

Give brief but complete answers to each of the following questions. Unless otherwise indicated, use complete sentences to write your answers.

1. Describe the basis of classification of the three types of mollusks.
2. What are two ways sea urchins protect their soft body parts?
3. To which group of arthropods do each of the following belong? lobster, tarantula, cockroach.
4. What are several reasons for the success of insects?
5. How does each of the following help make insects successful survivors?
   (a) protective coloration   (b) food sources   (c) social life style
   (d) rate of reproduction.
6. Define each of these terms, then state of which phylum it is characteristic: jointed legs, spiny skin, bivalve, tube feet.
7. How does the nervous system of an insect differ from that of a starfish?
8. How does each of the following animals obtain oxygen and remove carbon dioxide? clam, starfish, grasshopper.

9. What are some advantages and disadvantages of having an exoskeleton?
10. How are complete metamorphosis and incomplete metamorphosis alike? How are they different?
11. A starfish has an "open circulatory system." What is meant by this term, and how does it work?
12. Insects are the most complex animals discussed in this chapter. Why are they considered "complex"?

APPLYING SCIENCE

1. If a piece of sand gets into an oyster, the oyster secretes shell material around it to make a natural pearl. How do you think cultured pearls are made?

2. Brittle stars get their names because their legs break off easily, then regenerate. What advantage is this to the animal?

3. Some insects that lived with dinosaurs had wingspans up to 100 cm. Why aren't insects this big now?

4. Collect several insects from your neighborhood. Describe how their color and shape either help or hinder their ability to hide.

5. Collect a crayfish from a pond or stream. Set up a temporary aquarium, using water and other materials present in the crayfish's natural environment. Describe its physical traits and behavior. Determine the class of Phylum Arthropoda in which it belongs.

BIBLIOGRAPHY

Boraiko, Allen. "The Pesticide Dilemma." *National Geographic,* February 1980.

Hamner, William. "Krill, Untapped Bounty from the Sea." *National Geographic,* May 1984.

Jacobson, Morris, and William Emerson. *Wonders of the World of Shells: Sea, Land and Fresh Water.* New York: Dodd, Mead, 1971.

Jenkins, Marie. *The Curious Mollusks.* New York: Holiday House, 1972.

Hutchins, Ross. *Insects and Their Young.* New York: Dodd, Mead, 1975.

Mitchell, Robert, and Herbert S. Zim. *Butterflies and Moths.* New York: Golden Press, 1964.

Zim, Herbert S., and Clarence A. Cottam. *Insects.* New York: Golden Press, 1951.

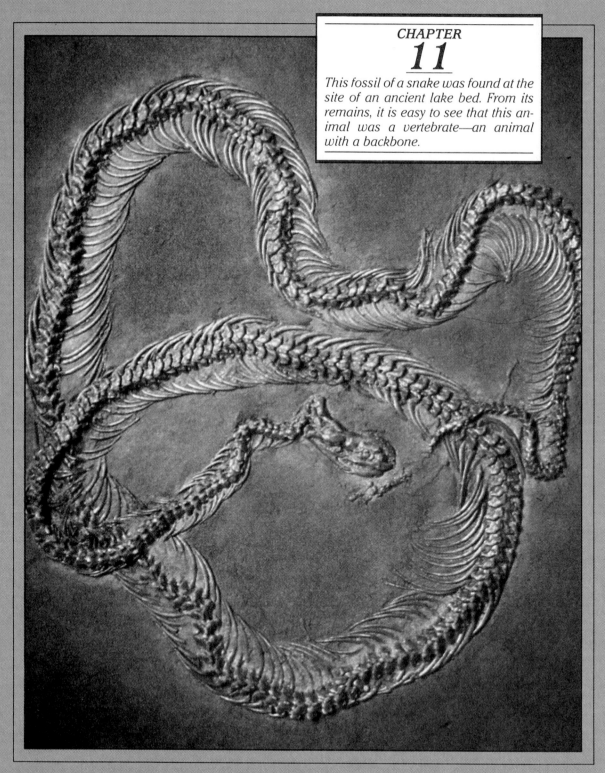

This fossil of a snake was found at the site of an ancient lake bed. From its remains, it is easy to see that this animal was a vertebrate—an animal with a backbone.

COLDBLOODED VERTEBRATES

CHAPTER GOALS

1. Identify the structures that all vertebrates have.
2. Name the classes of vertebrates and give examples of each.
3. Contrast body systems of coldblooded vertebrates.
4. Describe the life cycles of fishes, amphibians, and reptiles.

11-1. The Fish

At the end of this section you will be able to:

- List the characteristics of *chordates*.
- List the characteristics common to all *vertebrates*.
- Compare and contrast the three classes of fishes.
- Discuss the disadvantages of *external fertilization*.

The fossil in the photo on page 252 came from a site in West Germany. About 50 million years ago, the site was a lake. When the animal died, its body was carried into the lake by rivers. Thousands of other animal fossils in excellent condition have been recovered. Many are as complete as the one shown here. This site is very unusual, because so many of the fossils found here are of animals with backbones.

These fossil animals, as well as the scientists who found them, belong to Phylum Chordata (kor-**dah**-tah). So does the animal shown in Fig. 11–1. What can snakes and humans possibly have in common with these *sea squirts?* Scientists have compared these animals during their early stages of life.

At some stage of their lives, all **chordates** (**kor**-daytz) have a *stiff rod* of **cartilage** down their backs. *Chordates* also have *gill slits* in their throats at some stage. The gills of a fish are an example. Humans have both these features as embryos, long before birth. Another feature of all chordates is a *hollow nerve cord* down the back. In humans, this nerve cord develops into the brain and spinal cord.

Although they have these points in common, the animals in this phylum are very different. They are so different that several *subphyla* have been set up. We will study the largest

Chordate An animal having a nerve cord and gill slits at some time in its development.

Cartilage A firm but flexible tissue that gives shape and support to the bodies of some animals.

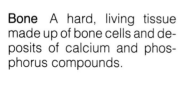

Fig. 11-1 These Light Bulb sea squirts are ocean animals, common along sea coasts. They are members of Phylum Chordata.

and most important subphylum, *Vertebrata* (vur-teh-**brah**-tah). This subphylum includes humans and the other kinds of animals that are most familiar to you.

VERTEBRATES

Vertebrate An animal with a backbone.

Endoskeleton An internal support structure in vertebrates, made of cartilage and/or bone.

Bone A hard, living tissue made up of bone cells and deposits of calcium and phosphorus compounds.

All **vertebrates** (**vur**-teh-braytz) have a backbone. This backbone surrounds or replaces the *cartilage* rod found in the embryo. The backbone is part of an internal skeleton called an **endoskeleton** (**en**-doe-skeh-leh-tun). See Fig. 11–2. The *endoskeleton* is an internal bone framework with muscles on the outside. Unlike the exoskeleton of the arthropods, an endoskeleton does not limit the growth of an animal. It is made up of **bone** and cartilage, both of which can grow as the animal grows. *Bone* and cartilage are similar types of tissues. However, cartilage is not as hard or brittle as bone.

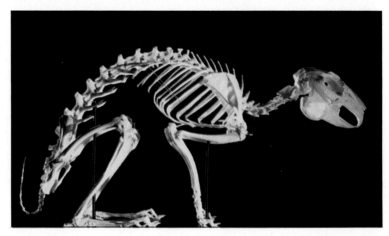

Fig. 11–2 All vertebrates have a backbone, as part of an endoskeleton.

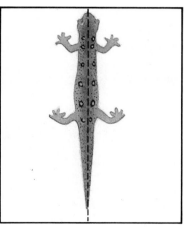

Fig. 11–3 In the salamander at left, the bone is stained pink and the cartilage blue. At right, vertebrates have a body plan in which one side of the body is a mirror image of the other side.

Cartilage is firm but flexible. You can bend your ears because they are made mostly of cartilage. See Fig. 11–3 (left).

Vertebrate animals have a head, neck, body, and usually a tail. Vertebrates have *bilateral symmetry*—that is, if you divide a vertebrate animal in half along its backbone, the two halves will be almost identical. See Fig. 11–3 (right). There are usually two pairs of limbs. Vertebrates have highly specialized systems of organs. For example, the nervous system has a brain and well-developed sense organs.

Vertebrates are divided into seven classes. These classes are: (1) *jawless fishes,* (2) *cartilage fishes,* (3) *bony fishes,* (4) *amphibians,* (5) *reptiles,* (6) *birds,* and (7) *mammals.* We will look at the first three classes in this section.

Fishes, amphibians, and reptiles are **coldblooded vertebrates.** This means that their body temperature changes with the temperature of their environment. The body temperature of a fish is very close to that of the surrounding water. *Coldbloodedness* has a disadvantage. Extreme temperatures of heat and cold can damage living cells. Thus, coldblooded animals cannot tolerate very high or very low temperatures.

Coldblooded vertebrate A vertebrate whose body temperature changes according to the temperature of the environment.

CLASSES OF FISHES

The sea lamprey is a good example of a *jawless fish.* Its mouth is round with sharp teeth. The lamprey has no jaws to open and close its mouth. This fish feeds by attaching its suckerlike mouth to another type of fish. The lamprey then sucks out the blood and body fluids. This weakens and often kills the other fish. See Fig. 11–4 on page 256.

Fig. 11–4 The sea lamprey is a jawless fish. It feeds by attaching its round mouth to the body of another fish.

The second class of vertebrates is made up of the *cartilage fishes.* Sharks, rays, and skates are cartilage fishes. Their skeletons are made entirely of cartilage. See Fig. 11–5. Most sharks feed on other ocean animals. Only 9 of the 250 species of sharks have been known to attack humans. Not all of the largest types of sharks are dangerous to humans. The whale shark may reach over 15 m in length. However, it feeds only on tiny ocean animals.

Skates and rays have broad, flat bodies with long, sharp tails. They almost seem to be flying through the water as they beat their large fins in a waving motion.

The most common vertebrate class is the *bony fishes.* There are more than 25,000 species, and some are found in almost every natural body of water.

Although *bony fishes* vary greatly in size and shape, most follow the same general body plan. In general, these fishes are streamlined in shape, tapered at both head and tail areas. See Fig. 11–6. A bony fish's body has two sets of paired fins. These can be compared to the two sets of limbs on other

Fig. 11–5 A Galapagos shark (left) *and an Eagle ray* (right) *are both cartilage fishes.*

Fig. 11–6 Bony fish live in both fresh and salt water.

vertebrates. There are often several unpaired fins. These may include one or two on the back, one on the bottom side, and one on the tail. Together, the body shape and fins allow the bony fish to move easily through the water.

LIFE PROCESSES OF FISHES

Fishes have many specialized organs to help them respond to their environment. Their senses of balance and of taste are very good. They can hear noises, sense pressure changes, and feel vibrations. However, the fish's eye is not well developed because there is little light in deep water.

One difference between fishes and arthropods is that fishes have a more advanced circulatory system. Arthropods have an open circulatory system. The circulatory system of a fish is a closed system. This means that the fish's blood stays in the blood vessels all the time. A second difference is the structure of the heart. The heart of a fish has two chambers. The first chamber receives blood from all parts of the body. It contracts and sends blood into the second chamber. When this chamber contracts, blood is pushed out of the heart. See Fig. 11–7 on page 258.

Like many invertebrates, fishes reproduce by **external fertilization.** In some species, nests are scooped out of a stream bottom or built from weeds or twigs. The female lays thousands of eggs into the water. The male then releases sperm to fertilize the eggs.

There are disadvantages to this method of reproduction. Many eggs do not get fertilized. Others are eaten by predators

External fertilization The joining of egg and sperm outside the female's body.

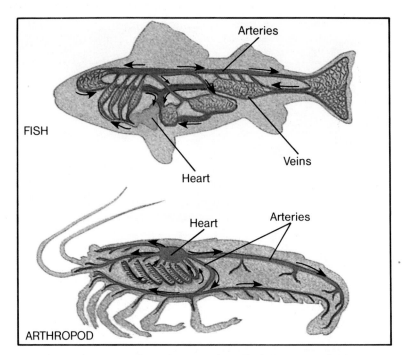

Fig. 11–7 The circulatory system of a fish is closed. In arthropods, blood moves away from the heart in arteries, but returns without the presence of blood vessels. This is an open system.

or other animals before they can hatch. Still more of the young fish will be eaten or die before reaching adulthood. However, since so many eggs have been produced, at least some will survive.

SUMMARY

Animals in Phylum Chordata share several important traits. However, their differences separate them into several subphyla. Chordates with a backbone are placed in subphylum Vertebrata. Fishes form three of the classes of vertebrates.

QUESTIONS

Use complete sentences to write your answers.

1. Why are sea squirts and humans both classified in Phylum Chordata?
2. How is your skeleton different from that of an arthropod?
3. How does the skeleton of a shark differ from that of a bony fish?
4. Why must animals that reproduce by external fertilization lay thousands of eggs?
5. Why is a fish called a coldblooded animal?

INVESTIGATION

OBSERVING FISH

PURPOSE: To observe the traits of a typical bony fish.

MATERIALS:

battery jar   fish food
aged water   thermometer
goldfish or guppy

PROCEDURE:

A. Sketch the general outline of the fish. Label these parts: *head, body, tail, paired fins, single fins, mouth, eye, gill cover, scales.*

   1. Why do most fish have a tapered shape?

   2. How many pairs of paired fins does this fish have?

   3. How many unpaired fins does it have?

   4. How does the fish push itself through the water?

   5. How does the fish turn?

   6. Can a fish back up? How?

   7. Which of these sense organs are found on the fish's head: eyes, ears, nostrils, tongue?

B. Observe the scales on the fish's body.

   8. In what way are the scales similar to shingles on a roof?

   9. What might be the function of these scales?

   10. The fish's body is covered with a slippery material. What do you think its function might be?

   11. Is the fish the same color all over its body?

   12. If yes, is it the same shade of this color all over?

   13. If no, where are the darker colors located on the body?

   14. What might be the reasons for this color pattern?

C. Observe the mouth and gills carefully.

   15. How are the movements of the mouth and gills related?

   16. How might this movement assist the fish in obtaining oxygen?

D. Use the thermometer to determine the temperature of the water in the container.

   17. What is the water temperature?

   18. What is the approximate body temperature of the fish?

   19. Why are you able to know this?

   20. If the aquarium were left in the sun, and the water temperature rose 8°C, what would happen to the fish's body temperature? Why is this so?

E. Add a pinch of food to the water.

   21. How does the fish know where the food is? What senses might be used to locate food?

   22. Are any teeth visible in the fish's mouth? If so, what can this tell you about the type of food it eats?

CONCLUSIONS:

   1. Why is a fish classified as a vertebrate animal?

   2. What is the function of the fins of a fish?

   3. Why are fishes called coldblooded?

11-2. The Amphibians

At the end of this section you will be able to:

- ☐ Describe the common traits of *amphibians.*
- ☐ Explain why amphibians are neither land nor water animals.
- ☐ Describe the stages of metamorphosis in the life cycle of a frog.

Have you ever seen a "fish out of water"? It flips, jumps, and wriggles about, trying to return to the water. It cannot live long, because it cannot take oxygen directly from the air. If it is lucky, it may wriggle its way back to the water before it dies.

THE DEVELOPMENT OF AMPHIBIANS

Scientists believe that many millions of years ago, there developed new types of fishes that could survive out of water for longer periods. How might this have happened? Perhaps fishes living in ponds or lakes were trapped during periods of drought. If these fishes had strong fins, they may have been able to flip and wriggle to the next pond. In this way they could find water and survive. In some fishes, structures other than gills became able to absorb water directly from the air. These adaptations would enable these fishes to survive out of water longer. Such fishes exist today. The lungfish are examples. See Fig. 11–8.

Living on land, even for short periods, has both problems and rewards. First, the bones and muscles of a land animal

Fig. 11–8 Lungfish are found in Australia, Africa, and South America.

must be stronger than those of a fish. This is because the surrounding water helps support a fish more than air does a land animal. Next, a waterproof body covering is necessary to prevent drying out. Also, gills do not work well out of water. They need to be kept moist. A *lung* or other body surface capable of absorbing oxygen from the air is needed.

There were also advantages for ancient fishes to move onto land. Animals that did so millions of years ago were safer because they had escaped from their enemies. On land, there was an abundance of plants and insects for food. Air contains plenty of oxygen to carry out cell activities. All of these factors probably played a role in the development of amphibians.

AMPHIBIANS TODAY

Amphibians are not land animals, and they are not water animals. The name *amphibian* means "double life." Most must live in a moist environment because their smooth skins do lose some water. They also take in some of their oxygen through their moist skins. Most important is the fact that amphibian eggs must be laid in a moist environment. They do not have shells or other coverings to keep them from drying out. The eggs must be laid in water or in a very moist location.

Frogs, toads, and *salamanders* are included in the amphibian class. See Fig. 11–9. As a group, amphibians have a thin, smooth, moist skin. Their feet and toes are soft, webbed, and without claws. Salamanders, unlike frogs and toads, retain their tails as adults.

Fig. 11–9 A leopard frog (left), *spadefoot toad* (middle), *and a red salamander* (right) *are all amphibians.*

LIFE CYCLES

Most amphibians reproduce by *external fertilization*. Eggs and sperm are released into the water, where they join. The fertilized eggs develop and hatch into larvae called *tadpoles*. See Fig. 11–10. Tadpoles have a tail but no legs. They are able to swim underwater. Most tadpoles breathe oxygen from water through their gills. In this stage, they are much like small fish. The tadpoles are usually plant eaters. They also have a two-chambered heart like the fishes.

Slowly, several changes take place in the tadpole. Like insects, amphibians undergo metamorphosis during their life cycle. The gills disappear, and lungs develop. Front and hind legs begin to develop. The tail disappears in frogs and toads. Once the lungs have formed, the tadpole must come to the surface of the water to take in air. Soon the tadpole moves onto land. The young amphibian eventually reaches the adult stage. Leopard frogs take about three months to complete this process. Bullfrogs take almost two years.

Fig. 11–10 The life cycle of a frog showing metamorphosis.

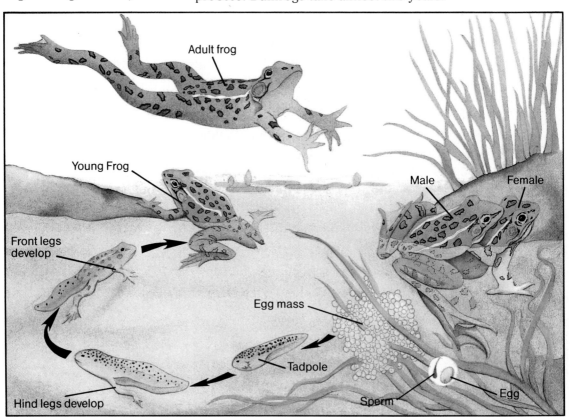

THE AMPHIBIAN HEART

The amphibian heart is different from that of all animals presented so far. See Fig. 11–11. Fishes have a simple, two-chambered heart. The larvae of amphibians also have this two-chambered heart. However, during metamorphosis, a third chamber develops. Two chambers receive blood from the body. One receives oxygen-poor blood from the body. The other receives oxygen-rich blood from the lungs. The two types of blood are mixed as they enter the third chamber. The third chamber forces some of the combined blood out to the cells and some to the lungs. This three-part structure speeds up the delivery of oxygen to the cells. This faster blood circulation is an advantage to the amphibian.

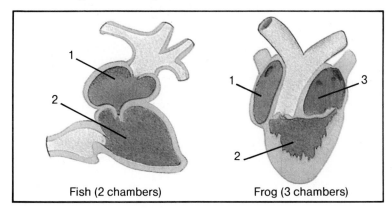

Fish (2 chambers)   Frog (3 chambers)

Fig. 11–11 The three-chambered heart of an amphibian is more efficient than the two-chambered heart of a fish.

SUMMARY

Amphibians represent a stage between water-dwelling animals and land animals. Their body structures and the stages of their metamorphosis reflect this change. Many body features, especially the structure of the heart, represent improvements over the fish.

QUESTIONS

Use complete sentences to write your answers.

1. What traits of amphibians set them apart from fishes?
2. How does the heart structure of an adult amphibian differ from that of a fish?
3. Describe three major changes that occur as a tadpole becomes a frog.

SKILL-BUILDING ACTIVITY

THE ANATOMY OF A FROG

PURPOSE: To *observe* the structure of a frog and gain skill in *using dissecting tools.*

MATERIALS:

preserved frog	probe
dissecting tray	dissecting pins
forceps	diagrams of frog structure
scissors	colored pencils

PROCEDURE:

A. Place the frog in the dissecting tray with its backside up. Use diagram A (supplied by your teacher) to locate the external features. Your teacher will give you the diagrams needed for this work. Check off each part as you identify it.

1. Describe the feel of the skin.

2. What sense organs are on the head?

3. Compare the size and structure of the forelegs, hind legs, and feet.

4. Why are the hind feet webbed?

5. Why is the upper surface darker colored than the lower surface?

B. Cut the jaws at the hinges so the mouth can be opened. See Fig. 11–12. Locate the structures shown in Diagram B. Check off each as you locate it.

6. How is the tongue attached to the floor of the mouth?

7. A frog's tongue is sticky. How does this help in food getting?

8. Which jaw holds the teeth? Teeth are not for chewing. What might they do?

9. Why is the glottis a muscular ring that can be opened and closed?

10. Push the probe down the gullet. Where does this tube lead?

C. With the frog on its back, pin the limbs down as in Fig. 11–12. Grasp the skin at point 1 with the forceps. Lift the skin and make a small cut. Insert the tip of the scissors and cut the skin along line 1. Do not cut too deeply or you may damage internal organs. Cut along lines 2, 3, 4, and 5. Pin back the skin flaps. Observe the muscle layers beneath the skin.

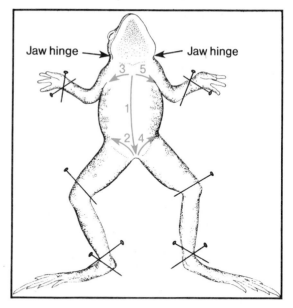

Jaw hinge   Jaw hinge

Fig. 11–12

11. Are blood vessels visible in the muscles or skin? What color are they?

D. Cut through the muscles in the same way. You will have to cut through the breastbone between the forelegs. Pin the flaps down as before.

E. If the body cavity is filled with small, dark eggs, your frog is female.

   12. What sex is your frog?

F. Use Diagram C to find the internal organs of the frog. Check off each part as you locate it.

   13. What is the large, dark, three-part organ in the upper body cavity?

   14. What is the green sac below it?

G. Cut out these organs. The heart should be visible in the upper chest area.

   15. What is the shape of the heart?

   16. How many chambers are there in this heart?

   17. How many blood vessels entering or leaving the heart can you find?

H. Follow the blood vessels from the heart to the two small lungs. Lungs provide only part of the frog's oxygen needs.

   18. How does the frog obtain the rest of the oxygen it needs?

I. Remove the heart. The digestive system is now visible. The small tube at the top leads from the gullet in the mouth to the large J-shaped organ.

   19. What is the J-shaped organ called?

   20. Cut this organ open. Had your frog eaten before it died?

J. Follow the small and large intestines from the stomach to the external body opening, the anus.

   21. How are the stomach and intestines held in place in the body cavity?

K. Snip the tube at the top of the stomach

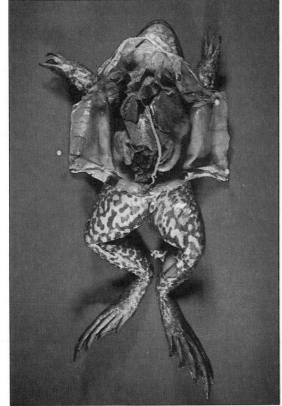

Fig. 11–13

and the bottom of the large intestine. Remove the entire digestive system.

L. Color the following structures on the diagram: the heart and blood vessels, light red; kidneys, dark red; liver, brown; gall bladder, green; lungs, blue; digestive system, yellow.

CONCLUSIONS:

   1. How are a frog's hind legs adapted for its type of life?

   2. How does the frog obtain oxygen?

   3. How does the circulatory system of a frog differ from that of a fish?

Fig. 11–14 Many reptile young, such as this painted turtle, hatch from shells laid on land.

11-3. The Reptiles

At the end of this section you will be able to:

- Contrast reptiles and amphibians.
- Discuss the advantages of *internal fertilization* and eggs with shells.
- Compare the three orders of reptiles.

Millions of years ago, *reptiles* were the dominant type of life on earth. The biggest were over 30 m in length. That made them the largest animals ever to live on land. And then they disappeared. No one knows exactly why, although there are several theories. Today, there are 6,000 species of reptiles left. These include turtles, alligators, crocodiles, lizards, and snakes.

WHAT IS A REPTILE?

Reptiles are different from amphibians in several important ways. The skin of a reptile is made of hard plates or scales. This covering prevents loss of water by evaporation. Although this is an advantage, it also means that reptiles cannot absorb oxygen through their skins. Therefore, their lungs are well developed to provide oxygen. Although many species of reptiles live in water environments, each comes to the surface to breathe.

Unlike amphibians, the limbs of reptiles have toes with claws. They use the clawed toes for climbing, digging, and moving about on land. Reptiles, like amphibians, have three-chambered hearts. However, in reptiles the third chamber is partially divided. This means there is less mixing of the oxygen-rich blood from the lungs with the oxygen-poor blood from the rest of the body. As a result, even more oxygen gets to the body cells.

Reptiles are coldblooded like fishes and amphibians. Their blood is not actually cold, but they cannot make their own body heat. They control their body temperature somewhat by moving in and out of warm sunlight. This helps explain why reptiles are commonly found in warm climates.

The most important difference between reptiles and amphibians is their way of reproducing. The male reptile deposits sperm inside the female's body to fertilize the eggs. This is called **internal fertilization.** *Internal fertilization* can take place on land. Since there is a greater chance the eggs will be fertilized, fewer eggs need to be produced. The *external* fertilization of fishes and amphibians is much less successful.

Another difference is the egg itself. See Fig. 11–14. Reptile eggs are covered and protected by thin *membranes* and tough, leathery shells. These shells allow oxygen and carbon dioxide to pass through, but not water. The shell prevents the embryo from drying out. Reptile eggs are laid on dry land.

The reptile egg contains a large supply of food for the developing embryo. This allows the reptile embryo to stay inside the shell a long time. When they hatch, reptiles are more completely developed than amphibians. They look like small adults. They do not go through metamorphosis. Some reptiles, such as garter snakes, keep the fertilized eggs inside the female's body until they hatch.

Internal fertilization The joining of egg and sperm inside the female's body.

TYPES OF REPTILES

Turtles are a group of reptiles that are easily recognized by the hard shells that protect them. See Fig. 11–15 (left). Some turtles can pull their limbs inside the shell for protection. However, on land this extra safety of the shell has its price. The shell is heavy and rigid. Land turtles move slowly and awkwardly. Turtles that live in the oceans have feet shaped like flippers and can swim very well.

Fig. 11–15 The land tortoise (left) *and crocodile* (right) *represent two of the three reptile orders.*

Fig. 11–16 The San Francisco garter snake (left) *is an endangered species. At right is a leopard lizard.*

Crocodiles and alligators spend much of their time in water. See Fig. 11–15 (right) on page 267. Their wide jaws and sharp teeth make them fierce hunters. They often drown their victims by dragging them underwater. The attacker keeps its own nostrils above water so it can breathe. These reptiles live in warm climates, including the southern United States.

Snakes and lizards form the third group of reptiles. See Fig. 11–16. They are so alike in their general structure that they are classified in the same group. There are, however, obvious differences between them. Lizards have legs, snakes do not. Lizards can close their eyes, snakes cannot.

SUMMARY

The structure of reptiles shows several important advances over the amphibians. Most important is the development of an egg with a shell. This egg, along with other changes, made reptiles the dominant land animal at one time. Reptile species today include turtles, lizards, snakes, and crocodiles.

QUESTIONS

Use complete sentences to write your answers.

1. How do the skins of reptiles and amphibians differ?
2. Compare the eggs of reptiles and amphibians. How are reptile eggs suited to the land environment?
3. How are snakes different from lizards?
4. Why is internal fertilization important to land animals?

TECHNOLOGY

ARTIFICIAL GILLS

Scientists are working on a way in which humans can learn how to breathe through gills! In the future, artificial gills may replace the heavy air tanks human divers now use to breathe underwater.

Artificial gills would do the same job a real fish gill does—take oxygen out of sea water. Bony fish usually have two gill chambers on each side of the head. Each gill is a rounded piece of bone with hundreds of tiny projections sticking out.

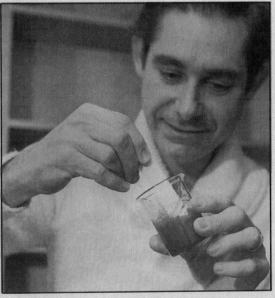

If you've ever watched a fish take in water through its mouth and then release it through the gills, you've probably glimpsed the gills' pinkish-red color. The red color comes from the many capillaries close to each gill's surface. Blood is red owing to a substance called hemoglobin. When hemoglobin encounters the gas oxygen, the oxygen molecules bond with it and "hitch a ride" wherever it goes. This is how a gill works. As the gills open and close, water is forced past them. The oxygen dissolved in the water leaves the water and bonds with the hemoglobin in the nearby capillaries. Then, this oxygen travels through the blood vessels to all parts of the fish.

The artificial gill works in a similar way, except that instead of the blood flowing to a place where it meets sea water as it does in fish, water is forced to flow through immovable gills. An artificial gill is made of fine plastic sponge, which feels like seat cushion foam. Hemoglobin, which keeps its special properties even when it's not inside a living thing, is added to the sponge. A sponge, like a real gill, has many projections around which the oxygen from sea water can be taken up by the hemoglobin. This artificial gill, known as a hemosponge, could be used by divers. In the photo, a model of the hemosponge is shown by its inventor Dr. Joseph Bonaventura.

Another design is for a hemosponge almost 1 meter across and 3 meters long. It would produce enough oxygen for 150 people to breathe. This hemosponge is placed in a container underwater. As oxygen is collected from sea water, a pump sends it into many smaller hemosponges, which people could wear as backpacks. Instead of a heavy tank, divers could wear one of these lightweight plastic sponges.

Having a way to take oxygen out of sea water might have other uses. One idea being tested might provide Hawaii with more electric power. Fuel, mixed with oxygen, is often burned at power plants to generate electricity. Why not carry on fuel-burning in enclosed power plants underwater? Vast amounts of oxygen from sea water could be pumped from artificial gills into combustion chambers and mixed with fuel. The result could be clean energy that is 300 times more efficient than the energy obtained from batteries. Perhaps someday people could actually live underwater!

CHAPTER REVIEW

VOCABULARY

On a separate piece of paper, match each term with the number of the statement that best explains it. Use each term only once.

internal fertilization   external fertilization   reptiles   bone

vertebrates   amphibians   cartilage   chordate

fishes   endoskeleton   coldblooded

1. Internal support structure of bone or cartilage.
2. Egg and sperm join outside the female's body.
3. An animal whose body temperature changes according to the temperature of the environment.
4. A group of animals that are coldblooded, have smooth skin, lay eggs in water, have gills in young, lungs in adults.
5. Hard, living tissue containing calcium and phosphorus compounds.
6. Animals with backbones as part of an endoskeleton.
7. A firm but flexible support tissue.
8. Coldblooded, scales, gills, lay eggs in water.
9. Egg and sperm join inside the female's body.
10. Coldblooded, scales, lungs, eggs with tough shells.
11. An animal having a nerve cord and gill slits during part of its development.

QUESTIONS

Give brief but complete answers to each of the following questions. Unless otherwise indicated, use complete sentences to write your answers.

1. List three traits common to all vertebrates.
2. What are the five classes of coldblooded vertebrates? Give an example of each class.
3. How did amphibians get their name, which means "double lives"?
4. How does the structure of a fish's body help it to swim?
5. In what ways are reptiles suited for life out of water?
6. How is the skeleton of a shark different from the skeletons of other vertebrates?
7. What is meant by the term coldblooded? Why can being coldblooded be a disadvantage?

8. Describe the metamorphosis of a frog.
9. Why is the three-chambered heart of amphibians and reptiles better than the two-chambered heart found in fishes?
10. Distinguish between internal and external fertilization. What advantages does internal fertilization have?
11. How are the eggs of reptiles adapted to exist on dry land?
12. What characteristics of amphibians led scientists to hypothesize that their ancestors were the first land animals?

APPLYING SCIENCE
1. What could happen to a reptile that could not find any shade?
2. Why are the top sides of fishes darker in color than the bottom sides?
3. Why is the tongue of a frog attached at the front of its mouth instead of toward the back?
4. Apply the dissection skill learned in the Skill-building Activity to the dissection of a typical bony fish, such as a perch. Compare and contrast the body systems of the fish with those of the frog.
5. Collect egg masses of frogs or toads during the spring. Set up an aquarium filled with pond water to contain them. Observe the development of the tadpoles into young frogs or toads.

BIBLIOGRAPHY

Blumberg, Rhoda. *Sharks.* New York: Franklin Watts, 1976.

Brownlee, Shannon, et al. "The Great Dyings." *Discover,* May 1984.

Fichter, George S. *Poisonous Snakes.* New York: Franklin Watts, 1982.

Gibbons, Whit. *Their Blood Runs Cold: Adventures with Reptiles and Amphibians.* University, AL: University of Alabama Press, 1983.

Jacobson, S. K. "Frog Feats." *International Wildlife,* May–June, 1984.

Minton, Sherman and Madge. *Giant Reptiles.* New York: Scribner's, 1983.

Patent, Dorothy. *Frogs, Toads, and Salamanders and How They Reproduce.* New York: Holiday House, 1975.

A Great Egret, perched above a swamp, preens its feathers. This keeps them clean of dirt and free of parasites. Feathers are what make birds different from other animals.

WARMBLOODED VERTEBRATES

CHAPTER GOALS:

1. Discuss the advantages of warmbloodedness to an animal.
2. List and discuss the principal characteristics of warm-blooded vertebrates.
3. Compare and contrast the body structures of birds and mammals with other vertebrates.
4. Describe the reproductive patterns of birds and mammals.

12-1. The Birds

At the end of this section you will be able to:

- ☐ List the advantages of being *warmblooded.*
- ☐ Describe the structures that allow a bird to fly.
- ☐ Describe how the organ systems of birds differ from those of other vertebrates.
- ☐ Describe the reproductive patterns of birds.

Birds are the masters of the air. For centuries, humans wanted to fly like most birds do. Some people even tried building wings and flapping them to imitate the ways of birds. But they failed. The human body is not meant to fly. It isn't built for flight.

Birds A group of vertebrate animals that has feathers.

WARMBLOODEDNESS

Only birds and mammals are **warmblooded.** This means that their internal body temperature is not dependent on their surroundings. Usually, a warmblooded animal's body temperature is warmer than the air around it. The warmth is produced by the release of energy from its food and its activity. Flying requires a lot of energy. Birds must eat and then break down a lot of food to produce that energy. The heat given off by these processes is used to keep a constant internal temperature. Down feathers and a layer of fat under the bird's skin help keep this heat from escaping out of its body.

This constant internal temperature is the best condition for carrying on cell activities. No matter how hot or cold the

Warmblooded Refers to those vertebrates whose body temperature remains constant despite temperature changes in the environment.

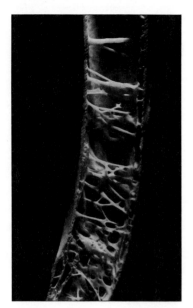

Fig. 12–1 *The bones of birds are strong and hollow, with reinforcing crosspieces.*

weather is, the cells can function well. For this reason, birds and mammals can be active and live in a wide range of temperatures and environments.

FIT FOR FLYING

Birds have developed many structures that allow them to fly. The bones of a bird are either thin or hollow. They have internal supports that make them strong but also very lightweight. See Fig. 12–1. By comparison, the bones of other vertebrates are solid and heavy.

The structure of a bird's wing is very much like the structure of your own arm. See Fig. 12–2. Both are jointed for flexibility. Yet, you cannot fly. This is because you do not have all the body structures that make flight possible.

A second adaptation to flight is the powerful breast muscles of a bird. These muscles are anchored to the large breastbone. Their function is to pull the wing downward and backward. This is the power stroke of the wing beat. A different set of muscles moves the wing upward and forward to repeat the stroke. Some birds must beat their wings as often as 50 times per second to stay up in the air.

Feathers are another adaptation for flight. Feathers are formed only by vertebrates of the Class Aves—the birds. This trait sets birds apart from all other animals. The feathers of a bird develop in the same way as the scales of a reptile. In fact, birds, like reptiles, have scales on their legs and feet. This is one reason scientists believe that birds are related to ancient reptiles. The feather is, in reality, a scale that has been modified or changed.

Fig. 12–2 *A bird's wing has many of the same bones as a human's arm.*

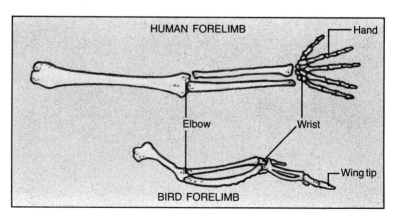

There are two types of feathers. See Fig. 12–3. *Down feathers* are small and fluffy. They act as insulation for the bird. The fluffy down traps a layer of air close to the bird's body. This layer helps prevent heat loss. That is why we use down feathers in winter clothing. *Contour feathers* cover the body and wings. They give the bird its streamlined shape. Large contour feathers on the wings and tail are used for flying and balance.

To release the energy needed for flying, the bird's cells must receive a lot of oxygen. In addition to its *lungs,* the bird also has a system of *air sacs* throughout its body. See Fig. 12–4. These sacs are connected to the lungs and provide them with a steady flow of oxygen-rich air. Thus, the bird is able to supply the cells with all of the oxygen they need.

To move the oxygen and food to the cells quickly requires a rapid heartbeat. The hearts of some birds beat over 500 times per minute. By comparison, a human heart beats only about 72 times per minute.

The structure of the bird's heart is an improvement over a reptile's three-chambered heart. It can deliver more oxygen to the cells. There are four chambers in a bird's heart, two on each side. Blood returns from the body to one side of the

Fig. 12–3 Birds have two kinds of feathers—contour and down.

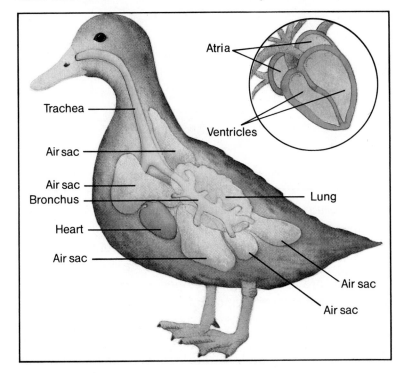

Fig. 12–4 A four-chambered heart and respiratory system with air sacs work together to supply a bird's high oxygen needs.

INCUBATION PERIODS	
Bluebird	12 days
Chicken	21 days
Duck	28 days
Kiwi	80 days

Table 12–1

heart. The two chambers on this side pump the blood to the lungs. Here, waste carbon dioxide is given off and fresh oxygen is taken in. The oxygen-rich blood returns to the other side of the heart. The two chambers on this side pump the blood out to the body cells. There is no mixing of the oxygen-poor and the oxygen-rich blood. In this way, the maximum amount of oxygen gets to the body cells.

REPRODUCTIVE PATTERNS

Like reptiles, birds reproduce by internal fertilization. A shell develops around the fertilized egg in the female's body. Birds' eggs have brittle shells instead of the tough, leathery shells of reptiles. The shell structure of birds' eggs allows oxygen to pass through into the egg. The shell also prevents water loss and protects the *embryo*. Within the egg are the *embryo,* the *yolk,* and the *albumen,* or "white" of the egg. See Fig. 12–5. The albumin and the yolk are the food supply for the developing embryo.

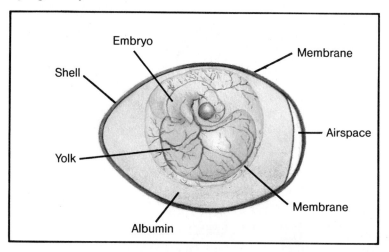

Fig. 12–5 The egg white, membrane, and shell form around the fertilized egg within the female's reproductive system.

Incubate To keep eggs warm so that they can mature.

To develop properly, the embryo must be kept warm. Birds must stay with their eggs to **incubate** them. Different species hatch at different stages of development. Some birds, like ducks, have long *incubation* periods. See Table 12–1. The ducklings are almost fully developed when they hatch. They are immediately active, finding food and following their parents. The ducklings are covered with down for warmth. Other birds have shorter incubation periods. When they hatch, the young are helpless and totally dependent on their parents for

food and warmth. These young birds are often unable to see and are almost featherless. They have to grow and develop more before they can care for themselves. See Fig. 12–6.

Fig. 12–6 *Young geese can swim, walk, and feed as soon as they hatch* (left). *Warbler babies* (right) *may need weeks of care before they can move out of the nest.*

TYPES OF BIRDS

There are about 8,700 different species of birds. One way to classify them is by the differences in their beaks and feet. There are *perching birds, water birds, birds of prey,* and *flightless birds.* The type of beak a bird has can tell you what a bird eats. See Fig. 12–7 on page 278. There are beaks for eating flesh, catching insects, cracking seeds, sipping nectar, digging in mud, or pecking through wood. The type of foot a bird has can tell you in what environment the bird lives. See Fig. 12–7. Birds' feet are shaped for paddling, running, grasping, wading, or perching. Birds are able to live in many different environments and eat many different foods. This has made them one of the most successful classes of vertebrates.

Because of their energy and food requirements, birds do not hibernate in cold weather. Many birds **migrate** with the coming of winter. Some species travel thousands of kilometers to find other sources of food. The following spring, they return again to build nests and raise their young. How birds find their way on these long trips is still a mystery. Scientists are investigating this question and others, such as what stimuli trigger the birds to begin their migrations.

Migrate Refers to seasonal movement of animals from one environment to another, usually in search of food.

Labels in figure:
OWL — Tearing, Grasping
BROWN PELICAN — Straining, Swimming
Spearing — GREAT BLUE HERON — Wading
Seed-eating — BLACK-HEADED GROSBEAK — Perching
Insect-eating — WILD TURKEY — Running

Fig. 12–7 There is great varia-tion in the forms of birds' beaks and feet.

SUMMARY

The ability to keep a constant body temperature is an important trait in birds and mammals. It enables them to survive in many environments and to keep active throughout the year. The skeleton, muscles, and body covering of birds are adapted for flying. Birds lay eggs with brittle shells, which must be kept warm while the embryo develops.

QUESTIONS

Use complete sentences to write your answers.

1. How do warmblooded animals differ from coldblooded animals?
2. What are three adaptations birds have for flying?
3. What structures help birds to get a lot of oxygen to cells?
4. Describe the eggs of birds. How do birds differ from reptiles in their care of the young?

MAINTAINING BODY TEMPERATURE

PURPOSE: To demonstrate one of the ways in which warmblooded animals maintain body temperature.

MATERIALS:

2 large beakers
overflow tray
2 thermometers
(plastic backed)

refrigerated water
towel
clock with second
hand

PROCEDURE:

A. Warmblooded animals must maintain a fairly constant internal temperature. The blood distributes heat to all body parts. Excess heat is released through body surfaces. To conserve heat, blood flow to limbs and body surfaces is reduced.

B. Place one thermometer on the desk. Grasp the other so that the bulb is touching the palm of your hand. CAUTION: Do not squeeze hard. The thermometer could break. Allow the thermometers one minute to adjust.

1. What is the air temperature?

2. What is the temperature of your palm?

C. Fill each beaker half full of the cold water. Place the beakers in the tray to catch any overflow.

D. Place a thermometer in each beaker. Allow time for the temperature to adjust.

Time	Beaker #1	Beaker #2
BT		
30 sec		

Table 12–2

E. Copy Table 12–2 onto a separate piece of paper. Record the water temperature as the beginning temperature (BT) for each beaker.

F. Keep your writing hand free so you can record data. Insert your other hand into beaker #1. Arrange the thermometer so the plastic backing is between the bulb and your hand. You should be able to read the temperature through the glass. Move your fingers often to stir the water.

3. What is the purpose of the second beaker and thermometer?

4. Why should the water in beaker #1 be stirred often?

G. Record the water temperature in each beaker every 30 seconds for at least five minutes.

H. Measure the temperature of the palm of your hand immediately after removing it from the water.

5. How has its temperature changed?

6. How much did the temperature of beaker #1 increase?

7. Where did the heat to cause this change come from?

8. How much did the temperature of beaker #2 increase?

9. Where did the heat to cause this change come from?

CONCLUSION:

1. How do humans control the amount of heat lost through the surfaces of the body?

SCIENCE INPUT

Organisms belonging to the animal Phylum Chordata, Subphylum Vertebrata, all have a major characteristic in common: Their skeletal systems are composed of bone, cartilage, and muscle tissue, which provide protection and allow body movement.

However, even through the organisms are similar in basic structure, they differ in characteristics, including environmental habitat, body temperature, body-covering composition, movement structures, and type of respiratory system.

Due to these and other different characteristics, VERTEBRATES are further classified into more specific groups, based on similar characteristics, known as CLASSES. In other words, the divisions in a classification system help to specify important differences within the overall group.

COMPUTER INPUT

After you've entered your data, the computer will choose 7 definitions at random from the 15 data statements. The computer's ability to select random numbers is built into the electronic circuitry. The basic idea in any random selection is that every item in the group must have an equal chance of being selected. A lottery, to be fair, is supposed to be random. No number is supposed to have more of a chance to win than any other number. A computer can make a random selection of any data that has been entered into its memory. In statement 130 of Program Classification, the computer is directed to select one of the 15 definitions at random that you entered as data statements.

WHAT TO DO

Review the characteristics of the five common vertebrate groups: mammals, amphibians, fishes, reptiles, and birds. Make a data chart similar to the one below and fill in the definitions and the class column. Define each class in three different ways. Each definition should fit that class only. An example is given for each group. The definitions and classes in the chart will be used for data statements 200 to 340.

Enter Program Classification, store it, and then run it. Take your time to enter the program correctly. The longer the program, the more chance there is for mistakes when it is run. An incorrectly entered program may not run, or it may run but not do what it was intended to do.

The program will use this data to choose 7 definitions at random. You will be asked which class of vertebrates they define. When entering the 15 data statements, use the format shown in the examples, lines 200 to 240. Do not put commas within your definitions. The comma must go after the definition and before the class name.

Data Chart

Class	Description/Definition
Fish	Lives in water and is cold-blooded
Bird	Has feathers and dry skin
Reptile	Has dry plates or scales for skin
Mammal	Has legs and feet or flippers and flukes
Amphibian	Has moist skin and gills or lungs

GLOSSARY

RND   a command in BASIC that instructs the computer to select a random number or random piece of data.

GOSUB   a statement directing the computer to go to another statement and follow the steps there, after which the computer will return to the next statement following the place where the GO-SUB appears. It indicates what is called a "loop," a smaller routine within the larger program. It is different from a GOTO statement, which does not instruct the computer to return to the original statement.

PROGRAM

```
100   REM CLASSIFICATION
110   FOR CS = 1 TO 24: PRINT : NEXT CS: DIM
      CH$(15),V$(15),R$(15)
120   FOR X = 1 TO 15: READ CH$(X),V$(X): NEXT
      X:K = 0
130   FOR Q = 1 TO 7:N = INT (15 * RND (1) + 1)
140   PRINT : PRINT : PRINT "     WHICH OF THE
      FOLLOWING: BIRDS, ": PRINT "AMPHIBIANS,
      FISH, REPTILES, MAMMALS,": PRINT
      CH$(N)"?"
150   PRINT : PRINT : INPUT " ";R$(N)
160   IF R$(N) < > V$(N) THEN GOSUB 400
170   IF R$(N) = V$(N) THEN GOSUB 500:K = K + 1
180   FOR CS = 1 TO 24: PRINT : NEXT CS: NEXT Q
190   P = INT ((K / 7) * 100): PRINT P"% OF 7
      QUESTIONS WERE CORRECT!":END
200   DATA     LIVES IN WATER & IS COLD-
      BLOODED, FISH
210   DATA     HAS FEATHERS AND DRY SKIN,
      BIRDS
220   DATA     HAS DRY PLATES OR SCALES FOR
      SKIN, REPTILES
230   DATA     HAS LEGS & FEET OR FLIPPERS &
      FLUKES, MAMMALS
240   DATA     HAS MOIST SKIN & GILLS OR
      LUNGS, AMPHIBIAN
250   DATA . . . (continue entering data statements)
400   FOR CS = 1 TO 24: PRINT : NEXT CS
410   A = INT (4 * RND (1) + 1): ON A GOTO
      420,430,440,450
420   PRINT : PRINT : PRINT : PRINT ,"UGH!",
430   PRINT : PRINT : PRINT : PRINT ,"SORRY!",
440   PRINT : PRINT : PRINT : PRINT ,"KEEP
      TRYING!",
450   PRINT : PRINT : PRINT : PRINT ,"WHOOPS!",
460   PRINT : PRINT " YOUR ANSWER IS WRONG!"
465   FOR TD = 1 TO 2000: NEXT TD: GOTO 600
470   FOR CS = 1 TO 24: PRINT : NEXT CS: RETURN
500   FOR CS = 1 TO 24: PRINT : NEXT CS
510   B = INT (4 * RND (1) + 1): ON B GOTO
      520,530,540,550
520   PRINT : PRINT : PRINT : PRINT ,"GREAT!",
530   PRINT : PRINT : PRINT : PRINT
      ,"WONDERFUL!",
540   PRINT : PRINT : PRINT : PRINT ,"YOU'RE A
      WHIZ!",
550   PRINT : PRINT : PRINT : PRINT ,"GOOD JOB!",
560   PRINT : PRINT "YOUR ANSWER IS CORRECT.
      CONTINUE!"
570   FOR TD = 1 TO 2000: NEXT TD: FOR CS = 1
      TO 24: PRINT : NEXT CS: RETURN
600   FOR CS = 1 TO 24: PRINT : NEXT CS
620   PRINT : PRINT "THE CORRECT ANSWER IS:
      ":PRINT : PRINT V$(N)": "CH$(N): FOR TD = 1
      TO 3000: NEXT TD: GOTO 470
630   END
```

BITS OF INFORMATION

Computers may prove to be a big help in locating schools of catchable fishes. A professor at the University of Southern California has designed a program that uses information received from NASA Weather Satellites. Large fish tend to be found in the boundaries between warm and colder water. These fish feed on the smaller fish and plant life found there. The fish-finding program uses the computer to analyze information from the satellites in orbit around the earth to locate these boundary areas.

12-2. The Mammals

At the end of this section you will be able to:

- ☐ Describe the characteristics of mammals.
- ☐ Compare and contrast reproductive patterns among mammals.
- ☐ Discuss some of the structural characteristics by which the mammals are classified.

Mammal An animal that has hair and feeds its young milk.

The oldest known fossils of **mammals** come from the time of the dinosaurs. They were tiny, shrewlike animals. See Fig. 12–8. After dinosaurs became extinct, mammals developed rapidly and became the dominant animals on earth.

Fig. 12–8 Fossil evidence suggests that this early mammal lived at the same time as the dinosaurs.

WHAT IS A MAMMAL?

Mammals are the most complex class of vertebrates. They can be found in every type of environment, although most are found on land. Mammals are warmblooded, have two pairs of limbs, have a four-chambered heart, and breathe air through lungs all their lives. Almost all give birth to live young.

They also have two characteristics that distinguish them from other classes of vertebrates. Mammals have a body covered with hair. This body hair acts as insulation to help regulate the internal temperature. After birth, mammals feed their offspring with milk from *mammary glands.* Both males and females have mammary glands, but they function only in females. These glands are the source of the class name *mammals.* See Fig. 12–9.

Mammals provide their young with more parental care than any other type of animal. This is because the young mammal

Fig. 12–9 A California sea lion nurses her pup.

takes a relatively long time to reach maturity. During this time of parental care, the young learn many types of behavior, including food getting and methods of defense.

The nervous system of mammals is far more developed than in other vertebrates. Mammals have large eyes and very keen senses. More importantly, the part of the brain involved with the thinking processes is much larger. Because of this, mammals are the most intelligent animals.

HOW MAMMALS REPRODUCE

The 5,000 species of mammals are classified into about 21 orders, the most common of which are listed in Table 12–3 on page 284. This classification is partly based on how the mammal reproduces. One of these orders has two species of mammals that lay eggs. These are the platypus and the spiny anteaters. See Fig. 12–10. After the eggs hatch, the young are fed milk that is produced by the female's body.

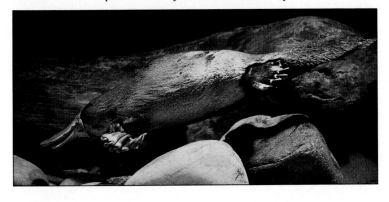

Fig. 12–10 Egg-laying mammals, such as this duckbill platypus, are the most primitive mammals.

Fig. 12–11 Newborn opossums inside their mother's pouch (left). When they are more mature, the babies cling to the mother's fur.

A second order is the *pouched mammals*. They give birth to live young that are very undeveloped. These young crawl into a pouch on the female's body. See Fig. 12–11. Here they are fed on milk from mammary glands until they complete their development. The best known of these pouched mammals are the kangaroo of Australia and the opossum of North America.

In the remaining 19 orders, the young are carried within the female's body until their early development is completed. After

SOME COMMON ORDERS OF MAMMALS		
Order	Common Name	Examples
Monotremata	Egg-laying mammals	platypus, spiny anteaters
Marsupialia	Pouched mammals	kangaroos, opossums, koala
Chiroptera	Bats	brown bat, vampire bat
Carnivora	Carnivores	cats, dogs, seals
Cetacea	Whales	whales, dolphin, porpoises
Rodentia	Rodents	beavers, rats, squirrels
Lagomorpha	Rabbits and hares	jack rabbits, cottontails
Artiodactyla	Even-toed, hoofed mammals	deer, cattle
Perissodactyla	Odd-toed, hoofed mammals	horses, rhinoceros
Proboscidea	Elephants	Indian and African elephants
Primata	Primates	apes, humans

Table 12–3

birth, the young are fed milk for a time until they are capable of eating other foods. Further classification of these 19 orders is based on many other traits. Two of these traits are the types of teeth and the structure of the foot.

TOOTH AND CLAW

There are three basic types of teeth in mammals. See Fig. 12–12. Each is related to the type of food the animal eats. One type of tooth is *chisel-shaped* and is used for cutting and gnawing. The large front teeth of a beaver are a good example. The second type of tooth is *long and pointed.* These teeth are used for ripping and tearing chunks of food. The fangs of a tiger are an example of this type. The third type are *mostly flat* and are used for grinding the food before swallowing. The back teeth of a horse are grinding teeth. Some animals, such as humans, have all three types of teeth. Can you guess why?

Differences in foot structure allow mammals to live in different kinds of environments and do different things. See Fig. 12–13. Some mammal feet are adapted for swimming. The flippers on seals and dolphins are examples. Feet with claws allow a mammal to grasp, rip, dig, and climb. The members of the cat family can do all of these things. Mammals with claws often run on their toes with the rest of the foot raised. This allows for great speed. The cheetah, for example, is capable of spurts of up to 80 km per hour. When it catches an antelope, it uses its paws to knock it down. The claws are used to hold the victim while the cheetah kills it with a bite to the throat.

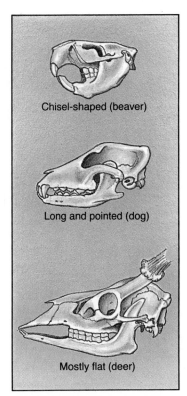

Chisel-shaped (beaver)

Long and pointed (dog)

Mostly flat (deer)

Fig. 12–12 The three types of mammal teeth.

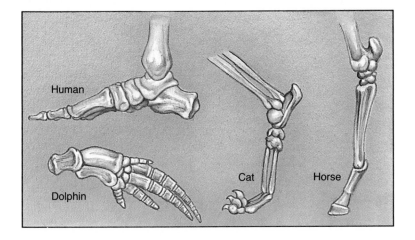

Human

Dolphin

Cat

Horse

Fig. 12–13 The shape of a mammal's foot is related to the way it moves.

Still another type of foot is the hoof. Hoofs are actually toenails, and the animal is actually "running on tiptoe." It does not hurt a horse to have horseshoes nailed on because the nails are driven into the toenail material, which is not living tissue.

Other mammals, such as humans and bears, walk on the entire foot from heel to toe. Because of this trait, these mammals move rather slowly.

KINDS OF MAMMALS

Most of the animals you are familiar with are mammals. *Bats* are in the only order of mammals that can fly. See Fig. 12–15. The fingers of their front limbs form the structure of the wings. Skin stretched between the fingers makes the wing surface. There are 900 or so different species of bats in the world.

Carnivores are animals that eat meat. They have teeth shaped for ripping and tearing. See Fig. 12–14. The order includes cats, dogs, bears, wolves, foxes, weasels, raccoons, skunks, walruses, and seals.

Mammals of two other orders are found in the sea. *Whales,* dolphins, and porpoises are in the Order *Cetacea.* Like all mammals, they have hair and feed their young milk. Adult whales have whiskers. These animals have fishlike bodies with fins and flippers. The blue whale is the largest animal that has ever lived. Some blue whales grow to 30 m long and may weigh 135 metric tons. (A metric ton equals 1,000 kg.) Scientists believe these ocean mammals first developed on land but slowly adapted to living in the sea because of a better food supply there.

The *rodents* are the largest order of mammals. There are over 3,000 different species in the order. In numbers of individual animals, rodents probably outnumber all the other orders combined. The beaver is one of the largest rodents. Like the beaver, all rodents have sharp, chisel-shaped teeth for cutting and gnawing. The group includes squirrels, chipmunks, rats, mice, gophers, and porcupines. The rodents' generally small size and rapid rate of reproduction have contributed to their success.

Hares and *rabbits* are sometimes called the rodentlike order

Fig. 12–14 Carnivores are meat-eating mammals. Their bodies are adapted for seeking their prey.

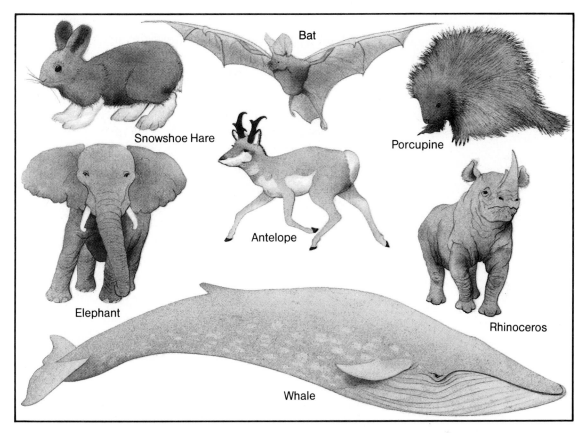

Fig. 12–15 All of these animals are mammals.

of animals. One difference from rodents is that they have two pairs of cutting teeth in the upper jaw while rodents only have one pair. The cottontail, jack rabbit, and snowshoe hare are in this group.

There are two orders of hoofed mammals. Members of one group, to which horses and rhinoceroses belong, have an *odd number of toes* on each foot. The members of the other group have an *even number of toes*. They include cattle, pigs, sheep, antelopes, giraffes, and hippopotamuses.

Only two species of the order containing *elephants* remain today. Elephants' long, muscular trunk and large tusks are easily recognized traits. At one time, there were up to thirty species of these animals. Today, all but two are extinct.

THE PRIMATES

Primates are the most highly developed order of the Animal Kingdom. The *primates* include monkeys, chimpanzees,

Fig. 12–16 Notice how this macaque uses its opposable thumbs.

Hominidae In the classification system, the family to which humans belong.

gorillas, and humans. Several traits in combination set this group apart. Primates are adapted for living in trees. Primates have hands with movable fingers shaped for grasping. The thumb can touch each of the other fingers. See Fig. 12–16. The ability to pick up things between the fingertips has given these animals an advantage over other mammals. The second trait is three-dimensional vision. The eyes of the primate are set on the front of the skull so that they look forward. This enables primates to perceive depth, which aids in judging distances. The parts of the brain used for sight and thinking are more developed in primates than other mammals.

We are primates, yet we are different from the apes and chimpanzees. We walk erect all the time. Apes and chimps use both their front and hind limbs for walking. Our larger brain provides us with better control of our fingers and hands. It is also better developed in those areas related to thinking and remembering. All humans are in the same genus and species. We are all classified as *Homo sapiens.* Together with other "humanlike" species that existed in the past, we make up the classification family **Hominidae** (haw-**mih**-nih-day).

SUMMARY

All mammals have hair at some point in their lives and feed their young milk from mammary glands. Mammals are classified into about 21 orders based on their reproductive pattern, tooth structure, and foot structure. Their body systems are all highly advanced.

QUESTIONS

Use complete sentences to write your answers.

1. What traits separate mammals from other vertebrates?
2. What are the three reproductive patterns found among mammals? Name one animal for each type of pattern.
3. What is the function of each of the three types of teeth found in mammals?
4. What are some of the different types of foot structure found among mammals? How does each type help the animal in its environment?

CAREERS IN SCIENCE

VETERINARIAN

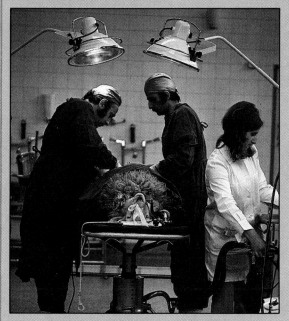

A pet that needs medical care can receive it from a veterinarian. Some veterinarians specialize in treating small animals—dogs and cats. Others care for large animals like horses, cattle, and sheep. Veterinarians provide medicines and vaccinations to prevent animal diseases. They also perform many different types of surgery, repairing organs, tissues, or broken bones. A few veterinarians work as inspectors of meat and poultry.

People who intend to become veterinarians must obtain a degree in veterinary medicine. The majority of veterinarians work for themselves in private practice, although many organizations employ veterinarians, including the police, the armed services, circuses, and zoos. Veterinarians often run their own animal hospitals, which may include a staff of assistants.

For further information, contact: American Veterinary Medical Association, 930 North Meacham Road, Schaumburg, IL 60196.

PET STORE RETAILER

If you have an interest in pets and want to own a business someday, you might consider becoming a pet supply retailer. These store owners sell a variety of items for pets, such as food, toys, leashes and collars, beds and cages, and an assortment of grooming tools. Some stores even sell the pets themselves. Usually, these include dogs, cats, fish, gerbils, and hamsters, although a few store owners specialize in more exotic animals.

There are no specific qualifications to become a pet store retailer. But you should have some knowledge of business, including merchandising, advertising, and accounting. It also helps to learn as much as you can about pets in order to sell healthy animals, judge the value of pet products, and be able to offer sound advice to your customers.

For further information, write: National Retail Pet Store and Groomers Association, P.O. Box 265, Danville, CA 94526.

CHAPTER REVIEW

VOCABULARY

On a separate piece of paper, write TRUE next to the number of each statement that is true. Next to the number of each false statement, write FALSE, then rewrite the statement to make it true, keeping the underlined term.

1. Birds are animals with fur that feed their young milk.
2. A warmblooded animal's internal body temperature is always changing.
3. Mammals have feathers covering their bodies.
4. Humans are in the classification family known as Hominidae.
5. Migrate means to lay eggs with hard shells.
6. Birds incubate their eggs by keeping them cold.

QUESTIONS

Give brief but complete answers to each of the following questions. Unless otherwise indicated, use complete sentences to write your answers.

1. What are the parts and the functions of a bird's egg?
2. How do birds' eggs differ from reptiles' eggs?
3. An ostrich is a *bird*. A bat is a *mammal*. What traits of each are used for this classification?
4. What traits do a bird and a mammal share?
5. What structure does each of the following use to obtain oxygen: fish, tadpole, snake, hawk, horse?
6. Give three examples of how birds' feet are structured for the type of life they live.
7. What are the three types of teeth found in mammals? What is the function of each type?
8. List three ways in which mammals reproduce. Identify animals that are examples of each.
9. How do the body coverings of birds and mammals differ? What are the functions of these coverings?
10. What are two characteristics that make primates different from the other mammals? How are these characteristics important to humans?

11. How does the structure of the heart in fish, amphibians, reptiles, and birds differ?
12. Why is being warmblooded an advantage to animals?

APPLYING SCIENCE

1. You are a living organism. Classify yourself into each of the following categories: kingdom, phylum, subphylum, class, order, family, genus, species. Give your reasons for your choices.
2. Why are birds and mammals the most common organisms in both the Arctic and Antarctic?
3. List ways in which birds are both useful and destructive to humans. In what ways are the hoofed mammals useful?
4. Obtain samples of various types of feathers. Use a scalpel to dissect them. Examine them under a hand lens. Sketch what you see. What makes flight feathers so sturdy?
5. Obtain, dissect, and compare the structure of the hearts of each class of vertebrates (for example, fish, frog, turtle, chicken, sheep).

BIBLIOGRAPHY

Hoyt, Erich. "The Whales Called 'Killer.' " *National Geographic Magazine,* August 1984.

Krapu, Gary L., and Jan Eldridge. "Crane River." *Natural History,* January 1984.

Madson, John. "A Lot of Trouble and a Few Triumphs for North American Waterfowl." *National Geographic Magazine,* November 1984.

McLoughlin, John C. *The Canine Clan: A New Look at Man's Best Friend.* New York: Viking, 1983.

Ricciuti, Edward. *They Work With Wildlife: Jobs for People Who Want to Work With Animals.* New York: Harper, 1983.

Robbins, Chandler S., Bertel Brunn, and Herbert S. Zim. *Birds of North America: A Guide to Field Identifications.* Jamestown, Ohio: Western Publishers, 1966.

Rue, Leonard Lee, III. *Furbearing Animals of North America.* New York: Crown, 1981.

CHAPTER
13

Imagine propelling your body 5 meters up into the air with only the aid of a pole. Which muscles and bones does the pole-vaulter use in this sport? Can you tell from this photograph?

SUPPORT AND MOVEMENT

CHAPTER GOALS

1. Describe the organization of the human body, including the major tissue groups, systems, body regions and their corresponding functions.
2. Identify the structures and functions of the skeletal system.
3. Compare and contrast the structure and functions of the different types of muscles in the muscular system.

13-1. Body Frameworks

At the end of this section you will be able to:

☐ Describe the four groups of tissues in the human body.

☐ List the human body systems and give the function of each.

☐ Name and identify the body regions of a human.

Humans have varying characteristics. Among these are different facial features, body size and shape, skin color, and hair color. These differences make each individual unique and special. However, every human being has the same biological characteristics. As a result, there is only one known genus and species of human alive today, *Homo sapiens*.

TYPES OF TISSUES

Some people do not like to think of themselves as members of the Animal Kingdom. Yet, like all animals and all living things, we must carry out life processes in order to survive. We must get food and oxygen for energy, growth, and repair. We must get rid of waste products. We must respond to changes in our environment. And, through reproduction, we must continue our own kind.

All the parts of the human body are made up of *cells* and the products of cells. The body of an average human adult contains about 40,000,000,000,000 (40 trillion) cells. Many of these cells are specialized to perform certain functions. For example, some groups of cells are specialized for movement.

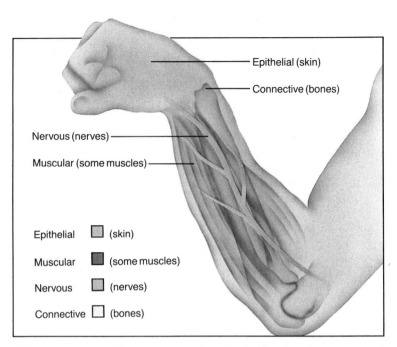

Epithelial (skin)

Connective (bones)

Nervous (nerves)

Muscular (some muscles)

Epithelial ☐ (skin)

Muscular ■ (some muscles)

Nervous ☐ (nerves)

Connective ☐ (bones)

Fig. 13–1 The four main tissue types in a human arm.

Epithelial tissue A type of tissue that covers all the body surfaces.

Muscular tissue A type of tissue that provides for movement.

Nervous tissue A type of tissue that helps an organism respond to stimuli in the environment.

Connective tissue A type of tissue that supports, joins, and binds various parts of the body.

Other cells provide protection for the body. Another group of cells supports the structure of the body. There are specialized cells that are only used for reproduction. Groups of cells that look alike and perform similar functions make up a *tissue* of the human body.

The human body is made up of four main groups of tissues. They are epithelial, muscular, nervous, and connective tissues. See Fig. 13–1. **Epithelial** (ep-ih-**thee**-lee-al) **tissues** cover and protect both the internal (inside) and external (outside) surfaces of the body. The lining of the nose and throat, the covering of the heart and lungs, and the outer portion of the skin are all *epithelial tissues.* In addition, some types of epithelial tissue are able to either secrete or absorb materials. For example, the stomach lining not only protects but also secretes digestive juices.

Muscular tissues provide movement. **Nervous tissues** help an organism respond to stimuli in the environment. *Nervous tissues* also help to coordinate body activities and body movement. In addition, they help to keep us informed about our surroundings.

Connective tissues are the most widely distributed tissues in the human body. *Connective tissues* support, join, and bind together the various other tissues. Some types of connective tissue also help distribute nutrients. Blood and the fluid surrounding cells are examples of this type of connective tissue. Connective tissues also include fat, bone, and cartilage.

In many parts of the body, all four of these groups of tissues work together to form *organs*. Some organs you are familiar with include your eyes, heart, lungs, liver, and stomach. Each of these organs performs a specific job or jobs to keep the body working properly. In addition, many organs work together to do a specific job. Organs that work together to perform a specific function make up a *system*. The body systems are listed in Table 13–1. Each system will be covered in the sections and chapters that follow. Most body activities involve many of these systems working together.

HUMAN BODY SYSTEMS	
System	Function
Skeletal	Provides form and support
Muscular	Produces movement
Digestive	Changes food into simpler compounds that can be used by the cells
Circulatory	Transports materials throughout the body
Respiratory	Exchanges gases between outside and inside the body
Excretory	Eliminates wastes from the body
Nervous	Receives, coordinates, and acts upon information from the environment
Endocrine	Chemical control system that helps keep the body's internal environment in balance
Reproductive	Allows humans to continue their own kind by producing more humans

Table 13–1

BODY REGIONS

The body has four main regions: the *head, neck, trunk,* and *limbs*. See Fig. 13–2 on page 298. The limbs are both the arms and the legs. The head contains many of the sense organs, such as the eyes and ears. It also contains the brain. The neck connects organs in the head to the rest of the body. It contains the windpipe, food pipe, nerves, and blood vessels.

The trunk includes two areas commonly referred to as the chest and abdomen. See Fig. 13–2. Within the chest area are the heart and lungs. The abdomen is below the lowest rib that you can feel. Many organs of the digestive, excretory, and endocrine systems are located in this area. The abdomen of a female also contains her reproductive organs.

Fig. 13–2 The four main body regions.

SUMMARY

As in other animals, human cells are organized into tissues. There are four types of tissue that make up the human body. In most parts of the body, these four types of tissues are organized into organs and systems. The organs and systems are found in the four main regions of the body.

QUESTIONS

Use complete sentences to write your answers.

1. Name the four types of tissues. What is the function of each?
2. How are most human tissues organized?
3. Make a list of the systems in the human body. Briefly describe the function of each system.
4. What are the four body regions? What are some of the organs and/or systems found in each?

13-2. The Skeletal System

At the end of this section you will be able to:

☐ List and explain the functions of a skeleton.

☐ Describe the structure of a typical bone.

☐ Describe the structure and give examples of major types of joints.

Many small organisms function well without a rigid framework. These include protozoa, jellyfish, earthworms, algae, fungi, and mosses. Larger land and water organisms need a strong, rigid framework. Trees and other large plants are supported by the hard walls of the xylem tubes. The material that we call *wood* is really old xylem built up over the years.

THE SKELETON

The framework of animals is called a *skeleton*. It gives shape and support to the bodies of larger animals. A skeleton also helps to protect and move parts of an animal's body. Some animals, such as insects, spiders, and crabs, have a hard, outerbody covering called an *exoskeleton*. See Fig. 13–3.

Humans, however, like all vertebrates, have an *endoskeleton*, which is made of bone and cartilage. The skeleton is one of the many human body systems. An endoskeleton provides an internal framework. Muscles are attached to the bones of the human skeleton.

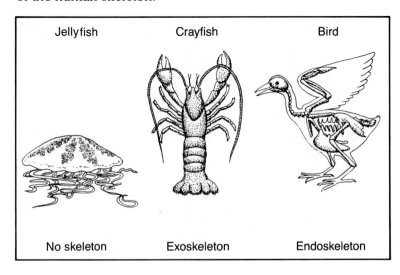

Jellyfish	Crayfish	Bird
No skeleton	Exoskeleton	Endoskeleton

Fig. 13–3 Which of these animals has a skeleton most like a human's?

THE SKELETAL SYSTEM

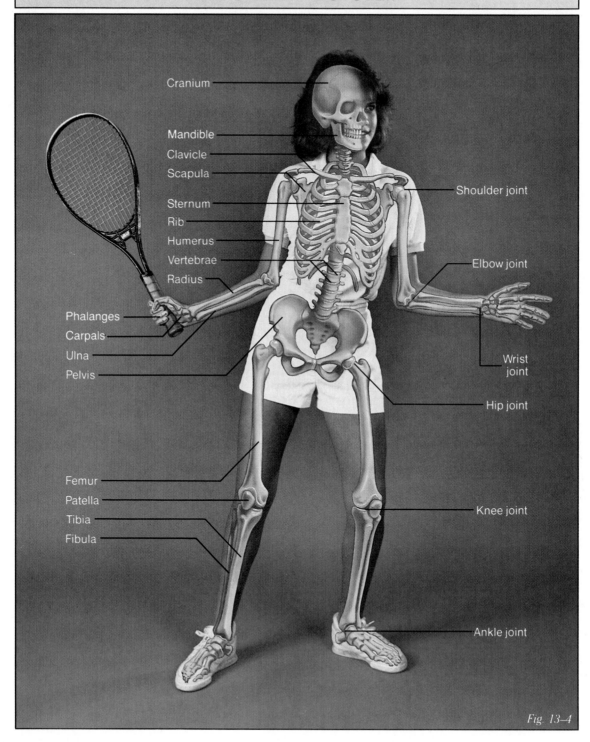

Cranium

Mandible

Clavicle

Scapula

Sternum

Rib

Humerus

Vertebrae

Radius

Phalanges

Carpals

Ulna

Pelvis

Shoulder joint

Elbow joint

Wrist joint

Hip joint

Femur

Patella

Tibia

Fibula

Knee joint

Ankle joint

Fig. 13–4

BONES

There are 206 separate bones in the human body. These bones make up only 18 percent of a person's body weight. Thus, the bones of a person weighing 100 kg would weigh 18 kg. Why is the hardest part of our body so light? It is because most bones are hollow inside. This makes them weigh less than if they were solid.

The parts of a bone are shown in Fig. 13–5. A tough membrane surrounds the bone. This membrane helps repair bone injuries. It also helps to attach muscles to bones.

Beneath the membrane is a hard bony layer. This layer contains live bone cells arranged in regular patterns. These live bone cells release the calcium and phosphorus compounds that form deposits and makes bones hard.

The hollow center of many bones is filled with soft, fatty tissue called **marrow**. The *marrow* contains many nerves and blood vessels. New blood cells for the body are made in the marrow.

Bones that have the same functions are constructed in a similar way. Look again at the diagram of the human skeleton in Fig. 13–4. Bones that protect delicate organs are strong and solid. The *skull* protects the brain. The *sternum* (breastbone) and *ribs* protect the heart and lungs. The *vertebrae* that make up the backbone protect the spinal cord. Can you see any other bones that look similar and have the same function?

Many of our bones have more than one function. Your backbone not only protects the spinal cord but also supports the

Marrow Soft, fatty tissue in the center of some bones that produces special cells found in blood.

Fig. 13–5 The three layers of a bone (left). *A photo of the bony layer* (right) *was taken through a microscope. The tiny, dark spots are bone cells.*

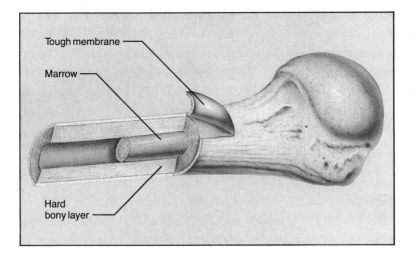

Tough membrane

Marrow

Hard bony layer

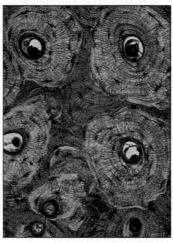

upper body and head. Leg bones provide support and also help the body to move.

JOINTS

Some bones are connected to other bones by tough strands of elastic tissue called **ligaments**. This tissue allows these bones to move freely. *Ligaments* are located in the movable **joints**. There are two main types of movable joints. A *hinge joint* can move in only one direction, like the hinge on a door. A hinge joint is very powerful because it cannot be twisted. The elbow is an example of a hinge joint. See Fig. 13–6.

Ligaments Tough strands of elastic tissue that connect bones at movable joints.

Joint A place where two bones meet.

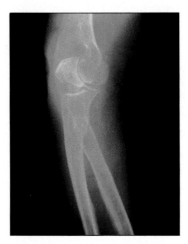

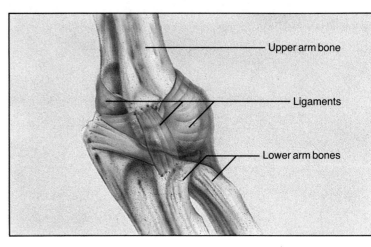

Fig. 13–6 An elbow, shown in an X-ray (left), *and a drawing* (right), *is a hinge joint.*

The second type of movable joint can be twisted. It is called a *ball-and-socket joint.* Look at Fig. 13–7. The joint at the shoulder and the upper arm is an example of a ball-and-socket joint. The end of the upper arm bone is round like a ball. It fits into a socket in the shoulder bone. This type of joint allows movement in many directions. It also allows rotating movements, such as those involved in throwing.

Exercise helps to keep movable joints loose by stretching the ligaments. Loose joints help you to move more easily. Even people whose arms or legs are paralyzed must have regular exercise. Without exercise, joints become tight and painful.

Some joints in the body are only *partially movable*. Your ribs are attached to your backbone by partially movable joints. See Fig. 13–7. In fact, the vertebrae of the backbone form partially movable joints called *gliding joints*. Your ribs are also attached to your breastbone. Some of the ribs are attached to

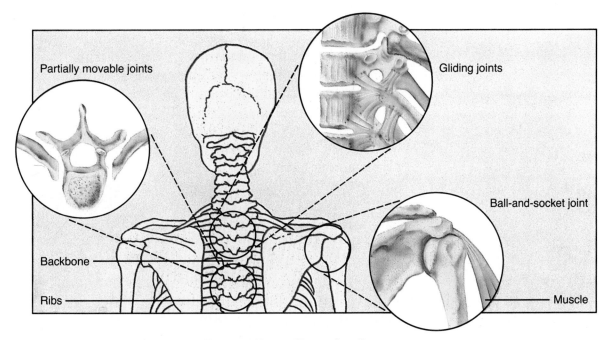

Partially movable joints

Gliding joints

Ball-and-socket joint

Backbone

Ribs

Muscle

the breastbone by cartilage. This cartilage allows the rib cage to expand when you breathe.

Other joints in the body are immovable. The cranial bones of the skull are immovable. However, these bones in a newborn baby are not joined together. See Fig. 13–8. As babies grow bigger, the cranial bones gradually join and the top of the skull becomes solid. This slow process allows room for the brain to grow and develop.

Cartilage covers the inside surfaces of most joints. It acts as a cushion between the bones. Joints also contain a special fluid. This fluid lubricates the joints. It also reduces the rubbing together of the two bones.

Fig. 13–7 Three types of joints.

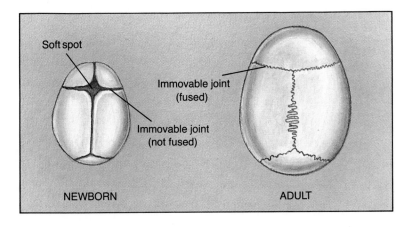

Soft spot

Immovable joint (fused)

Immovable joint (not fused)

NEWBORN

ADULT

Fig. 13–8 Compare the joints in the skull of a baby and an adult.

If joints are suddenly twisted or pulled, a *sprain* can result. A sprained ankle or wrist becomes swollen and painful. This injury results from stretched or torn ligaments and *tendons*. Therefore, it is important to see a physician if you have a severe sprain.

Breaks to bones may or may not be completely through the bone. In either case, swelling and pain occur and medical treatment is necessary. See Fig. 13–9.

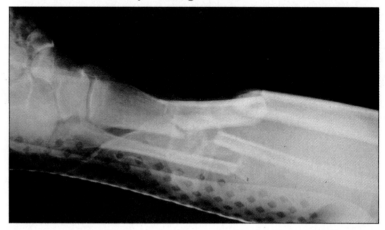

Fig. 13–9 An X-ray of broken bones in a person's lower leg.

SUMMARY

The human skeleton supports the body. It also houses and protects many organs and tissues. The human body contains 206 bones of different shapes and sizes. A joint is where two or more bones meet. Cartilage and a special fluid cushion bones at most joints. Some of these joints allow a wide variety of movements.

QUESTIONS

Use complete sentences to write your answers.

1. What are four functions of the skeleton?
2. Which bones protect the organs listed: lungs, brain, spinal cord, heart?
3. What are two main types of movable joints? Give examples of each.
4. What do ligaments and cartilage have to do with joints?
5. What structures in insects and trees have the same function as the human skeleton?

SKILL-BUILDING ACTIVITY

COMPARING AND CONTRASTING

PURPOSE: To *compare and contrast* some parts of the human skeleton and to identify joints by examining bone movement.

MATERIALS:
Fig. 13–4 or model
of human skeleton

PROCEDURE:

A. Compare the arm bones and leg bones.

   1. How are they similar?

   2. How are they different?

B. Gently bend your ear and wiggle the end of your nose with your hand. Compare these two movements.

   3. What type of tissue forms your ears and the end of your nose?

   4. Where else is this firm but flexible tissue found?

C. On a separate sheet of paper, copy Table 13–2. Use it to record your answers to some of the questions that follow.

D. Stand up. Keeping one leg straight, raise it forward, backward, and to each side.

Joint	Bones Connected	Type of Movement

Table 13–2

Still keeping your leg straight, swing it around in a *small* circle.

   5. What type of joint connects the upper leg bone to the hip? Write your answer on your chart.

   6. How do you know? Record the type of movement on your chart.

   7. Where else on your body do you find this same type of joint? Fill in the next line of your chart.

E. While sitting, move your lower leg forward and backward at the knee. Then move your lower leg forward so that your leg is straight out in front of you. Now try to move your *lower* leg (from the knee down) from side to side.

   8. Can you do it?

   9. What type of joint is a knee? Write this on your chart.

   10. What movement does this joint allow? Write this on your chart.

   11. Where else is this type of joint located? Fill in the next line of your chart.

CONCLUSIONS:

   1. How many ball-and-socket joints did you locate? Name them.

   2. How many hinge joints did you find? Name them.

   3. In the backbone is firm, flexible tissue like that forming your ears and the tip of your nose. This same tissue is located between each vertebra and is disc-shaped. What is its function in this location?

13-3. Muscles

At the end of this section you will be able to:

- ☐ Describe three types of muscles.
- ☐ Explain how muscles work.
- ☐ Compare involuntary and voluntary muscles.

In contrast to the 18 percent taken up by bones, muscles make up 40 percent of your total body weight.

SMOOTH MUSCLE

Smooth muscles Involuntary muscles made up of cells that are long, thin, and have one nucleus.

Involuntary muscles Muscles that cannot be controlled at will.

There are three types of muscle cells: smooth, skeletal, and cardiac. **Smooth muscle** forms the walls of the digestive system. It is also found in blood vessels and other organs. *Smooth muscles* are **involuntary**. This means you usually cannot consciously control these muscles. Smooth muscle cells are thin, long, and pointed at each end. Each cell has one nucleus, usually near its center. See Fig. 13–10. Smooth muscle cells are joined to form sheets of muscle tissue.

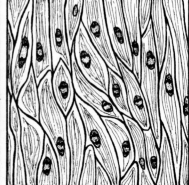

Fig. 13–10 Smooth muscle cells.

SKELETAL MUSCLE

Skeletal muscles Voluntary muscles made up of cells that are shaped like long cylinders and have many nuclei.

Voluntary muscles Muscles that can be controlled at will.

One of the most abundant tissues in the human body is **skeletal muscle**. *Skeletal muscles* are **voluntary**. You can consciously control these muscles. Each skeletal muscle cell is like a long cylinder with many nuclei. It looks striped because dark and light bands run across it. See Fig. 13–11. Skeletal muscle cells are joined together in bundles instead of sheets. These bundles are clustered to form the muscles that you can feel in your arms and legs, for example.

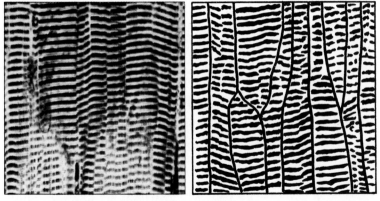

Fig. 13–11 Skeletal muscle cells.

Some skeletal muscles are attached to bones by **tendons** (**ten**-duns). *Tendons* are tough, nonelastic tissue. They act like the strings of a marionette. When the muscle contracts, or tightens, the tendon pulls on the bone. See Fig. 13–12. Other skeletal muscles are attached directly to bones. Still others, like the tongue and lips, are attached to other muscles.

Tendons Tough, nonelastic tissue that attaches some skeletal muscles to bones.

Since muscles work only by contracting, working muscles get shorter and thicker. As you move, you can feel which voluntary muscles are contracted by feeling how firm or relaxed each one is. Many skeletal muscles work in pairs. One skeletal muscle pulls on a tendon to bend a joint. A muscle that bends a joint is called a **flexor** (**flek**-sor). The other skeletal muscle pulls on a different tendon to straighten the same joint. A muscle that straightens a joint is called an **extensor** (ek-**sten**-sor). *Flexors* and *extensors* are used in almost every movement of the body. See Fig. 13–12.

Flexor A muscle that bends a joint.

Extensor A muscle that straightens, or extends, a joint.

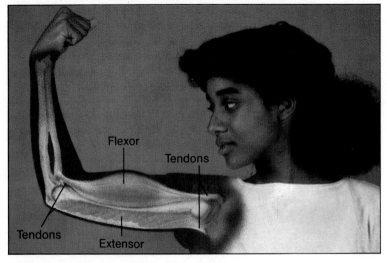

Fig. 13–12 Tendons attach muscles to bones. Which muscle is used to bend the elbow—the flexor or the extensor?

Skeletal muscles must be used often if you want to keep them strong and firm. When you do not use muscles, they become smaller, weaker, and flabby. Bulging muscles are not always the strongest muscles. Many outstanding athletes do not have big, bulging muscles.

Often after playing, exercising, or working very hard, a person's muscles will become sore. This is caused by the buildup of chemical wastes in the muscles. These wastes cause the muscle to ache. Often a hot bath or shower and a good rest will bring relief.

Sometimes muscles will *cramp*, which is painful. Cramps occur when a muscle suddenly and involuntarily contracts and will not relax. A temporary lack of food or oxygen to the muscle may cause a cramp. Rubbing the muscle will help.

A muscle can be overstretched. When this happens, the muscle *strains* and tears itself or the tendon. A sudden and intense pain can be felt immediately. Within another few seconds, a hard, tender swelling can be felt. Many muscle injuries can be avoided by doing warm-up exercises.

CARDIAC MUSCLE

Cardiac muscle is a third type of muscle. The heart is made of *cardiac muscle*. Like smooth muscle, cardiac muscle is involuntary. The cells of cardiac muscles have dark and light bands, just as the cells of skeletal muscles do. These cardiac cells are woven together so tightly that it's hard to tell that they are separate cells. See Fig. 13–13.

Unlike other types of muscle, cardiac muscle cells are able to contract without a nerve impulse. However, nerve signals

Cardiac muscle An involuntary muscle found only in the heart and made up of tightly woven cells.

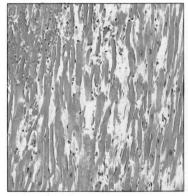 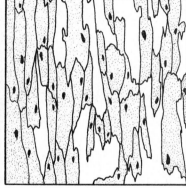

Fig. 13–13 Cardiac muscle cells, magnified 100 times.

coordinate and help regulate this rhythmic contraction that we know as the heartbeat. Cardiac muscle works 24 hours a day. It rests only between heartbeats. Like skeletal muscle, the heart is strengthened with exercise. The stronger the heart is, the more blood it can pump with each beat. Thus, it can beat more slowly and rest more often.

HOW MUSCLES WORK

Muscles work only by contracting. Muscle cells contract, or get smaller, when a message from a nerve signal tells them to do so. Muscle cells need energy to contract.

When a muscle contracts, it becomes shorter and firmer. The muscle relaxes when the nerve signal stops. Some muscles never completely relax. They remain partially contracted all the time. This is called muscle tone. See Fig. 13–14. These muscles help hold up the body. If these muscles relaxed, the body would collapse. Does this mean these muscles work 24 hours a day? In these support muscles, some groups of muscle cells contract while other groups rest. That way, no single group of cells get tired and overworked.

Fig. 13–14 Exercising muscles for strength or endurance results in good muscle tone.

SUMMARY

Muscles are the movement specialists of the body. You cannot control smooth and cardiac muscles. You can control skeletal muscles to move many parts of your body. Skeletal muscles are usually connected to bones and work in pairs to move the bones.

QUESTIONS

Use complete sentences to write your answers.

1. Make a chart to compare the three types of muscle cells. Indicate their structure, location, and function.

2. Explain what this statement means: "Muscles never push, they always pull."

3. What is the difference between voluntary and involuntary muscles?

4. Classify the muscles of the following as voluntary or involuntary: stomach walls, leg, heart, blood vessels, arm, hand.

INVESTIGATION

SKELETAL MUSCLES

PURPOSE: To locate pairs of skeletal muscles in the arms and legs.

MATERIALS:

desk or table   chair

PROCEDURE:

A. Sit straight in your seat and place one hand, palm up, under a desk or table top. Try to lift it with this hand. Use the other hand to feel the front and back of your upper arm as you press up against the desk or table.

    1. Which muscle is harder, the one in front or the one in back?

    2. Which muscle pulls your arm up toward your body, the front or the back muscle of the upper arm?

B. Next, place your hand, palm up, on top of your desk. Press down on your desk. Use your other hand again to feel the front and back of your upper arm as you do this.

    3. Which muscle is harder, the front one or the back one?

    4. Which muscle pulls your arm away from your body?

C. Sit on the edge of your chair and place both of your feet flat against the floor. Place one hand under the middle of your thigh. Place your other hand on top of the same thigh. While pushing against the floor with your foot, slide your foot slowly forward. See Fig. 13–15.

    5. Which muscle is harder, the top or the bottom muscle?

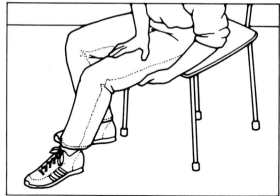

Fig. 13–15

6. Which muscle extends your leg?

D. Hold the same position as in C. Try to pull in your lower leg while still pushing down with your foot.

    7. Which muscle pulls the lower leg toward the body?

E. You are feeling skeletal muscles that work in pairs. A flexor bends a joint and an extensor straightens it. You may wish to explore other muscles of your body to see if you can identify other pairs.

    8. Where is the flexor muscle that bends the foot? Where is the extensor muscle that extends the foot?

    9. Swing one arm back and forth. What is the name of the muscle that swings it backward? Which muscle swings it forward?

CONCLUSIONS:

1. How can you determine if a muscle is working?

2. In your own words, explain why two different muscles are needed to bend and extend limbs.

EXOSKELETAL ROBOTS

Who could overturn a small car with just a flick of the wrist? The answer is not a science-fiction character. It is a real robot named Hardiman. This is no cuddly character with flashing lights out of a movie. Hardiman is an exoskeleton, a robot shell that is worn over the real skeleton of a human being. Exoskeletal robots have arms and legs and can usually walk. They are designed for different purposes. One purpose, like that of Hardiman, is to multiply human force. The person inside Hardiman is attached to the robot at the feet, forearms, and waist. When the person lifts something, his or her own strength is multiplied several times, enough to lift up to 680 kilograms. Hardiman is used underwater to salvage objects from sunken ships. Similar exoskeletons have been used in space.

Another design for exoskeletons is a walking machine. This machine can be used by people who have lost the ability to move their arms and legs. The walking shell is made up of 27 kilograms of tubing, motors, and artificial joints, which are activated by a computer. This exoskeleton can walk forward and backward, turn, and go up stairs.

Amazing as it is that robots can be made to walk up stairs and pick up heavy objects, they still cannot do things that a human hand does—like peel an apple—nor move with the freedom of the human arm. Scientists describe the human arm as having 42 degrees of freedom—that is, it has movement in 42 directions. The most advanced artificial arm has no more than 10 degrees of freedom.

Nevertheless, artificial arms available today allow people to comb their hair, eat with utensils, and even scratch their backs. In one design, a strap across the back holds the arm in position. As the strap is moved by other parts of the body (such as the leg), it sends signals to a computer, which activates the arm electronically.

Another type of artificial arm can be controlled almost like a real one. The messages from the brain are communicated to the remaining muscles near the artificial arm. There, the nerve signals are changed into electrical signals and sent to the artificial arm, commanding it to move. The speed and force of the movement are related to the amount of electricity in the signal. Artificial arms are usually powered with long-life batteries and computer chips to relay commands.

Work is now being done to make artificial limbs more sensitive to different textures and temperatures. In Japan, where much factory work is done by assembly-line robots, researchers have developed artificial hands with hundreds of contact points that can be stimulated by different textures. Some hands can "recognize" objects. Can you think of other ways in which exoskeletons might prove useful in the home and in the workplace?

CHAPTER REVIEW

VOCABULARY

On a separate piece of paper, match the number of each blank with the term that best completes each statement. Use each term only once.

cardiac	extensors	ligaments	smooth
connective	flexors	muscular	tendons
epithelial	joints	skeletal	

Aerobic dance is good exercise that is fun and doesn't require much time. Some professional ice hockey teams have even added it to their regular training programs. Aerobic dance thoroughly works the large voluntary muscles called ___1___ muscles. It also gives the heart or ___2___ muscle a vigorous workout. The involuntary or ___3___, muscles benefit indirectly from aerobic exercise. Therefore, the entire ___4___ system benefits.

An aerobic routine is "danced" to popular music. It generally starts with three or four minutes of easy bends and stretches. This warms up the muscles and loosens the group of tissues that connect muscles and bones called ___5___ tissue. Tough strands of elastic tissue called ___6___ connect bones at movable ___7___. The tough, nonelastic tissue that connects some skeletal muscles to bones, called ___8___, are also stretched. This helps a person to become more flexible.

After the warm-up comes a series of jog steps, arm swings, hops, jumps, and bends. During these movements, the arms and legs alternately bend and extend. The muscles that bend a limb are called ___9___, and the muscles that extend a limb are called ___10___. These muscles work in pairs, and both are exercised.

The last few minutes of the routine are for cooling down and easy stretching again. This helps to prevent the muscles from becoming tight. After aerobic exercise the skin has a healthy glow. The skin is part of one of the major tissue groups in the body called ___11___ tissue. Aerobic dance can be an enjoyable way to stay healthy and fit.

QUESTIONS

Give brief but complete answers to each of the following questions. Unless otherwise indicated, use complete sentences to write your answers.

1. Briefly describe the functions of each of the four groups of tissue in the body.
2. What are some examples of connective tissue?
3. List the four main regions of the human body.
4. How is the skeleton of a human different from the skeleton of a crab?

5. What are the functions of the human skeleton? How do the ribs perform these functions?
6. Describe the structure and give examples of the following joints: hinge, ball-and-socket, immovable, partially movable.
7. What are the three types of muscle tissue? Give an example of each.
8. How are the tendons of a person like the strings of a marionette?
9. Why must some muscles work in pairs to move parts of the body?
10. What are the systems of the body? What is the function of each system?

APPLYING SCIENCE

1. When a cast is removed from a broken arm or leg, that limb is often smaller compared to the other limb. Why does this happen? What can be done to correct this?
2. Why is an exercise program important to your heart?
3. Classify each of the following as a tissue or an organ. Explain how you reached your decision in each case: skin, teeth, heart, fingernails, lining of the stomach, kneecap, hair, lungs, eyes, ears.
4. Which do you think is stronger, the skeletal muscles in your hand or the cardiac muscles of your heart? To test this, squeeze a rubber ball in your hand about 70 times each minute. This is about the same number of times your heart beats each minute. Each squeeze is somewhat like the beat of your heart. Continue to do this until you are too tired to do it any longer. Record your results. Remember that the heart rests only between beats.
5. Assuming your heart beats about 70 times per minute, calculate how many times the heart beats in an hour. In a day. In a week. In a month.

BIBLIOGRAPHY

Allen, Oliver E., and The Editors of Time-Life Books. *Building Sound Bones and Muscles*. New York: Time-Life Books, 1981.

Rahn, Joan Elma. *Grocery Store Zoology: Bones and Muscles*. New York: Atheneum, 1977.

Showers, Paul. *You Can't Make a Move Without Your Muscles*. New York: Crowell, 1982.

Silverstein, Alvin and Virginia. *The Muscular System: How Living Creatures Move*. Englewood Cliffs, NJ: Prentice-Hall, 1972.

NUTRITION AND DIGESTION

CHAPTER GOALS

1. Name the basic nutrients found in food and describe their importance to body processes.
2. Identify major organs of the digestive system.
3. Describe the sequence of events in mechanical digestion.
4. Describe how chemical digestion takes place.

14-1. Nutrition

At the end of this section you will be able to:

☐ Identify six basic *nutrients* found in food.

☐ Explain how processing, additives, and cooking affect the nutritional quality of food.

☐ Discuss the *calorie* content of foods.

"You are what you eat." You may have heard this expression and wondered what it meant. If you think about it, you may realize that you and the food you eat are made up of some of the same compounds. Through *digestion,* the food is changed into substances that your body uses to work, grow, and carry out life processes. The food you eat really does become part of you.

What is food? Food is any substance that provides your body with **nutrients.** *Nutrients* furnish energy and build, repair, maintain, and regulate body processes. Nutrients are divided into six groups: carbohydrates, fats, proteins, water, minerals, and vitamins.

Nutrient Substance that does one or more of the following: furnishes energy, builds, repairs, and maintains the body; regulates body processes.

CARBOHYDRATES

Plants combine carbon, oxygen, and water during photosynthesis to make **carbohydrates** (car-boe-**hye**-draytz). *Carbohydrates* are organic compounds. Organic compounds contain carbon. Carbohydrates are also energy providers. When carbohydrate molecules are broken apart, the stored energy that held them together is released. This energy is used by an organism to perform its many activities.

Carbohydrates Organic compounds, such as sugars and starches, that are sources of energy.

Fig. 14–1 Simple carbo-
hydrates (left) and complex
carbohydrates (right).

Carbohydrates can be divided into two groups: simple and complex. *Simple carbohydrates* are sugars. *Complex carbohydrates* include starches, which are in foods such as breads, potatoes, and cereals. See Fig. 14–1.

FATS

Fats Organic compounds that are storage compounds. They provide twice as much energy as carbohydrates.

Fats are another group of organic compounds. Most *fats* are storage compounds. They store large amounts of energy for future use. When you eat more food than your body needs, the extra food is changed to body fat. Body fat insulates the body and cushions the organs. Fats provide more than twice as much energy as carbohydrates. Foods containing fats stay in the stomach longer. This means that meals containing some fats keep you from feeling hungry longer than meals without fats.

Fats can be divided into two groups. See Fig. 14–2. Fats that are solid at room temperature are called *saturated fats.* Animal fats, dairy products, coconut and palm oils, and solid vegetable shortenings contain large amounts of saturated fats. Fats that are liquid at room temperature are called *unsaturated*, or polyunsaturated, fats. Oils made from corn, sunflower seeds, safflower seeds, and soybeans are examples of unsaturated fats.

Another word that is often used when discussing fats is *cholesterol* (koe-**les**-teh-rawl). Cholesterol is found in saturated fats. It is also produced by the body and is needed for normal body functioning. However, in some people cholesterol collects on the inside of blood vessels. This blocks the

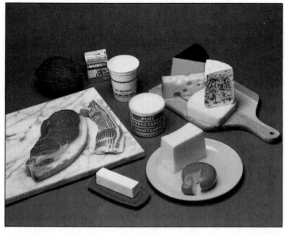

flow of blood and could be a factor in causing heart attacks. See Fig. 14–3. Does this mean you should limit the amount of saturated fats you eat? This question has been difficult for scientists to answer. It is known that eating too many fats can be a cause of overweight. Thus, it is generally recommended that people limit the amount of fats in their diets.

Fig. 14–2 Foods containing saturated fats (left) *and unsaturated fats* (right).

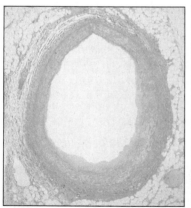 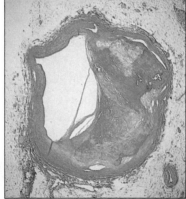

Fig. 14–3 A cross-section of a healthy artery (left) *and a partly clogged artery* (right), *a condition that is called arteriosclerosis.*

PROTEINS

Proteins are the most common organic compounds in your body. They make up about 75 percent of the solid matter of the body. *Proteins* build, repair, and maintain the body. Proteins can also provide energy. However, proteins are used to meet the other body needs first. Extra protein is then either used for energy or is stored as fat. The diets of infants, teenagers, and pregnant and nursing mothers need more protein. More protein is also needed during an illness or injury.

Proteins Complex organic compounds that are the building blocks of organisms.

Amino acids About 20 different organic compounds that form proteins.

Fig. 14–4 Combining incomplete proteins can provide all the essential amino acids.

Proteins are made up of smaller organic compounds called **amino** (ah-**mee**-noe) **acids**. A single protein is many *amino acids* joined together. There are over 20 different amino acids that form proteins. Each type of protein differs in the number, kind, and order of these amino acids.

Plants can make all of the amino acids they need. Animals cannot. There are about nine amino acids that animals, including humans, need but cannot make in their bodies. These must be supplied by the food that is eaten. They are called *essential amino acids.* Foods that supply all of these essential amino acids are called *complete proteins.* Dairy products, eggs, meat, fish, and poultry provide complete proteins.

Foods that do not supply all the essential amino acids are called *incomplete proteins.* Whole grains, beans, peas, lentils, nuts, and seeds contain incomplete proteins. However, combining two or more incomplete proteins can provide complete proteins. See Fig. 14–4.

WATER

Water is necessary for life. You can live without food much longer than you can live without water. About 60 to 75 percent of your body is water. Water is needed for protoplasm, tissue fluid, and blood. Water is also needed for dissolving substances so they can diffuse into and out of the cells. Water is needed for every body process.

Half of the water in your body is lost and replaced every ten days. Water is lost through perspiration and the removal of

Fig. 14–5 Competing athletes require large amounts of water to replace that lost by perspiration. These marathoners must actually drink during the race.

waste products from the body. Much of the water lost from the body is replaced by the foods you eat. However, it is also important to replace water lost from the body by drinking five or six glasses of water daily. See Fig. 14–5.

MINERALS AND VITAMINS

Minerals and vitamins are taken into the body in smaller amounts than the other nutrients. Minerals are elements used to form bones, teeth, and other parts of the body. Both vitamins and minerals are necessary to the growth, repair, and maintenance of every living cell. These two nutrient groups are also important in the functioning of enzymes. Enzymes are chemicals that regulate chemical reactions in a living cell. For the most part, your body cannot make vitamins and minerals. Therefore, they must be supplied. A well-balanced diet should provide all the vitamins and minerals you need. A severe lack of one vitamin in the diet can lead to a *deficiency disease.* See Table 14–1 below and Table 14–2 on page 320.

How food is prepared and cooked affects whether the nutrients it contains are destroyed or not. To prevent vitamins and minerals from being lost during cooking, use a minimum amount of water. It is best to cook food at a higher temperature for a shorter amount of time. Cook with fresh foods, rather than canned or frozen, and serve the food as soon as it is prepared.

MINERALS	NEEDED FOR	SOURCES
Sodium	blood and body tissues	table salt, vegetables
Calcium	strong bones and teeth; heart and nerve action; clotting of blood	meat, vegetables, fruit, milk, and bread
Phosphorus	strong bones and teeth; making ATP	meat, vegetables, fruit, milk, bread, and cereal
Magnesium	muscle and nerve action	vegetables, whole grains
Potassium	blood and cell activities; growth	vegetables and fruits
Iron	making red blood cells	liver, lean meats, shellfish, whole grains, green leafy vegetables
Iodine	proper functioning of thyroid gland	iodized salt and seafood
Zinc	important part of many enzymes	protein foods

Table 14–1

VITAMIN	NEEDED FOR	SOURCES	DEFICIENCY DISEASE
A	growth; healthy eyes and skin	liver, eggs, fortified milk, dark-green vegetables	night blindness, poor growth
B_1	growth; use of carbohydrates; healthy heart and nerves	pork, liver, oysters, whole grains	beriberi (incomplete digestion and nervous disorders), poor growth
B_2	growth; healthy skin and mouth, eye function	milk, meat, eggs, vegetables, whole grains	skin disorders, poor growth
B_3 (niacin)	use of carbohydrates, proteins, and fats; normal body growth and energy supplies	meat, tuna, eggs, whole grains, beans	pellagra (mental disorders, digestive problems, skin disorders)
B_6	healthy skin, nervous and digestive system functioning	whole grains, liver, spinach, bananas, nuts, potatoes	skin disorders, anemia, poor white blood cell functioning
B_{12}	making red blood cells	all foods from animals	pernicious anemia (reduced red blood cells)
C	repairing blood vessels; healthy teeth, gums	citrus fruits, melon, green pepper, potatoes	scurvy (tendency to bruise), sore gums
D	growth; building bones and teeth	fortified milk, eggs, made by your skin when exposed to sun	rickets (soft bones, poor development of teeth)
E	growth; reproduction; response to stress; digestion; use of polyunsaturated fats	vegetable oil, whole grains, vegetables	not known
K	blood clotting; liver function	green leafy vegetables, peas, potatoes	bleeding problems

Table 14–2

FOOD ADDITIVES

Today, many foods are processed into products that are different from traditional foods. Some examples include frozen dinners, breakfast cereals, and canned stews. Many essential nutrients are lost during processing. It is sometimes difficult to tell exactly what these foods contain.

Processed foods usually contain many additives, such as salt and sugar. Too much salt may contribute to high blood pressure. Sugar in the mouth can cause tooth decay. Many other additives used today have been created in the laboratory. Some additives once used have now been found to be unsafe.

Why are food additives so widely used? There are four basic reasons. First, foods are easier to process and prepare with food additives. This makes it cheaper to produce food products. Second, attractive food colors and flavors make food more appealing to customers. Third, preservatives reduce spoilage and increase shelf life. Fourth, additives are used to enrich and fortify foods. Minerals and vitamins are added to food products to replace nutrients lost during processing. Sometimes they are used to fortify foods by adding nutrients that may be lacking in the diet.

CALORIES

Many food products provide energy with little or no nutrients. These "junk foods" are said to have "empty calories." A **food calorie** is a measure of energy that foods provide.

Food calorie A measure of energy that foods provide.

How many *calories* do you need daily? This depends upon your age, your size, and how active you are. In general, the average person needs from about 2,500 to 3,500 food calories a day. See Fig. 14–6. Remember, this will vary from person to person. Also, along with these calories, a person should be getting the necessary nutrients. If you take in more calories than your cells can use up, the extra food will be changed to fat. This results in weight gain.

What foods should we eat to keep healthy? In other words, what is good nutrition? No one food contains all the essential nutrients. Good nutrition means eating a variety of foods.

SUMMARY

Food contains six essential nutrients. Food also provides energy for the body. The use of food processing and food additives sometimes makes it difficult to know what nutrients are in a food product. Eating a variety of foods is the best way to get the essential nutrients.

QUESTIONS

Use complete sentences to write your answers.

1. What six nutrients are provided by foods?
2. What is the difference between processed foods and traditional foods?
3. Why are food additives used?
4. What does the term "calorie" mean?

Fig. 14–6 An Olympic runner must get adequate nutrients and calories to compete.

SKILL-BUILDING ACTIVITY

PROBLEM SOLVING

PURPOSE: To determine the foods to eat by comparing and contrasting food labels.

MATERIALS:
food labels

PROCEDURE:

A. Look at the food labels in Fig. 14–7. One is for grape juice, the other is for grape drink. In addition to the ingredients, notice the nutrition information.

   1. Which product contains more ingredients? List the ones that are different.

B. Compare the ingredients of the two products. Ingredients are always listed in order of decreasing weight. Whatever ingredient is most abundant is listed first.

   2. What is the most abundant ingredient in grape drink?

   3. In most cases, grape drink is less expensive than grape juice. Suggest a reason why this might be true.

   4. How could you make grape juice less expensive yet more nutritious than grape drink?

C. Look at a food label from one of your favorite foods.

   5. What is the name of the product?

   6. List anything the product contains or provides that you may not have been aware of before you read the label.

   7. Is there a more nutritious alternative to this product? If so, what is it?

   8. What benefits would you get by eating the alternative?

CONCLUSIONS:

   1. Why is eating nutritious food vital?

   2. What types of information can you get from a food label?

   3. How can reading food labels help you to make better food choices?

Fig. 14–7

14-2. Mechanical Digestion

At the end of this section you will be able to:

- [ ] List the functions of the human digestive system.
- [ ] Describe the sequence of events that occur in *mechanical digestion.*
- [ ] Explain how food moves through the food tube.

Do you like riddles? Try this one. When is a tube of toothpaste like the *digestive system?* You will find the answer as you study this section.

DIGESTIVE SYSTEM FUNCTIONS

All animals need to get food. Food is a source of energy for living cells. Without energy, your body could not run, jump, read a book, or even breathe. Food must first be changed into a form that can be used by cells. Changing food into a simpler form that can be used by cells is called **digestion.**

In humans, *digestion* takes place in the digestive system. The human digestive system has two parts. One part is the digestive tube. The other part of the digestive system consists of the organs that help digestion. The digestive system has several functions:

1. Breaking down food into smaller pieces.
2. Pushing food along by a series of muscular contractions.
3. Producing special chemicals that help digestion.
4. Absorbing the digested food into the bloodstream.
5. Absorbing water.

Digestion The process that breaks down food into a simpler form so it can be used by cells.

THE MOUTH

Before your cells can use food, such as a cheeseburger, it must be broken down into smaller pieces. This process is called **mechanical digestion.** *Mechanical digestion* begins in the mouth where the teeth bite and chew the food. The tongue helps mash and push the food around during chewing. As it is chewed, the food is mixed with **saliva** (sah-**lie**-vah). *Saliva* is a watery liquid produced by the *salivary* (**sal**-ih-ver-ee) *glands.* See Fig. 14–8 on page 324. It moistens the food and makes it easier to swallow.

Mechanical digestion The process in which food is broken down into tiny pieces.

Saliva A liquid produced by the salivary glands in the mouth.

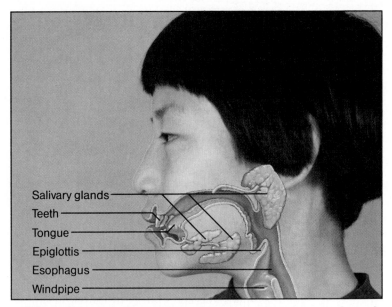

Fig. 14–8 Digestion begins in the mouth. Next, the food moves down the food tube to the stomach.

Salivary glands
Teeth
Tongue
Epiglottis
Esophagus
Windpipe

When chewing, the teeth and tongue grind, tear, and mash food into smaller pieces. Let's see what is happening as that piece of food travels down to your stomach.

There are two openings at the base of your throat. One opening is the windpipe, leading to the lungs. The other is the opening of the food pipe leading to the stomach. These two openings lie next to each other. See Fig. 14–8. When you swallow, a flap of tissue, called the *epiglottis* (ep-ih-**glot**-iss), covers the windpipe. This prevents food from going into your lungs, which would cause you to choke. When you talk, the epiglottis must be open. If you chew and talk at the same time, this increases the chances of choking.

Esophagus The food tube that connects the throat and the stomach.

The food enters the food pipe called the **esophagus** (eh-**sof**-ah-guss). The *esophagus* is a long tube that connects the throat to the stomach.

MOVEMENT OF FOOD

A series of circular muscles make up part of the wall of the esophagus. These muscles push the food downward toward the stomach. The muscles in front of the food relax while the muscles behind the food contract (squeeze). This causes the food to be pushed along. This series of muscular movements occurs so quickly that it has a wavelike motion. Think of how you get the last bit out of a tube of toothpaste. This is similar to how muscles move the food through the esophagus. See

Fig. 14–9. The muscles of the esophagus move automatically without your thinking about them. What other body muscles are involuntary like these?

The inside surface of the esophagus makes a liquid called *mucus* (**mew**-kuss). Mucus is also produced by the inside surface of the entire digestive tract. It provides a slippery surface to enable the food to travel along easily. This process is similar to coating your finger with soap to get a ring off. Muscular contractions are still needed to push the food along. The mucus simply makes the pushing easier.

Once the food leaves the esophagus, it enters the **stomach.** Where the esophagus meets the *stomach*, there is a special ring of muscle. This muscle is called a **sphincter** (**sfingk**-ter). When it contracts, the entrance to the stomach closes. (Pucker your lips. Your lips are a type of sphincter muscle.)

Stomach A muscular storage sac that helps break food down into smaller pieces.

Sphincter A ring of muscle that closes an opening or a tube.

Fig. 14–9 Rings of muscle and mucus move food along. Notice the locations of sphincter muscles throughout the digestive system.

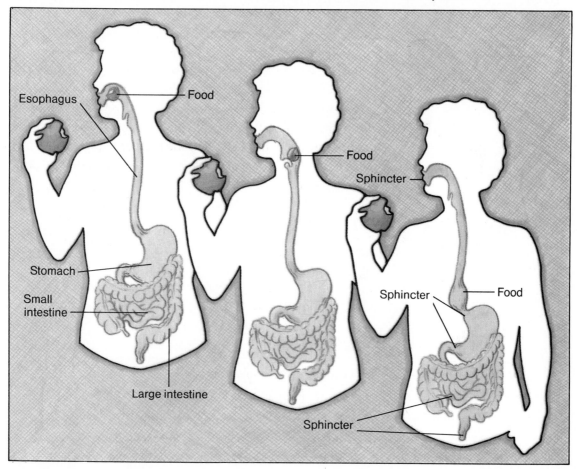

Esophagus — Food

Food

Sphincter

Stomach

Small intestine

Sphincter — Food

Large intestine

Sphincter

The *sphincter* at the base of the esophagus is normally closed. This prevents the contents of the stomach from moving back into the esophagus. The sphincter opens when food reaches it, allowing food to enter the stomach.

The stomach is a large, J-shaped, muscular sac that stores food. An adult's stomach can measure 20 cm from top to bottom and 10 cm across. When full, the stomach can stretch to be twice as long. Because the stomach can store food, a person does not have to eat all the time. When food is in the stomach, the muscular walls contract in a wavelike motion. This churning action breaks up the larger pieces of food. In this way, the stomach helps the mechanical digestion of foods.

Have you ever gotten a stomachache when you were feeling anxious, worried, or upset? Emotional stress can cause an increase in the muscular contractions of the stomach. These contractions result in the sensation of a "knot in the stomach" or a stomachache.

Food moves through the entire digestive tract the same way it moves through the esophagus. Mucus makes this movement of food easier. In addition, there are sphincters found along the entire digestive tract.

The stomach sphincter releases partially digested food in spoon-sized amounts into the small intestine. Little or no mechanical digestion takes place in either the small or large intestine. Even so, these are very important digestive organs. Their functions are described in the next section.

SUMMARY

The food you eat must be broken down into simpler forms before your body cells can use it. As the food moves along the digestive system, several functions do this task. Mechanical digestion occurs mainly in the mouth and the stomach.

QUESTIONS

Use complete sentences to write your answers.

1. What are five functions of the digestive system?
2. How is food moved through the digestive system?
3. Which steps of mechanical digestion take place in the mouth? In the stomach?

FOOD MOVEMENT MODEL

PURPOSE: To create a model for the movement of food in the digestive system.

MATERIALS:

dialysis tubing (40 cm long)	medicine dropper
	marble
water	clock with a second
1 mL vegetable oil	hand

PROCEDURE:

A. You will need a partner for this activity. Your partner will time how long it takes you to move a marble through a piece of dialysis tubing.

B. Wet a piece of tubing thoroughly. If it is dry, it may rip. Open it up by rubbing it between your fingers. Put a marble in one end of the tubing. Move the marble through the tubing by squeezing behind it. Do this until the marble comes out the other end. See Fig. 14–10.

  1. How many seconds did it take you to move the marble through the tubing?

  2. What caused the marble to move through the tubing?

C. Add one mL (or 10 drops) of oil to the inside of your piece of tubing. Flatten the tubing to spread the oil along the entire inside surface. Place the marble in one end of the tubing. Move it through the tubing as you did before, with your partner timing you.

  3. How many seconds did it take this time?

  4. Through which tube did the marble move faster?

  5. Did you notice any other differences this time? If so, describe them.

CONCLUSIONS:

1. How does the model you have just made compare with what happens to food in the esophagus?

2. What is produced along the lining of the digestive tract that acts like the oil?

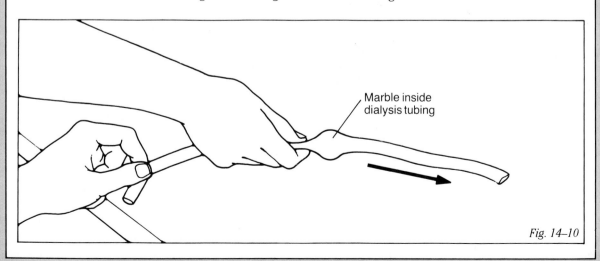

Marble inside dialysis tubing

Fig. 14–10

14-3. Chemical Digestion and Absorption

At the end of this section you will be able to:

- ☐ Contrast mechanical and chemical digestion.
- ☐ Name the end products after the chemical digestion of proteins, carbohydrates, and fats.
- ☐ Identify the parts of the digestive system.
- ☐ Describe the structure and functions of the small and large intestines.

In 1822, a young French-Canadian trapper was accidentally wounded in his left side by a gunshot. His name was Alexis St. Martin. See Fig. 14–11. Dr. William Beaumont, a U.S. Army doctor stationed nearby, saved St. Martin's life. However, he could not completely close the wound. St. Martin's stomach healed around an opening to the outside. This left a hole leading to St. Martin's stomach, which Dr. Beaumont used as a window to study the workings of the stomach. Much of what we know about the function of the stomach is based on his observations.

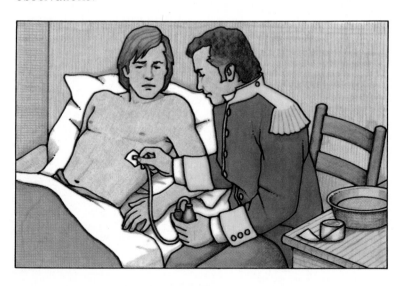

Fig. 14–11 Dr. Beaumont's research on Alexis St. Martin lasted almost 10 years. None of the experiments harmed the patient.

CHEMICAL DIGESTION

In the last section you learned that mechanical digestion breaks food down into small pieces. However, these small pieces are still too large to be used by your cells. They must be broken down even more. Special chemicals break down these small pieces of food into smaller and simpler compounds. Carbohydrates are broken down into simple sugars. Proteins are broken down into amino acids. Fats are broken down into *fatty acids* and *glycerol.* Fatty acids and glycerol are the building blocks of fats, just as amino acids are the building blocks of proteins. This chemical breakdown of food is called **chemical digestion.**

Let's follow the *chemical digestion* of a cheeseburger as it travels through the digestive system. The basic foods in a cheeseburger are illustrated in Fig. 14–12. As you chew the cheeseburger, saliva moistens it. Saliva also contains an enzyme that begins digesting carbohydrates. The bun is mostly carbohydrates. Some starch is broken down into sugar right in your mouth.

IN THE STOMACH

After the cheeseburger is swallowed, it travels down the esophagus to the stomach. See Fig. 14–13 on page 330. No digestion takes place in the esophagus. The food stays in the stomach for one to six hours. Chemical digestion of proteins begins here. **Glands** lining the stomach produce *gastric juice* and mucus. *Gastric juice* contains an acid and a protein-digesting enzyme. The acid helps the enzyme to work. This enzyme acts on the beef and the cheese. The rest of the cheeseburger remains chemically unchanged. As the stomach churns, the food is mixed with the gastric juice and becomes a thick, soupy mixture. This soupy mixture passes into the **small intestine** a little at a time.

SMALL INTESTINE

Very little chemical digestion takes place in the stomach. The *small intestine* is the main digestive organ. It is a coiled tube about 3 cm wide and 7 m long.

The first 25 cm of the small intestine is the *duodenum* (doo-oh-**dee**-num). Juice produced by a large *gland* called the **pancreas** (**pan**-kree-us) flows into the duodenum. The juice enters the duodenum through a small *duct,* or tube. Pancreatic

Fig. 14–12 A hamburger contains protein, carbohydrates, fat, and fiber.

Chemical digestion The process in which our foods are chemically broken down into simpler substances.

Gland An organ that secretes substances having special functions in the body.

Small intestine Part of the digestive tube where food is digested and absorbed.

Pancreas A gland whose secretions help digest food chemically.

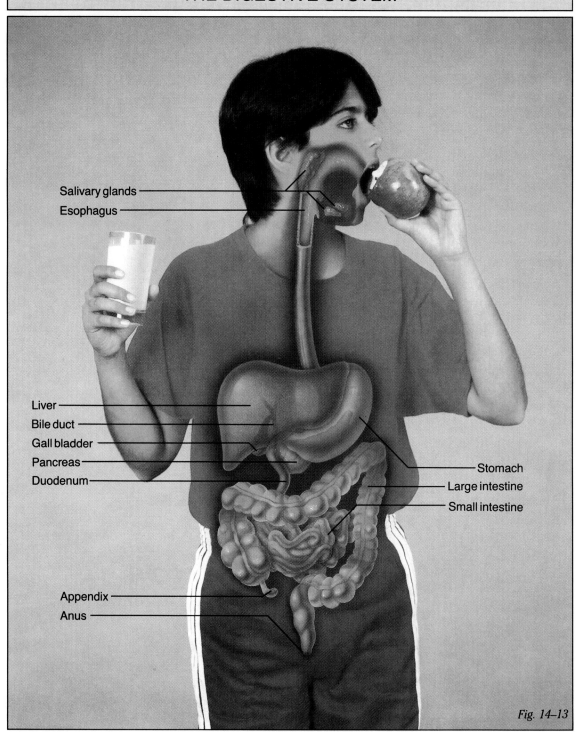

Salivary glands

Esophagus

Liver

Bile duct

Gall bladder

Pancreas

Duodenum

Stomach

Large intestine

Small intestine

Appendix

Anus

Fig. 14–13

juice contains enzymes that digest carbohydrates, proteins, and fats. The starches in your hamburger roll that weren't broken down in your mouth will now be broken down into double sugar molecules. Protein digestion, which began in the stomach, continues.

A yellow-green substance called *bile* flows into the duodenum through the bile duct. Bile is made by the **liver** and stored in the **gall bladder.** Look again at Fig. 14–13. Bile breaks up fats into tiny droplets. This is similar to the way detergents "cut" grease. These tiny fat droplets can then be easily digested by the enzymes in pancreatic juice. The digestion of all fat from this meal is now completed.

As the hamburger moves on through the small intestine, digestion will be completed. *Intestinal juice* is produced by the lining of the small intestine. One enzyme in this juice changes the double sugars into simple sugars. Another completes the digestion of proteins by breaking them down into amino acids. Simple sugars, amino acids, fatty acids, and glycerol are the end products of digestion. These substances are simple enough to be used by the body cells. Table 14–3 summarizes chemical digestion.

Liver The largest gland in the human body. One of its functions is to produce bile.

Gall bladder A sac in which bile from the liver is stored.

ORGAN	DIGESTIVE JUICE	GLAND WHERE JUICE IS MADE	GENERAL PROCESS
Mouth	Saliva	Salivary glands	Beginning of starch digestion
Stomach	Gastric juice	Glands in stomach wall	Beginning of protein digestion
Small intestine	Bile	Liver	Helps the fat-digesting pancreatic enzyme
	Pancreatic juice	Pancreas	Continuation of protein
			Continuation of starch digestion
			Fat digestion
	Intestinal juice	Glands in wall of the small intestine	Protein and carbohydrate digestion completed

Table 14–3

Lettuce, pickles, and onions come from plants. Although we get many nutrients from plants, we cannot digest all plant materials. Plants have cell walls made of *cellulose.* Some animals, such as deer and cows, can digest cellulose. These animals have bacteria living in their appendixes that break down the cellulose. Humans do not have these bacteria. Thus, we cannot digest cellulose. (Indeed, the human appendix no longer seems to have any function in the human digestive system.) Cellulose is the part of our food that we call "fiber."

ABSORPTION

Absorption The diffusion of water and dissolved materials into cells.

Once digestion is completed, the process of **absorption** begins. The digested food leaves the small intestine and moves into the bloodstream.

The small intestine is a very long tube. It is coiled back and forth inside your abdomen. Its inner lining has many tiny folds. Both the length and the folds increase the places where digested food can be absorbed.

Villi Very small, fingerlike projections on the lining of the small intestine that absorb digested foods. (Singular: *villus*).

Even more surface area is added by tiny, fingerlike structures called **villi** (**vill**-eye). *Villi* cover the folds lining the small intestine. They make the absorption of food even faster. They give the small intestine a velvety appearance. Each villus contains many tiny blood vessels. Digested food moves by diffusion through the villi and into the blood vessels. Blood then

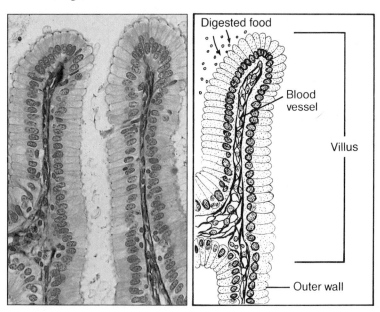

Fig. 14–14 Two villi, greatly magnified, from the lining of the small intestine.

carries the digested food to the cells throughout the body. See Fig. 14–14.

LARGE INTESTINE

After *absorption,* undigested foods are left in the small intestine. These materials then pass into the **large intestine** as a watery mixture. No digestion takes place in the *large intestine.* Most of the water in this mixture is absorbed by the large intestine. The water is returned to the body tissues. Plant "fiber" provides bulk to the watery mixture. With this bulk, the mixture can be easily moved through the large intestine. This easy movement usually prevents the removal of either too much or too little water. The removal of too much water may result in *constipation.* The removal of too little water may result in *diarrhea.*

Bacteria in the large intestine partially break down the undigested food. At this point, the watery liquid is more solid. It is now called *feces* (**fee**-seez). The feces are stored in the lower part of the large intestine. Later, they will be released from the body through the *anus* (**ay**-nus).

Large intestine Part of the digestive tube in which wastes are stored until they can be eliminated from the body.

SUMMARY

Chemical digestion changes food, through chemical reactions, into simpler molecules that can be used by the body cells. Enzymes help cause these chemical changes in food. Digested foods are absorbed into the bloodstream from the small intestine. Undigested food passes from the small intestine into the large intestine, where water is absorbed.

QUESTIONS

Use complete sentences to write your answers.

1. In what way is chemical digestion different from mechanical digestion?
2. What are the end products of carbohydrates, proteins, and fats after they are changed by chemical digestion?
3. List in order the organs that food passes through in the digestive system.
4. How is the surface area of the small intestine increased?
5. What is the function of the large intestine?

INVESTIGATION

DIGESTION OF STARCH

PURPOSE: To observe how saliva breaks down starch into sugar.

MATERIALS:

goggles	graduated cylinder
apron	water bath
iodine solution	crackers without
medicine dropper	sugar
paper towel	water
3 test tubes	test tube holder
Benedict's solution	test tube rack

PROCEDURE:

A. You will need a partner for this activity. CAUTION: Both you and your partner will need to wear goggles and an apron when heating liquids in this activity. Both of you will also need to copy Table 14–4 on a separate piece of paper.

	Starch Test	Sugar Test
Cracker		
Saliva		
Chewed Cracker		

Table 14–4

B. Break off a small piece of cracker. Place it on a paper towel. Test the piece of cracker for starch by putting two drops of iodine on it. A blue-black color indicates the presence of starch. CAUTION: Iodine may stain clothing.

1. What color did the cracker change to?

C. Record your results in column 1 of the table. Place a (+) in the box if the test was positive for starch. If the test was negative, place a (−) in the box. Throw the tested cracker away.

D. Crumble a new piece of cracker into a clean test tube. Cover the pieces of cracker with water. Test the cracker for sugar by adding 5 mL of Benedict's solution. Heat the test tube in a hot water bath until the solution boils. A red-orange color indicates the presence of sugar. CAUTION: Point the open end of the test tube away from you. Be careful not to heat the test tube too long. Record your results in the table as you did before.

2. What color did the solution change to? Was sugar present?

E. Place a small amount of your saliva into a clean test tube. Now test it for sugar as you did before. Record the results in the table.

F. The person who provided the saliva in step E should also chew the cracker for this test. Chew a piece of cracker for at least one minute. DO NOT SWALLOW IT. Place the chewed cracker in a test tube and test it for sugar. Record your results in the table.

CONCLUSIONS:

1. Why was saliva tested for sugar?
2. What did saliva do to the chewed cracker?
3. How does saliva help in digestion?
4. Why is it important to chew your food thoroughly?

CAREERS IN SCIENCE

CHEMICAL TECHNICIAN (FOOD INDUSTRY)

What foods will we be eating in the year 2000? How will these foods be processed or preserved? These questions directly concern chemical technicians who work in the food industry. These technicians assist food scientists in developing new foods. Technicians are involved in conducting some experiments on the nature of food and then recording the results. They help food scientists determine which foods are suitable for processing or freezing. After foods have been frozen, technicians examine them for freshness, flavor, and nutritional value.

Technicians can receive their training through two-year college programs. However, a Bachelor of Science degree, with a major in chemistry, would also provide a good background for entry into the field. Jobs are available in food manufacturing and processing companies.

For further information, write: Institute of Food Technologies, 221 North LaSalle Street, Suite 2120, Chicago, IL 60601.

DIETITIAN

Dietitians are responsible for developing meals that promote good health. Many dietitians work in hospitals, where they plan meals for the patients. Because of their illnesses, some patients must stay on special diets. Dietitians consult with doctors in designing these diets and then explain them to the patients.

Other dietitians plan meals for institutions such as schools, prisons, and corporations. They are involved with budgeting, purchasing food, and training food service workers. Also dietitians may deal with special areas of nutrition, such as the dietary needs of infants or the elderly, and they may serve as consultants wherever food is prepared and distributed.

To obtain a position as a dietitian, you'll need a bachelor's degree, usually with a major in food or nutrition.

For further information, write: American Dietetic Association, 430 North Michigan Avenue, Chicago, IL 60611.

CHAPTER REVIEW

VOCABULARY

On a separate piece of paper, match the number of each blank with the term that best completes each statement. Use each term only once.

absorbed   carbohydrates   small intestine   fats

stomach   glands   large intestine   proteins

enzymes   chemical digestion   mechanical digestion   villi

    Eating nutritious food is the basis for a healthy body. But before your cells can use it, the food must be digested. __1__ must be broken down into simple sugars. __2__ must be broken into amino acids. __3__ are broken into fatty acids and glycerol. Your grinding teeth and churning __4__ break apart food during a process called __5__ .

    __6__ is the process that breaks food down by chemical reactions. The chemicals that do this, called __7__ , are secreted by __8__ . Once the food is digested, it can be __9__ into the blood. This takes place in the __10__ . Tiny projections, called __11__ , create a large surface area for this to take place. The undigested part of food passes into the __12__ .

QUESTIONS

Give brief but complete answers to each of the following questions. Unless otherwise indicated, use complete sentences to write your answers.

1. What is good nutrition? Why is it so important?
2. What are the six basic nutrients in foods?
3. How does your body use carbohydrates, proteins, and fats?
4. Why is it recommended that people reduce the amount of fats in their diets?
5. Why are some foods enriched or fortified?
6. What is "junk food"? Why is it said to contain empty calories?
7. What happens to food when it is in the mouth? Why is it important to chew your food thoroughly?
8. How is food mechanically and chemically digested in the stomach?
9. How is food moved through the digestive system?
10. Name the organ(s) associated with each of the following functions. (a) A large gland that makes bile. (b) Sac that stores bile. (c) Chemical digestion of some proteins begins in this storage organ. (d) A gland that makes a digestive juice containing many enzymes. (e) Most chemical digestion takes place here.

11. Why is water considered a nutrient?

APPLYING SCIENCE

1. List the ingredients in a meal you have recently eaten. Find out the nutritional value of each of the foods in it.

2. How is it that people can eat while upside down or weightless in space?

3. You could still continue to live if you had to have your stomach removed. How do you think your life might change if this happened?

4. Do different fruit and/or vegetable juices contain different amounts of vitamin C? Is there a difference in the amount of vitamin C in fresh, frozen, and canned juices? To answer these questions, use 10 mL of 0.1 percent indophenol solution in a test tube for each juice tested. Using a medicine dropper, add one drop of juice at a time to the test tube. Gently swirl the test tube after each drop. Count and record the number of drops of juice needed to turn the solution from blue to colorless. The fewer drops of juice added, the more vitamin C the juice contains.

5. The Center for Science in the Public Interest, 1501 Sixteenth Street, N.W., Washington, DC 20036 is a nonprofit, tax-exempt organization that provides the public with information on the effects of science and technology on society. Its main focus has been on food and health. Write to the center for additional nutrition information.

BIBLIOGRAPHY

Allen, Oliver E., and Editors of Time-Life Books. *Secrets of Good Digestion.* Alexandria, VA: Time-Life Books, 1982.

Bershad, Carol, and Deborah Bernick. *Bodyworks: The Kid's Guide to Food and Physical Fitness.* New York: Random House, 1980.

Editors of Time-Life Books. *Wholesome Diet.* Alexandria, VA: Time-Life Books, 1981.

Grady, Denise and Sana Siwolop. "An Anti-Cancer Diet?" *Discover,* June 1984.

Hopson, J. "Carbohydrates: A Key to Health and Digestion." *Science Digest,* January 1984.

Langone, J. "Heart Attack and Cholesterol." *Discover,* March 1984.

Zim, Herbert S. *Your Stomach and Digestive Tract.* New York: Morrow, 1973.

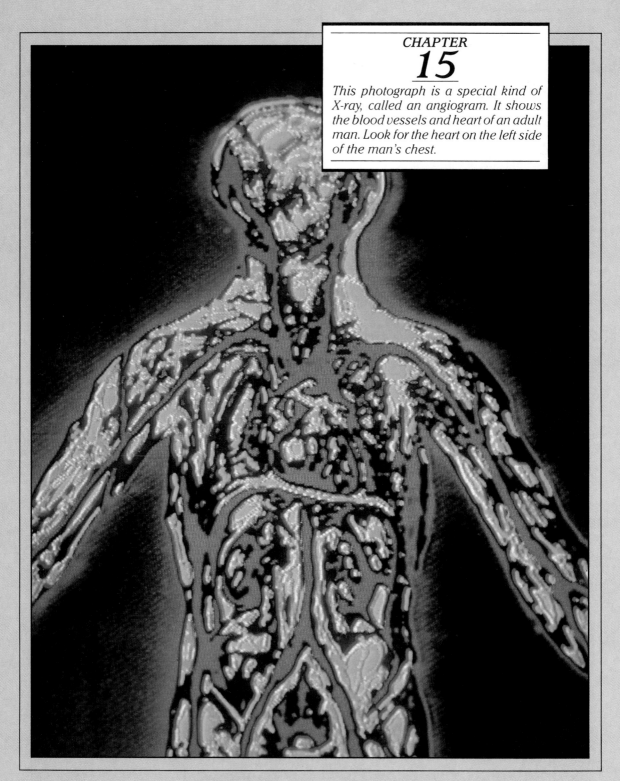

CIRCULATION

CHAPTER GOALS

1. Describe the parts of blood and the function of each part.
2. Name the blood vessels and describe their relationship to each other in the circulatory system.
3. Identify the parts of the heart and trace the path of blood through it.

15-1. River of Life: Blood

At the end of this section you will be able to:

- ☐ Identify the parts of blood and describe the function of each part.
- ☐ Explain blood types and their role in transfusions.

Throughout the ages of mankind, blood has been viewed with awe and mystery. Blood was thought to contain both good and bad spirits. It was closely associated with life forces. It was even thought to be partly responsible for intelligence. Terms and phrases such as "blood brother," "blue blood," "bad blood," "blood is thicker than water," "make one's blood boil," and "in cold blood" indicate the many other special characteristics that were given to blood. Today, blood is still viewed with awe, but some of its mysteries have been solved.

FUNCTIONS

The average person's body contains about 5.5 liters of blood. This is about 8 percent of a person's body weight. There are five basic jobs that blood carries out. Blood transports materials such as digested food, water, and oxygen to our body cells. It carries wastes from the cells to the parts of the body where the wastes are released. The blood carries chemical messengers between cells far away from each other. Also, special cells and proteins that help fight disease travel in the blood. In addition, blood helps regulate body temperature. The blood distributes heat from the tissues throughout the body. Excess heat is released when extra blood travels to the skin.

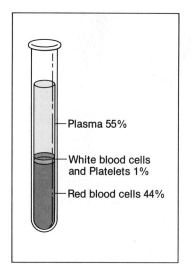

Fig. 15–1 *Blood is made up of four basic parts.*

Red blood cells Cells in the blood that contain hemoglobin and carry oxygen.

PARTS OF BLOOD

There are two main parts to whole blood—a liquid part and a solid part. See Fig. 15–1. About half of whole blood is a clear, yellowish-colored liquid called plasma. Most of plasma is water with dissolved nutrients and waste products in it. The rest consists of proteins. Digested food, wastes, and heat are carried by plasma.

Plasma proteins fight disease and help clot the blood. The disease-fighting plasma proteins are called *antibodies.* Each antibody fights a certain type of foreign substance, such as bacteria and viruses. Antibodies provide the power to resist disease; this is called *immunity.*

The solid part of the blood consists of red blood cells, white blood cells, and platelets. Blood looks red because it contains so many **red blood cells.** An average person has about 27 trillion (27,000,000,000,000) *red blood cells.* Each red blood cell lives only from 30 to 120 days. Then it breaks up. The spleen, a small organ in the upper abdomen, filters from the blood the remains of these dead red blood cells. Millions of red blood cells are always being made in the soft, red center of bones called *marrow.* See Fig. 15–2.

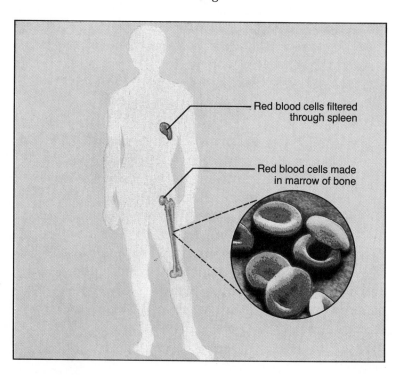

Fig. 15–2 *Red blood cells* (inset) *are produced in bone marrow and removed from the blood by the spleen.*

Mature red blood cells contain a special compound called **hemoglobin** (**hee**-moe-gloe-bin). *Hemoglobin* gives the red blood cells their red color. Oxygen from the lungs attaches to hemoglobin and is carried to the cells.

Hemoglobin A protein that combines with oxygen, carrying it to all cells in the body.

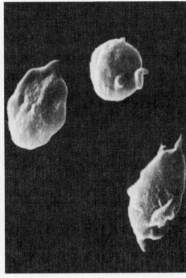

Fig. 15–3 White blood cells (left), *and platelets* (right), *photographed by a scanning electron microscope.*

Inhaling car exhaust fumes is very dangerous. Understanding how hemoglobin works makes it clear why. Exhaust fumes contain a poisonous gas called carbon monoxide. Hemoglobin combines more easily with carbon monoxide than with oxygen. Breathing in exhaust fumes means that body cells would be getting carbon monoxide instead of oxygen. As a result, a person can suffocate and die. For this reason, motor vehicles should not be left in a closed garage with the engine running.

White blood cells are larger than red blood cells. There are fewer *white blood cells* than red blood cells. For every white cell, there are about 600 red blood cells. White blood cells are also made by the bone marrow. There are several different kinds of white blood cells. Some release chemicals that poison bacteria. Others surround and eat the bacteria. White blood cells attack foreign invaders faster than the antibodies do. White blood cells can squeeze through the walls of the smallest blood vessels. They leave the bloodstream in order to reach an area of infection. During an infection, the bone marrow makes many extra white blood cells and releases them into the blood. See Fig. 15–3 (left).

White blood cells Cells in the blood that protect the body from infection.

Platelets Tiny particles in the blood that help the blood to clot.

Another solid part of the blood is the **platelets** (**plate**-letz). *Platelets* are tiny, colorless particles that help form blood clots. See Fig. 15–3 (right). Blood clots stop the bleeding from a cut or injury. The clot plugs the opening in the blood vessel. If your blood didn't clot, you could die from a small cut.

BLOOD TYPES

Blood is often called the "gift of life." Many lives are saved each year by blood transfusions. See Fig. 15–4. When a blood transfusion is given, the blood types of the donor and the patient receiving the blood must be *compatible.* This means that several characteristics of their blood must be the same.

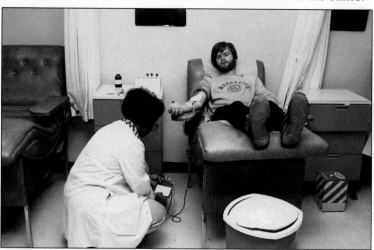

Fig. 15–4 A person who is in good health may donate a pint of blood at one time.

There are four main types of blood—named A, B, AB, and O. The *blood type* is determined by a protein on the surface of the red blood cells. Only two proteins, type A and type B, make up the four main blood types. People with type A protein have type A blood. People with type B protein have type B blood. Others, who have both proteins, have type AB blood. People with neither protein have type O blood. If two different blood types are mixed, the blood cells clump together and block the flow of blood inside the body.

Transfusions do not have to be in the form of whole blood. Whole blood can be separated into its parts—plasma, red cells, white cells, and platelets. Depending on the illness, a patient will receive only those parts of the blood that are needed. Transfusions may be given for some of the blood disorders listed in Table 15–1.

BLOOD DISORDERS		
Disease	Condition	Symptoms
Anemia	The number of red blood cells is low.	Weakness, always tired, breathlessness.
Sickle-cell anemia	An inherited disease in which red blood cells are sickle-shaped. These abnormal cells block small blood vessels; this prevents nearby tissues from getting oxygen.	Pains in hands, feet, arms, and legs. Easily tired and weak. Growth is often affected.
Leukemia	Too many white blood cells are formed. Many of these do not function properly. They interfere with many of the body functions.	Body's defense against disease lowered. Person gets infections easily.
Hemophilia	An inherited disease in which the blood does not clot.	Internal bleeding caused by even slight injuries.

Table 15–1

SUMMARY

Blood has four main parts that together transport nutrients and other materials to and from all the body cells, fight disease, and regulate body temperature. There are four basic blood types. Blood transfusions from one person to another must be with compatible types.

QUESTIONS

Use complete sentences to write your answers.

1. Identify the part of blood for each of the descriptions that follow: (a) helps clot blood; (b) transports oxygen from the lungs to the cells; (c) transports food and wastes; (d) fights disease.
2. What does it mean if a person has type O blood?
3. Why is it important for the blood types to match when giving a transfusion?

¡COMPUTE!

SCIENCE INPUT

The human being's circulatory system is composed of three basic components: the heart, various blood vessels, and blood. The circulatory system assists other body systems, such as the excretory and the digestive systems, to perform their tasks. At the same time, the nervous system and the respiratory system help the circulatory system to do its job.

This intertwining network carries a large share of responsibility for the health of our bodies.

COMPUTER INPUT

It has become quite common to use the computer as a learning tool. As we have seen so far, it can give immediate feedback as well as generate a random set of questions. Through random selection, feedback, and repetition, learning that requires that you memorize a great deal of information such as the parts and purposes of the circulatory system can be made a lot easier.

WHAT TO DO

Make a data chart, similar to the one below, listing 33 terms related to the human circulatory system, and brief definitions or descriptions of them. A few examples are given. Your terms and definitions will be the data for statements 230 to 262. Save Program Circulation on disc or tape and run it.

The program will use your definitions and descriptions in question form. For example, see the sample data chart below. If the computer reads statement 230, it would print out the question "WHAT ARE MICROSCOPIC BLOOD VESSELS?" If the computer reads statement 231, it would print out the question "WHAT DISEASE RESULTS IN NO BLOOD CLOTTING?" Program Circulation will ask you ten questions and then calculate what percentage of your answers was correct.

Data Chart

Term	Definition/Description
Capillaries	Are microscopic blood vessels
Hemophilia	Disease results in no blood clotting
etc.	

GLOSSARY

BIT   short for binary digit. A bit is a piece of computer-coded information. The coding process is done by the computer hardware, not the program. The hardware is designed to translate the data input into binary numbers, which can be stored electronically in the computer's memory.

BYTE   a collection of eight bits. The power of a computer's memory is often expressed as 8K or 16K. Each K represents 1,024 bytes of information. The more bytes in the computer's memory, the greater its capacity.

PROGRAM

```
100   REM PROGRAM CIRCULATION
110   DIM CH$(33),T$(33),R$(33),AM(2):AO
      = 0:AF = 0
120   FOR R = 1 TO 2: FOR CS = 1 TO 24:
      PRINT : NEXT CS
130   PRINT : PRINT : PRINT : PRINT "
      USE THE LISTING OF TERMS IN
      YOUR": PRINT : PRINT DATA
      CHART, TO ANSWER THE 10
      RANDOMLY": PRINT : PRINT:
      "COMPUTER-SELECTED QUIZ
      QUESTIONS."
140   FOR TD = 1 TO 3500: NEXT TD:
      FOR CS = 1 TO 24: PRINT : NEXT CS
150   FOR X = 1 TO 33: READ CH$(X),
      T$(X): NEXT X:K = 0
153   FOR Q = 1 TO 10
155   FOR X = 1 TO 34
157   X = INT (34 * RND (1) + 1)
160   IF X = 1 THEN NEXT X
165   N = (X - 1)
170   PRINT : PRINT : PRINT : PRINT Q;".
      WHAT ";CH$(N);"?"
180   PRINT : PRINT : INPUT "";R$(N)
190   IF R$(N) = T$(N) THEN PRINT :
      PRINT "THAT'S CORRECT! LET'S
      CONTINUE!": FOR TD = 1 TO 1500:
      NEXT TD:K = K + 1: FOR CS = 1
      TO 24: PRINT : NEXT CS: NEXT Q
200   IF R$(N) < > T$(N) THEN PRINT :
      PRINT : PRINT : PRINT "SORRY, THE
      CORRECT ANSWER IS:": PRINT :
      PRINT "--> ";T$(N): PRINT : PRINT :
      PRINT "COPY THE MISSED TERM
      ON YOUR WORKSHEET.": FOR TD =
      1 TO 3500: NEXT TD: FOR CS = 1
      TO 24: PRINT : NEXT CS: NEXT Q
210   AM(R) = INT ((K / 10) * 100): PRINT :
      PRINT AM(R);"% OF 10 QUESTIONS
      WERE CORRECT!": FOR TD = 1 TO
      2500: NEXT TD:AO = AO + AM(R):
      RESTORE: NEXT R
215   AF = (AO / 2)
220   FOR CS = 1 TO 24: PRINT : NEXT
      CS: PRINT : PRINT "YOUR
      COMBINED AVERAGE FOR BOTH
      QUIZZES": PRINT : PRINT "          IS
      ";AF;"%.": END
230   DATA     ARE MICROSCOPIC
      BLOOD VESSELS, CAPILLARIES
231   DATA     DISEASE HAS NO BLOOD
      CLOTTING, HEMOPHILIA
232   CONTINUE DATA STATEMENTS
263   END
```

BITS OF INFORMATION

There are many computer programs in use today assisting the medical profession. MYCIN, a program developed at Stanford University, diagnoses blood diseases and disorders after the patient's symptoms have been entered. The program also explains its reasoning process and recommends treatment. It is correct about 90 percent of the time.

Another type of program uses computer graphics. This program is being developed for medical students at New York University. Using three-dimensional graphics, the image of the skull can be viewed on the screen. Medical students "perform" brain surgery with electronic devices similar to the joy sticks used in video games. The students can learn and receive immediate feedback about their performance in a field of medicine where practicing on live patients is not possible, and where mistakes are especially dangerous.

INVESTIGATION

COMPATIBLE BLOOD TYPES

PURPOSE: To determine which blood types can be safely given for transfusions.

MATERIALS:

13 clear plastic or glass containers
grease pencil
red food coloring
blue food coloring
water

PROCEDURE:

A. Label four containers A, four containers B, four AB, and the last one O. Fill each container half full with water.

B. To each of the containers labeled A, add a few drops of red food coloring and stir. If the solution is not a bright red color, add more food coloring. These four containers will represent type A blood.

C. Repeat step B for the next four containers, this time using blue food coloring. These represent type B blood.

D. To each of the containers labeled AB add a few drops of red food coloring and the same amount of blue food coloring. These represent type AB blood.

E. Do not add coloring to the container labeled O. This represents type O blood.

F. Pour about one-fourth of the liquid from one container of each blood type into one container of each of the other blood types. These represent transfusions. If the liquid changes color, the transfusion is not safe. After each transfusion, set the container aside so you do not use it again.

G. Copy Table 15–2 on a separate piece of paper. Record the results by writing SAFE or UNSAFE in your table. Since it is always

safe to mix the same blood types, it is not necessary to test these transfusions. Also, any blood type (other than O) added to O will cause a color change. Therefore, it is not necessary to test these transfusions either.

Transfusions	Safe or Unsafe
Type A into A Type A into B Type A into AB Type A into O	
Type B into B Type B into A Type B into AB Type B into O	
Type AB into AB Type AB into A Type AB into B Type AB into O	
Type O into O Type O into A Type O into B Type O into AB	

Table 15–2

CONCLUSION:

On a separate piece of paper, copy Table 15–3 and use it to summarize your results.

Blood Type	Can Act as Donor to Types	Can Receive Blood from Types
O A B AB		

Table 15–3

15-2. Circulation

At the end of this section you will be able to:

- ☐ Identify the three types of blood vessels and the characteristics of each.
- ☐ Explain the functions of each type of blood vessel.
- ☐ Trace the path of blood through the circulatory system.

In 1628, a major turning point occurred in understanding the functions of the human body. In that year, William Harvey, an English physician, proved that blood moves through the blood vessels in a circle, pushed by the heart. Before that time, scientists said that blood oozed in all directions in the body and went no place in particular. Harvey outlined the basic parts of the circulatory system as we know them today without ever actually seeing the smallest blood vessels. See Fig. 15–5.

All living things must be able to transport food and oxygen to their cells. They also must be able to transport wastes away from their cells. In single-celled organisms, transport occurs through the cell membrane by diffusion. In larger animals, transport is more complicated. Materials are transported from cell to cell by a fluid. In some animals, a pump is needed to push this fluid. The earthworm has a special *circulatory system*. It has five simple pumps that push blood through blood vessels. Fish, amphibians, reptiles, birds, and mammals all have similar circulatory systems. They have a main pumping organ called the heart. The heart pumps blood through a system of blood vessels.

Fig. 15–5 A painting of William Harvey demonstrating his theory of circulation to the king of England.

CIRCULATORY SYSTEM

Let's study the parts of the human *circulatory system.* See Fig. 15–6. The heart, located in the middle of your chest, is a pear-shaped organ. It is made of special muscle tissue.

The heart is divided into four parts. The upper parts function like receiving rooms. The lower ones function like shipping rooms in a factory.

The blood vessels include arteries, veins, and capillaries. The structures and functions of these blood vessels are compared in Table 15–4. **Arteries** (**ahr**-teh-reez) carry blood from the heart to all parts of the body. The *arteries* branch into smaller arteries. These smaller arteries continue to branch into smaller arteries. Eventually, they will connect with the smallest blood vessels, the **capillaries** (**kap**-ih-leh-reez).

Capillaries are so small that they cannot be seen without a microscope. They are just wide enough for the red cells to pass through them in single file. See Fig. 15–7 on page 350. No cell in the body is more than two cells away from a capillary. The exchange of materials between the blood and the body cells takes place at these capillaries.

The capillaries join to form small **veins.** These small *veins* join to form larger veins. These larger veins are the blood vessels that carry blood back to the heart. Veins have valves

Artery A blood vessel that carries blood away from the heart. (Plural: *arteries.*)

Capillaries Small blood vessels connecting arteries to veins.

Vein A blood vessel that carries blood toward the heart.

COMPARING BLOOD VESSELS		
Artery	Vein	Capillary
Carries blood away from the heart.	Carries blood to the heart.	Connects small arteries and small veins.
Wall—thick, elastic, muscular.	Wall—thin, not elastic, little muscle.	Wall—one layer of cells; exchange of materials between blood and body cells occurs through it.
Does not have valves.	Has valves.	Does not have valves.

Table 15–4

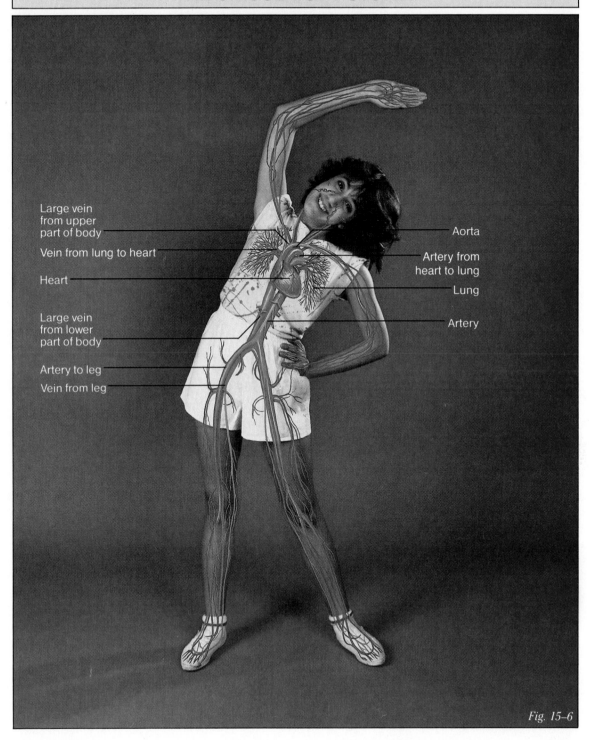

Large vein from upper part of body

Vein from lung to heart

Heart

Large vein from lower part of body

Artery to leg

Vein from leg

Aorta

Artery from heart to lung

Lung

Artery

Fig. 15–6

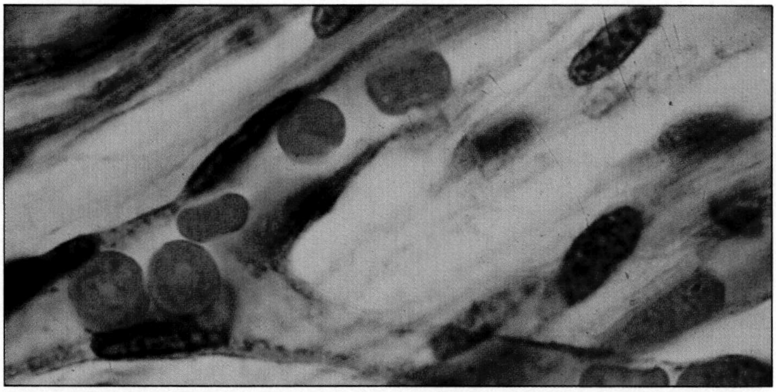

Fig. 15–7 Red blood cells move in single file through capillaries.

in them. These valves act like one-way doors. They prevent the blood from flowing backwards.

When you move, the muscles in your legs and arms squeeze the veins. This squeezing helps to push the blood back to the heart. This is like pushing water through a hose by squeezing it. Whenever you are motionless for a long time, the blood can collect in the veins. This is what happens when your arm or leg "goes to sleep." Stretching helps to get the blood moving again.

CIRCULATION

Let's follow the blood through the circulatory system. Use Fig. 15–6 to help you see what happens. Blood returns to the heart from all parts of the body. This blood contains very little oxygen and much carbon dioxide. The heart pushes the blood into special *arteries*. These arteries carry blood to the lungs.

In the lungs, the blood gets rid of carbon dioxide and picks up oxygen. This oxygen-rich blood is brought back to the heart by special *veins*. The heart pumps this blood into the *aorta* (ay-**or**-tah).

The aorta is the largest artery in the body. It is 2.5 cm in diameter. The aorta branches out to all parts of the body. The first branch carries blood to the heart itself. Another branch carries blood to the head, neck, and arms. Yet another one carries blood to the lower parts of the body. After the blood exchanges materials in the capillaries, it flows back to the heart through the veins.

As the heart beats, blood surges against the artery walls. This force is known as **blood pressure.** When the walls of arteries are narrow in diameter, more force is needed to push

Blood pressure The amount of force caused by blood pushing on the artery walls.

blood through them. This increased force is known as *high blood pressure* or *hypertension*. This is similar to what happens with a water hose and nozzle. The smaller the nozzle opening, the more pressure on the walls of the hose.

High blood pressure makes the heart work harder. In some people, fatty substances build up on the inside walls of arteries. Look back at Fig. 14–3 on page 317. Sometimes an artery can be completely blocked. If this happens to an artery bringing blood to the heart, part of the heart muscle does not get enough blood. The result is a *heart attack.* If an artery carrying blood to the brain is blocked, a *stroke* will occur.

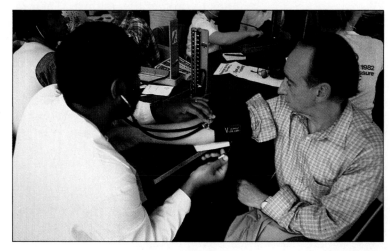

Fig. 15–8 Testing the blood pressure is a very simple procedure.

SUMMARY

Arteries, veins, and capillaries are blood vessels. They carry blood from the heart to all parts of the body and back again.

QUESTIONS

Use complete sentences to write your answers.

1. How is each type of blood vessel specially suited to carry out its functions?
2. What is the function of valves in the veins?
3. From the lungs, where does the blood go?
4. Where does the actual exchange of materials between the blood and the body cells take place?
5. What type of blood vessel is the aorta? What is special about it?

SKILL-BUILDING ACTIVITY

OBSERVING

PURPOSE: To observe blood vessels.

MATERIALS:

mirror (magnifying one if available)	absorbent cotton
	glass slide
good light source	eyedropper
goldfish	microscope
petri dish	beaker

PROCEDURE:
Part I

A. Using the mirror and the light, look at the underside of your tongue.

B. Use Fig. 15–9 to help you identify the blood vessels. Thick, bluish lines are veins. Thick, pinkish ones are arteries.

C. Draw what you observe.

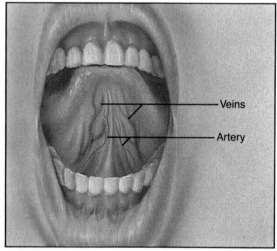

Fig. 15–9

Part II

A. Carefully wrap the body of a goldfish in cotton soaked with fish-tank water so that only the tail is uncovered. Place it in the petri dish with a small amount of water.

Place some water from the fish tank in the beaker. Use the eyedropper and a small amount of fish-tank water to keep the cotton and fish wet at all times.

B. Gently place a slide over the tail to hold it flat and spread out. See Fig. 15–10. Try to do the remainder of the steps in 15 minutes to protect the goldfish. Then return it to the aquarium.

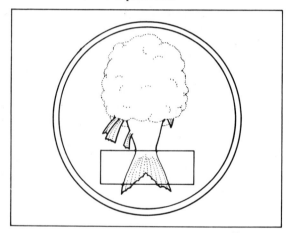

Fig. 15–10

C. Remove the clips from the stage of the microscope. Place the petri dish on the stage so that the fish's tail is over the hole.

D. Observe the tail under low power. You should see smaller arteries and veins as well as crisscrossing capillaries. You may see red blood cells flowing in single file through the capillaries. Quickly sketch what you see. Then label your sketch.

CONCLUSION:

1. How does knowing exactly what you are looking for help you to make better observations? What were you observing in Part I of this activity? Part II?

15-3. The Heart

At the end of this section you will be able to:

- [ ] Identify the parts of the heart and follow the blood flow through them.
- [ ] Explain what causes a heartbeat.
- [ ] Explain what causes a pulse.

On December 2, 1982, Barney Clark, a 61-year-old retired dentist from Seattle, Washington, became the first human being to receive an artificial heart. See Fig. 15–11. He survived for 16 weeks. His death was owing to problems caused by the heart disease he had before his natural heart was replaced with the artificial one.

The artificial heart works by puffs of air that are pumped into it from a large air-pumping machine. An artificial heart may seem amazing. Yet, it can't outdo your own natural heart.

Your own heart is a pear-shaped hollow muscle about the size of your two fists clenched together. It weighs about 340 grams. It is located near the center of your chest between your lungs, just beneath the breastbone. It is tipped a little to the left side of the chest. The heart can be most easily felt and heard here.

Every minute the heart pumps about 5 liters of blood. This amount, as you learned, is almost the entire supply of blood in your body. In one day, the heart beats over 100,000 times. In an average lifetime, it pumps enough blood to fill more than three large ocean-going tankers! It is easy to see why the heart is considered the hardest working organ in the body.

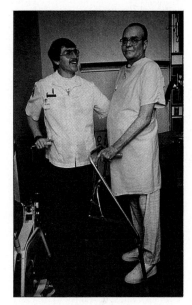

Fig. 15–11 Barney Clark, at right, was the first human to receive an artificial heart.

STRUCTURE

The heart is divided into four muscular rooms, or chambers. See Fig. 15–12 on page 354. The upper chambers that function like receiving rooms are called **atria** (**ay**-tree-uh). The lower chambers, called **ventricles** (**ven**-trih-kulz) are more muscular than the *atria*. They function like shipping rooms. Between the atria and ventricles are valves. The *heart valves* open and close like doors to keep the blood flowing in only one direction. A wall of muscle, called the *septum,* separates the right side of the heart from the left side. See Fig. 15–12.

Atria The two upper chambers of the heart that receive blood from the veins.

Ventricles The two lower chambers of the heart that pump blood to the arteries.

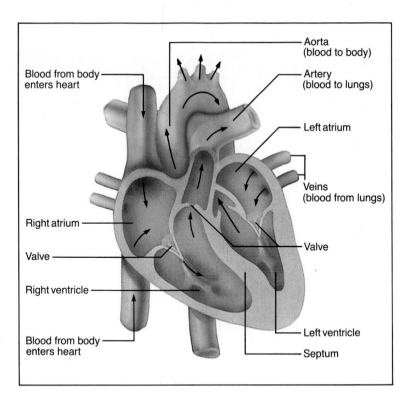

Fig. 15–12 The structure of the human heart. "Left" and "Right" refer to the left and right sides of the human body.

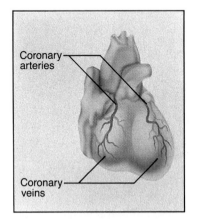

Fig. 15–13 If the coronary arteries become blocked, the heart will not get enough food and oxygen. This may cause a heart attack.

BLOOD FLOW THROUGH THE HEART

Oxygen-poor blood enters the heart at the *right atrium*. The right atrium contracts and pushes the blood through a *valve* into the *right ventricle*. The right ventricle contracts and pushes the blood out of the heart, sending it on its way to the lungs.

At the same time, oxygen-rich blood from the lungs empties into the *left atrium*. The left atrium pushes the blood through a valve into the *left ventricle*. The left ventricle then contracts and pumps the blood out to the body.

Although the heart is small, it uses about ten times the food and oxygen required by other organs. The heart does not get its food and oxygen from blood as it passes through its four chambers. Instead, it is fed blood by its own two *coronary arteries*. The coronary arteries are straw-sized tubes that branch out to reach all parts of the heart. See Fig. 15–13.

HEART SOUNDS AND PULSE

Have you listened to the sound of your heartbeat with a stethoscope? You can hear a "lub-DUB" sound repeated over and over. This sound is made by the closing of the valves in the heart. The "lub" sound is made by the closing of the valves between the atria and the ventricles. The "DUB" sound is made by the closing of the valves between the ventricles and the large arteries.

When resting, the heart of an adult beats about 70 times per minute. The only rest the heart gets is between beats. Then it relaxes for about a half a second. During exercise, the heartbeat may increase to 180 beats per minute. Exercise strengthens the heart. An athletic heart can pump more blood in each beat. Thus, the athletic heart gets to rest longer between beats.

The heartbeat is regulated by special tissue in the right atrium called the *pacemaker.* The pacemaker has direct nerve lines to the brain. It receives messages from the brain either to speed up or slow down the pace of the beat.

You can count the number of heartbeats without using a stethoscope. To count your own or another person's heartbeats you can take a **pulse.** The *pulse* is caused by the stretching and relaxing of the *arteries* as the blood spurts through them. When the ventricles of the heart contract, blood is forced through the arteries. This high-pressure blood flow causes the walls of the arteries to stretch. When the ventricles relax, the elastic walls of the arteries also relax. You can feel your pulse wherever an artery is close to the surface of the skin. One such place is the wrist. Other pulse points include the sides of the neck, armpits, inside of the thighs, inside of the ankles, and the temples. See Fig. 15–14.

Fig. 15–14 A runner takes his pulse by counting the number of pulses in a minute.

Pulse Regular stretching and relaxing of the artery walls caused by the beating of the heart.

SUMMARY

Every second of every day, your heart pumps blood to all the cells in your body. This amazing organ is a double pump. The right side of the heart pumps oxygen-poor blood to the lungs. At the same time, the left side of the heart pumps oxygen-rich blood to the body.

QUESTIONS

Use complete sentences to write your answers.

1. Outline the circulation of blood through the heart.
2. What causes the "lub-DUB" sound of a heartbeat?
3. By what is your pulse caused?

INVESTIGATION

TAKING A PULSE

PURPOSE: To measure a heartbeat rate by taking a pulse.

MATERIALS:

clock with a second hand · toothpick or wooden matchstick
ball of clay about the size of a dime

PROCEDURE:

A. Run in place for 30 seconds. Locate the pulse point on the inside of your wrist. Do this by lightly placing three fingers on the thumb side between the bone and tendon. See Fig. 15–15. Do not use your thumb. When you feel a regular throb, you have found your pulse.

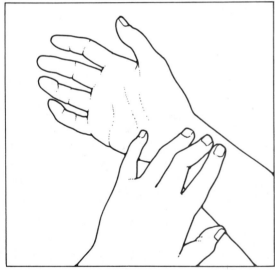

Fig. 15–15

B. You will need a partner for the remainder of this activity. Find your partner's pulse. Have your partner find your pulse.

C. Relax quietly for at least three minutes. During the last minute, your partner will count your pulse rate. The number of throbs in that minute will be your pulse rate "at rest." Create a data table on a separate piece of paper. Record this pulse rate in your table.

D. For the next three minutes, run in place again. Then quickly sit down. Now have your partner take your pulse rate. Record your "after exercise" pulse rate.

E. Repeat procedures C and D. This time you will be measuring your partner's pulse rate.

1. Was your "at rest" pulse rate the same as your partner's?

2. How did exercising affect your pulse rate?

F. You can make and use a pulse-watching device. Stick a toothpick into the ball of clay. Exercise vigorously for one minute. Place the clay with the toothpick on your wrist at the spot with the strongest throb.

G. Count the number of pulse beats you see using this pulse meter. Record this pulse rate.

3. Does using the pulse meter make it harder or easier to count your pulse? Why?

CONCLUSIONS:

1. What is the relationship between heartbeat rate and pulse rate?

2. Should everyone have the same pulse rate? Why? Why not?

3. Why does your pulse rate change when you exercise?

TECHNOLOGY

TAKING THE DOCTOR'S SOUND ADVICE

Using a stethoscope, the doctor listens to the sounds within your chest cavity. Now with the help of new technology, a doctor may "listen" to your body in a very different way. Using sound waves, called ultrasound, that bounce off arteries, a doctor may be able to detect conditions that could cause a stroke.

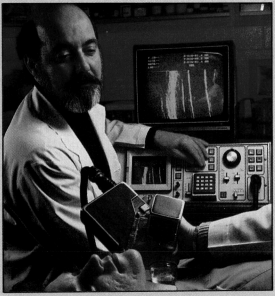

A stroke occurs when blood cannot reach a certain part of the brain. That portion of the brain cannot function without the oxygen supplied by the blood. If the part of the brain that controls muscle coordination is cut off from a fresh blood supply, the stroke victim usually becomes paralyzed. If the part that controls speech is affected, the stroke victim has difficulty speaking—or even loses speech entirely.

Eight out of ten strokes are caused by the clogging of the carotid (kah-**rawt**-id) arteries. The carotids are the large blood vessels on either side of your neck that supply blood to your head. In some people, these arteries can become clogged by plaque (**plak**) build-up. Plaque is a deposit of fat that collects beneath the cells that line the arteries. Plaque may be caused by a diet high in cholesterol, high blood pressure, smoking, or other factors. Over many years, plaque can build up inside the body's arteries. If it builds up in the coronary arteries, it can cause coronary heart disease resulting in a heart attack. If plaque buildup occurs in the carotid arteries, the result could be a mild or severe stroke.

Now, ultrasound may be used to detect plaque buildup before it causes a stroke. An ultrasound machine sends sound waves through the arteries. If blockages such as a plaque buildup are present, the sound waves are stopped. The pattern made by the sound waves goes through a computer and produces an image on a screen. The image shows the shape of the blockage. (Ultrasound is also used to show the image of a fetus when it is still inside its mother.)

Ultrasound can detect arterial disease in a safe way. There is no incision, there are no X-rays, and it is relatively inexpensive. Once people are warned of plaque buildup, they might take their doctor's advice more seriously, and change their diet or quit smoking.

If ultrasound can detect conditions that cause a stroke, why isn't it used to detect diseases of the coronary arteries? There are several reasons why. The heart is protected by the breastbone and the ribs. Ultrasound waves cannot travel through bone to reveal what is underneath. Instead they bounce back off the breastbone. The beating of the heart also interferes with the sound waves.

Ultrasound has brought medical science one step closer to its goal of preventing catastrophic illness.

CHAPTER REVIEW

VOCABULARY

On a separate piece of paper, match the number of each blank with the term that best explains it. Use each term only once.

red blood cells platelets capillaries atrium
hemoglobin   artery   transfusion ventricle
white blood cells vein(s)

 Blood from the body travels back to the heart in the ___1___. It enters the heart at the right ___2___. Then it is pumped through a valve into the right ___3___. From here the blood is pumped into the lungs. In the lungs, the ___4___ pick up oxygen and give off carbon dioxide. The oxygen attaches to the ___5___ in these cells. This exchange takes place in the ___6___, where the red blood cells travel in single file. From the lungs, the blood enters the heart again. The heart pumps the blood into the largest ___7___, the aorta, where it is sent to the rest of the body.

 Special cells in the blood, called ___8___, fight infection. Another solid part of blood, called ___9___, helps form blood clots. Blood clots stop bleeding. People that lose a lot of blood may require a ___10___.

QUESTIONS

Give brief but complete answers to each of the following questions. Unless otherwise indicated, use complete sentences to write your answers.

1. What are five functions of blood?
2. List the four parts of blood. What is the function of each part?
3. Why is breathing car exaust dangerous?
4. How is blood type determined?
5. Why do you have to know your blood type before receiving a transfusion?
6. List the three kinds of blood vessels. What is the function of each?
7. How are the three types of blood vessels connected in the circulatory system? How does the blood travel through them?
8. What is the function of the heart's atria? What is the function of the heart's ventricles?
9. Why is the heart considered a double pump?

10. What sound does the heart make as it beats? What causes this sound?
11. How are pulse and heartbeat related?
12. Why can you feel a pulse only at certain places on the body?

APPLYING SCIENCE

1. The brain cells have first priority for the blood supply. Knowing this, why might standing at attention for a long time cause a person to faint?
2. Some advertisements talk about iron-poor blood. Find out what part of your blood needs iron.
3. To avoid heart disease, why are people often encouraged to change their diet and life style, as well as to exercise?
4. It is important promptly to diagnose and treat a strep throat because it can develop into a disease called rheumatic fever. An end result of rheumatic fever can be heart damage, especially to the valves. Why could this be a serious problem?
5. Design experiments to test the following: Is it true that younger people have faster heartbeats? How are body size and heartbeat related? (You can include your pets in this.) How do the pulse rates of the male and female members of your family compare?

BIBLIOGRAPHY

Fisher, Arthur, and the Editors of Time-Life Books. *The Healthy Heart.* Arlington, VA: Time-Life Books, 1982.

Gilmore, C. P., and the Editors of Time-Life Books. *Exercising For Fitness.* Arlington, VA: Time-Life Books, 1982.

Silverstein, Alvin and Virginia B. *Heartbeats: Your Body, Your Heart.* New York: Harper, 1983.

Tully, Mary-Alice and Marianne. *Heart Disease.* New York: Watts, 1980.

Booklets and pamphlets from your local American Red Cross and the American Heart Association.

Humans have a system for obtaining oxygen and getting rid of carbon dioxide that is suited for life on land. What kinds of equipment allow these divers to breathe underwater?

RESPIRATION AND EXCRETION

CHAPTER GOALS

1. Describe the structures and functions of the respiratory system.
2. Describe the structures and functions of the excretory system.
3. List some functions of the liver and explain why they are so important.

16-1. The Respiratory System

At the end of this section you will be able to:

- ▣ Describe the function of respiratory structures.
- ▣ Identify the pathway air follows through the respiratory system and explain how gas exchange occurs.
- ▣ Explain the process of breathing.

Do you know why deep-sea divers must be careful not to rise to the surface too quickly? The deep-sea diver's body is put under enormous pressure when the diver is at great depths. The extra water pressure causes more nitrogen than usual from the air tank to enter the diver's blood and tissues. If the diver surfaces too quickly, the nitrogen bubbles out of the blood and tissues. This is similar to the bubbles formed when soda pop is poured. The nitrogen bubbles can cause pain in the muscles and joints, deafness, vomiting, paralysis, fainting, and sometimes death. This is called "the bends." If the diver returns to the surface slowly, the nitrogen is carried by the blood and eliminated by the *respiratory system.*

FUNCTIONS OF RESPIRATORY STRUCTURES

Respiration is the process that provides energy for most living things. To carry out this process, an organism must be able to obtain oxygen from the environment. It must also have a way to release carbon dioxide. Different organisms exchange these gases in different ways. But in all organisms, these gases must diffuse through a moist surface.

Many animals that live in water, such as fishes, obtain oxygen and release carbon dioxide through gills. All land vertebrates exchange gases through their lungs. The lungs of a mammal are very complicated. They are made up of thousands of small sacs and are found in a chest cavity.

THE HUMAN RESPIRATORY SYSTEM

In humans, air enters and leaves the lungs through a series of branching tubes. See Fig. 16–2. Together these tubes look like an upside-down tree. Air entering the nose or mouth passes down the **trachea** (**tray**-kee-ah). A common name for the *trachea* is the windpipe. At the top of the trachea is the *larynx,* or voice box. The larynx contains the vocal cords. The vocal cords are two leathery but flexible sheets of tissue that are stretched across the larynx. When you speak, muscles move the vocal cords close together. As air from the lungs streams past them, they vibrate, like the reed of an oboe, to make sound.

The trachea branches into two **bronchi** (**bron**-kee). The bronchi lead into each lung and branch into the **bronchial tubes.** In the lungs, the *bronchial tubes* divide into many smaller tubes, the *bronchioles* (**bron**-kee-olz). These smaller tubes lead into tiny *air sacs.* The lungs are made up of about 300 million of these tiny sacs. The wall of each air sac has many tiny pouches called **alveoli** (al-**vee**-oh-lye). The *alveoli* look like clusters of grapes and give the lungs its spongy appearance. If all your alveoli were spread out in a single layer, they would cover an area the size of a tennis court!

The alveoli have thin walls, usually only one cell thick. These thin, moist pouches are surrounded by a dense network of tiny blood vessels called capillaries. Oxygen entering the alveolus diffuses first through its walls and then through the capillary walls into the blood. Carbon dioxide diffuses from the blood into the alveoli. Some water from the blood is also released into the alveoli. See Fig. 16–1.

This diffusion of oxygen helps us to understand why it becomes harder to breathe at higher attitudes. As you move up from sea level, the air becomes thinner and contains less oxygen. The oxygen supplied by diffusion is not enough to meet the needs of the body.

Trachea A stiff, flexible breathing tube, also known as the windpipe.

Bronchi Two breathing tubes that branch out of the trachea and lead to the lungs (Singular: *bronchus*).

Bronchial tubes Large breathing tubes that branch out from the bronchi.

Alveoli Microscopic sacs in the lungs in which gas exchange takes place (Singular: *alveolus*).

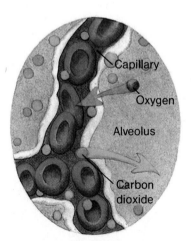

Fig. 16–1 Exchange of oxygen and carbon dioxide takes place where capillaries and alveoli meet.

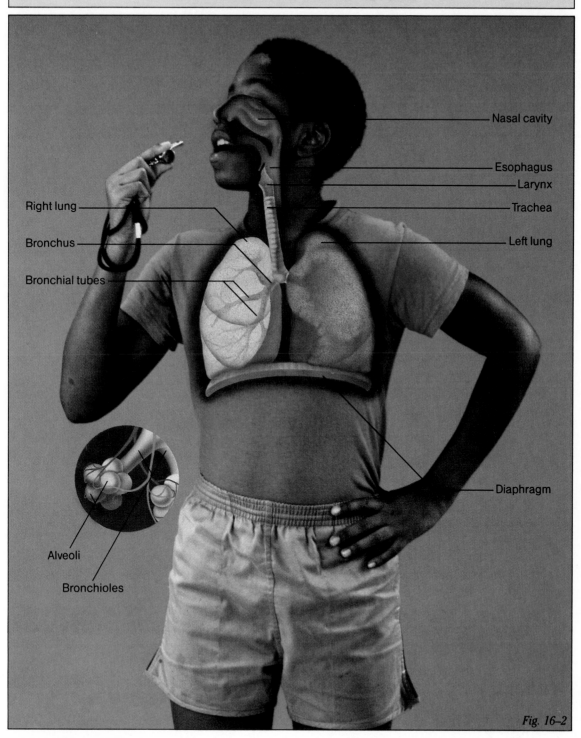

Nasal cavity

Esophagus

Larynx

Trachea

Left lung

Right lung

Bronchus

Bronchial tubes

Diaphragm

Alveoli

Bronchioles

Fig. 16–2

THE RESPIRATORY SYSTEM		
Body Part	Structure	Functions
Nose	Two nostrils lined with mucus and tiny hairs. Curvy passages.	The air pathway to the trachea. Tiny hairs and mucus clean, warm, and moisten the air.
Epiglottis	Flap of tissue.	Keeps food from entering the lungs.
Trachea	Large tube 2.5 cm in diameter. Supported by rings of cartilage and lined with mucus and tiny hairs.	Carries air to the bronchi. Moist membranes keep dust and bacteria out of the lungs.
Bronchi	Two tubes, similar in structure to the trachea, leading to the lungs.	Carry air from the trachea to the bronchial tubes.
Bronchial Tubes	Large air tubes in the lungs, reinforced with cartilage, and lined with mucus and tiny hairs.	Carry air from the bronchi to the bronchioles.
Bronchioles	Smallest air tubes in the lungs; some about one mm in diameter. No cartilage. Smooth muscle winds around them.	Carry air from the bronchial tubes to the alveoli.
Alveoli	Many small, moist, thin-walled pouches.	Exchange oxygen and carbon dioxide between blood and lungs.

Table 16–1

HOW WE BREATHE

Air moves into and out of the lungs when you breathe. At rest, you usually breathe about 14 times each minute. You take in about 0.5 L of air with each breath.

Your ribs and **diaphragm** help you to breathe. They act like a bellows. The *diaphragm* is a strong, dome-shaped sheet of muscle. It is located just below the lungs, forming the floor of the chest cavity. See Fig. 16–3. When the diaphragm contracts, it flattens a little and moves down about 1.5 cm. This increases the size of the chest cavity. As a result, air rushes in to fill the extra space when you *inhale.* When you *exhale*, the air is forced out because the diaphragm relaxes.

When you breathe deeply, the diaphragm may move down as much as 10 cm. At the same time, the muscles in between the ribs move the rib cage outward. Feel how your rib cage moves up and out when you inhale and returns to normal when you exhale. This causes the size of the chest cavity to increase and decrease even more. Thus, more air is drawn into and forced out of the lungs during deeper breathing.

Even after you forcibly exhale, there is still air in your lungs.

Diaphragm A strong sheet of muscle that forms the floor of the chest cavity.

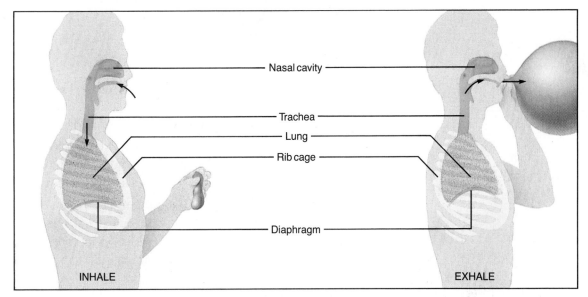

INHALE

EXHALE

This air is warm and helps to warm the incoming air quickly. The reserve air also prevents the lungs from collapsing.

Fig. 16–3 Breathing is controlled by the movement of the diaphragm.

Your lungs do not have any muscle tissue. Therefore, they cannot get bigger or smaller on their own. Your lungs are made up of elastic tissue. When they are filled with air, this elastic tissue is stretched. This is like a blown-up balloon. When you exhale, the elastic tissue acts like a deflating balloon.

People with the disease emphysema have lost the elasticity in their lungs. This makes it very difficult to breathe, especially to exhale. Emphysema is common among smokers.

SUMMARY

Organisms that get their energy by respiration must be able to exchange oxygen and carbon dioxide. This exchange of gases must occur across a moist surface. In humans, this moist surface is in the lungs. Air is brought into and sent out of the lungs by the process of breathing.

QUESTIONS

Use complete sentences to write your answers.

1. What is the purpose of the human respiratory system?
2. List the structures air passes through on its way to the lungs.
3. How does oxygen move from the lungs into the blood?
4. How do the ribs and diaphragm control breathing?

MEASURING LUNG CAPACITY

PURPOSE: To determine lung capacity by measuring the amount of exhaled air.

MATERIALS:

4-liter jug (marked   rubber tubing
   off in 100 mL   1 soda straw
   units)   pan
water   grease pencil

PROCEDURE:

A. Set up the equipment as shown in Fig. 16–4. Mark the water level with a grease pencil. Insert a straw into the free end of the tubing. Be sure to keep the opening of the jug underwater. Take a normal breath and hold it. Place your mouth on the straw and exhale normally into the straw. Hold your nose closed while you do this. The amount of water forced out of the bottle will equal the amount of air

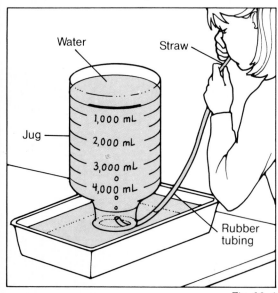

Water   Straw

Jug

1,000 mL
2,000 mL
3,000 mL
4,000 mL

Rubber tubing

Fig. 16–4

that you exhale. Mark the new level of the water.

1. Why must you hold your nose?

B. Record the volume of air you normally exhale in mL. This volume is the difference between the first and second marks.

C. Set up the equipment again. This time, take a deep breath and exhale into the straw. Hold your nose as before. Keep exhaling until you cannot exhale any longer. Mark and measure the amount of water that was forced out of the bottle. Record the amount of air you forcibly exhaled. Now record the difference between the amount of air you normally exhale and the amount of air you forcibly exhale.

2. Is there a difference? What might the reason be for this difference?

D. With the help of your teacher, make a list of the names of the students in your class. Next to each name, write the amount of air he or she forcibly exhaled. Now, analyze this data to answer the following questions.

3. What is the range of lung capacities in the class?

4. What is the most common lung capacity among the students?

5. What is the average lung capacity?

CONCLUSIONS:

1. What type of person would you expect to have a large lung capacity? Why?

2. What factors other than size might affect lung capacity?

16-2. The Excretory System

Fig. 16–5 Mammals have to be able to get rid of excess heat.

At the end of this section you will be able to:

☐ Identify some metabolic wastes and explain the importance of excretion.

☐ Describe how the kidneys and urinary system function together.

☐ List the functions of the skin.

☐ Explain how the skin excretes wastes.

Pigs like to roll themselves in mud during warm weather. It is not because they like dirt. Actually, pigs have cleaner habits than most other farm animals. It is because they cannot sweat. To keep cool, they try to cover themselves with mud or water. Is the common phrase "sweat like a pig" accurate?

EXCRETION

Fortunately, humans *can* sweat, or perspire. Perspiration not only helps to cool us but it also contain wastes. Where do these wastes come from?

All life processes require chemical activities. As a result, they all give off wastes. The sum of all the chemical processes that take place in an organism is called **metabolism** (meh-**tab**-oh-liz-um). Wastes that are produced by these processes are called *metabolic wastes*. These wastes include carbon dioxide, water, nitrogen wastes, inorganic salts, and heat.

Metabolism The sum of all the chemical processes that take place in an organism.

If wastes build up in the body, they become poisonous to the organism. Too many wastes prevent food and oxygen from entering the cells. Therefore, organisms must get rid of these harmful wastes. The process by which organisms get rid of wastes is called **excretion.**

Excretion is simple to explain in organisms such as protozoa. Wastes diffuse through the cell membrane into the surrounding water. Water balance is maintained by osmosis and, in some cases, special structures. Higher organisms have specialized organs for both getting rid of wastes and maintaining water balance.

Excretion The process by which an organism gets rid of metabolic wastes.

In humans, the excretory organs include:
1. Lungs that excrete carbon dioxide and water.

2. Kidneys that excrete nitrogen wastes, inorganic salts, and water in the form of *urine.*

3. Skin that excretes nitrogen wastes, inorganic salts, and water helps regulate body temperature.

THE KIDNEYS AND URINARY SYSTEM

Kidneys Excretory organs that remove wastes from the blood.

Your **kidneys** are bean-shaped organs. Each one is about the size of your fist. They are located in back of the stomach area on either side of the spine. They are well protected by your bottommost ribs and by layers of fat. See Fig. 16–6.

Each minute, more than a liter of blood passes through your *kidneys.* As it circulates, the blood carries wastes away from the body cells. Blood rushes into each kidney from large arteries. It flows into smaller and smaller arteries until it reaches the capillaries. The capillaries come in close contact with millions of tiny tubes in the kidneys. These tubes are called **nephrons** (**nef**-ronz). The liquid part of the blood (plasma) is *filtered* (separated) by the *nephrons.* Wastes, water, sugar, and minerals pass into the nephrons. Blood cells and large protein molecules remain in the blood.

Nephrons Special tubes in the kidneys that filter the blood and return essential substances to it.

If all the materials filtered by the nephrons were excreted, many essential substances would be lost. A person would have to drink about 170 L of water each day just to replace them. Each nephron returns needed water, sugar, and minerals to the blood. Only wastes and excess water are held. The wastes in the nephrons are called urine. Urine consists mostly of water, small amounts of nitrogen waste called *urea,* and inorganic salts. Urine also contains small amounts of usually useful substances that are in excess in the blood. These may include small amounts of sugars and vitamins.

The cleansed blood passes out of the kidneys through special veins. The veins connect the kidneys to a larger vein leading back to the heart. The cleansed blood is carried back to the heart for circulation again.

The urine passes from the nephrons into larger collecting tubes in the center of the kidneys. From here, a tube called the *ureter* (yoo-**reet**-er) carries the urine to a muscular sac. This sac is called the **urinary bladder.** The *urinary bladder* stores the urine until it is removed from the body.

Urinary bladder A muscular sac at the end of the ureters that stores urine.

As it fills, the urinary bladder stretches. Special nerve cells

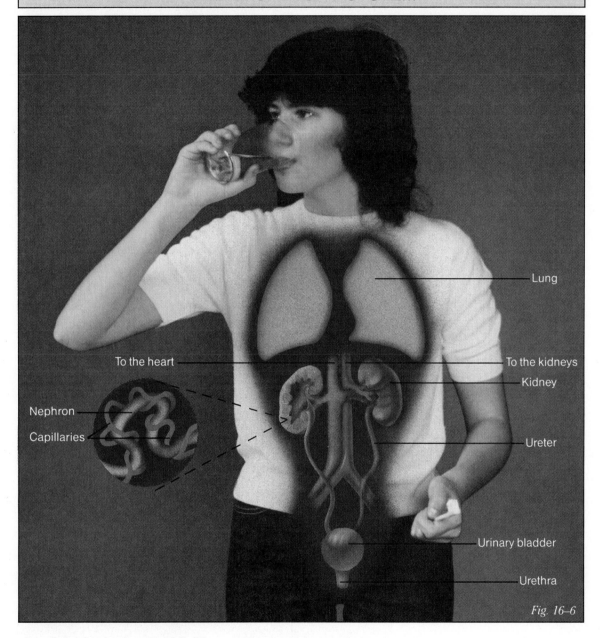

To the heart

To the kidneys

Kidney

Nephron

Capillaries

Lung

Ureter

Urinary bladder

Urethra

Fig. 16–6

signal the brain that the bladder needs to be emptied. At an appropriate time, the brain sends a signal to relax the sphincter muscles that hold the bladder shut. The bladder then contracts and pushes the urine into another tube called the *urethra* (yoo-**ree**-thruh). The urethra carries the urine out of the body.

THE SKIN

The skin is another excretory organ. It is the largest organ of the human body. It is attached to the rest of the body by a fatty layer. Skin is a combination of different cells and tissues. See Fig. 16–7. These different types of cells and tissues are organized into two general layers.

Epidermis The outer layer of skin.

The outer layer of skin is called the **epidermis** (ep-ih-**der**-miss). It is made up of layer upon layer of broad, flat cells. Some of these cells are not living. These cells are on the upper surface of the *epidermis*. They form a tough, waterproof shield that most germs cannot penetrate. These dead, hardened cells are constantly being rubbed off. You may see this happen when you dry yourself with a towel. Every time you rub your hands together, you rub off hundreds of cells. These dead cells are constantly replaced by the living cells underneath them.

Dermis The thick, active layer of skin underneath the thinner epidermis.

Below the epidermis is the inner layer of skin. It is called the **dermis.** All of the *dermis* is living. It is a tough layer containing blood vessels, nerve endings, sweat and oil glands, hair follicles, and fat cells. These are shown in Fig. 16–7.

Sweat glands in the dermis are coiled tubes that open at the surface of the skin. These openings are called sweat pores. Perspiration is released through these pores. It is mostly water with some dissolved wastes in it. These wastes include inorganic salts and urea. One of the skin's most important functions is the release of perspiration.

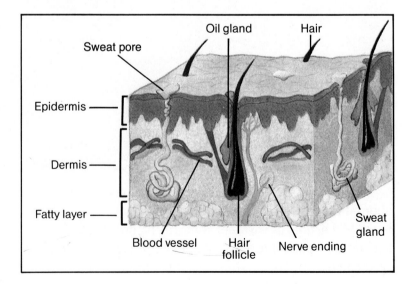

Fig. 16–7 The structures in a section of skin.

Perspiration on the surface of the skin evaporates. Even in cold weather, some perspiration evaporates from the body surface. During hot weather, more perspiration is released. As perspiration evaporates, it cools the body. In this way, perspiration also helps to regulate your body temperature. After perspiring, the skin feels sticky. This is due to the wastes remaining on the skin's surface. It is important to wash the skin to remove these wastes. In addition, washing prevents dirt from clogging the pores. It also prevents body odors from becoming offensive. All of the functions of the skin are listed in Table 16–2.

Fig. 16–8 Vigorous exercise increases the body's rate of perspiration.

FUNCTIONS OF THE SKIN
1. Excretes wastes.
2. Provides protection from germs and injury.
3. Waterproof covering.
4. Keeps body tissues from drying out.
5. Helps regulate body temperature.
6. Contains senses for heat, cold, touch, and pain.
7. Stores excess nutrients in the form of fats.
8. Produces Vitamin D in the presence of sunlight.

Table 16–2

SUMMARY

Living things give off wastes. These wastes are the result of the metabolic activities of an organism. Organisms get rid of metabolic wastes by excretion. The human excretory organs include the lungs, the kidneys, and the skin.

QUESTIONS

Use complete sentences to write your answers.

1. Why is excretion an important process for living organisms?
2. What are the main excretory organs of humans? What wastes does each excrete?
3. How do the kidneys clean the blood?
4. What is urine? How is it removed from the body?
5. How are wastes excreted by the skin? What are four other functions performed by the skin?

DISSECTING A KIDNEY

PURPOSE: To observe the structure of a kidney similar to that of a human.

MATERIALS:

lamb kidney	carving or butcher knife
dissecting pan	probe
scalpel	hand lens

PROCEDURE:

A. Using the knife, slice the kidney in half lengthwise. (This may already have been done for you.) Draw and label as many parts as you can, using Fig. 16–9 and other illustrations in Section 16–2.

B. Using the hand lens, examine the outer part of one half of the kidney. If you look closely, you will see the tiny, threadlike nephrons. On a separate piece of paper, draw what you see through the hand lens.

   1. To where do all the nephrons point?

   2. What is sent here?

C. Using your probe, lift the lighter-colored band in the center part of the kidney. These are the larger collecting tubes. Label these in your drawing.

   3. What is the function of these tubes?

   4. What are these tubes connected to in the kidney?

   5. What is the name of the tube they are connected to that leads to the bladder? Is this tube included as part of your kidney specimen?

CONCLUSIONS:

1. On your drawing trace the path that urine follows in the kidney.

2. What parts of the urinary system work with the kidneys to remove urine from the body? What is the function of each part?

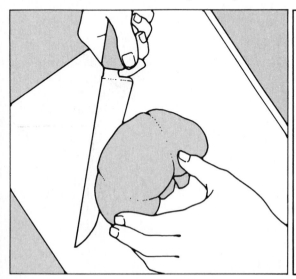

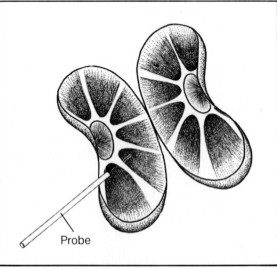

Probe

Fig. 16–9

SKILL-BUILDING ACTIVITY

CAUSE AND EFFECT RELATIONSHIPS

PURPOSE: To determine how the rate of evaporation affects the surface temperature of the skin.

MATERIALS:

hand lens   cotton
skin model   water
 (optional)   rubbing alcohol

PROCEDURE:

A. Closely examine the surface of your skin with a hand lens.

    1. Does the skin look smooth?

    2. What layer of skin are you looking at?

    3. What kind of cells make up this layer?

B. Find the very small sweat pores in between the hairs in your skin.

    4. What is a sweat pore?

    5. What is the function of the sweat pores?

C. You will need a partner to do this step. Get two pieces of cotton. Wet one piece with water and the other with alcohol. Have your partner rub the cotton with water on the inside of one of your wrists, and the cotton with alcohol on the inside of the other wrist. See Fig. 16–10.

    6. Which is evaporating faster, the water or the alcohol?

    7. Which wrist feels cooler?

D. Repeat step C for your partner.

CONCLUSIONS:

    1. What are the functions of the outer layer of skin?

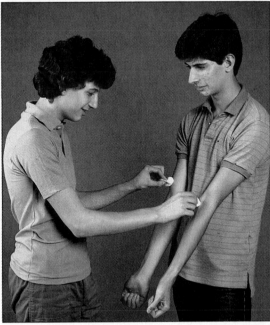

Fig. 16–10

    2. When you rubbed water or alcohol on your wrists, why did they feel cooler than the rest of your body?

    3. Why did one wrist feel cooler than the other one?

    4. Complete the following cause and effect statement: The faster _____ evaporates from the skin, the _____ the temperature of the skin becomes.

    5. When it is hot outside, what type of day can make you feel cool and comfortable? Why?

    6. What makes air feel sticky? Why can you feel uncomfortable when it is hot, sticky, and still outside?

    7. Why can adding moisture to the heated air in your house make you feel warmer when the weather is cold?

16-3. The Liver

At the end of this section you will be able to:

☐ List several functions of the liver.

☐ Describe how the liver controls the blood sugar level.

☐ Explain how the liver functions as an excretory organ.

Proper functioning of the liver is important. Serious liver disease or damage usually causes rapid death. There is very little that can be done. As yet, there is no artificial liver machine, and liver transplants need to be perfected. What makes the liver such a vital organ?

The **liver** is the most complex organ in the human body. It weighs about 1,500 g and is located just below the diaphragm. Most of the *liver* lies on the right side of the body. Part of it extends to the left side and partly covers the stomach. See Fig. 16–11.

Liver The largest gland in the body; it performs many important functions.

Fig. 16–11 The structure and location of the liver.

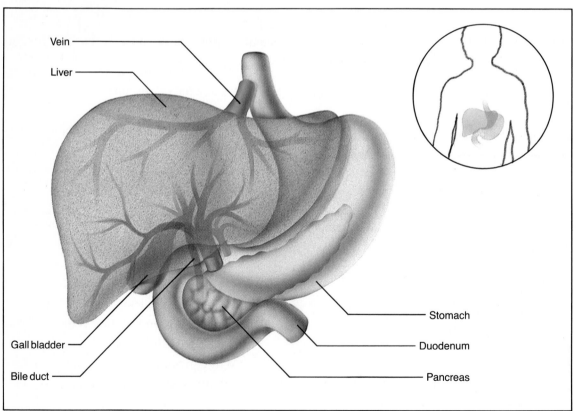

Vein

Liver

Gall bladder

Bile duct

Stomach

Duodenum

Pancreas

LIVER FUNCTIONS

The liver and kidneys together have the most control over what stays in and what is removed from the bloodstream. This is the reason that death comes so quickly when the liver stops working or does not work properly. The liver is remarkable because it has so many functions. See Table 16–3.

Let's look at some of these functions in more detail. The liver was first introduced in this text as a digestive gland. As a part of the digestive system, it makes bile. Bile also contains wastes from the breakdown of red blood cells. These wastes are removed from the body with the bile. From the small intestine, bile goes with the undigested food into the large intestine. There, they are eventually eliminated as feces.

REGULATING BLOOD SUGAR LEVEL

Digested food from the small intestine travels in the blood directly to the liver. See Fig. 16–12. Food is removed and stored in the liver. The food is then gradually released back into the blood at a controlled rate. For example, the liver removes sugar from the blood and stores it as a substance called **glycogen** (**glie**-koh-jin). When the amount of sugar in the blood gets low, the liver breaks down *glycogen* and releases sugar into the blood.

When the liver cannot store any more glycogen, the excess sugars are changed into fat. This fat is sent to various parts of the body, where it is stored.

FUNCTIONS OF THE LIVER
1. Stores digested food, vitamins, and iron.
2. Regulates the amount of sugar in the blood.
3. Acts like a chemical factory in which fats are changed and amino acids are split.
4. Makes plasma proteins.
5. Breaks down worn-out red blood cells.
6. Breaks down poisonous substances.
7. Makes bile, a digestive juice.

Table 16–3

Glycogen A starch that is made and stored in the liver.

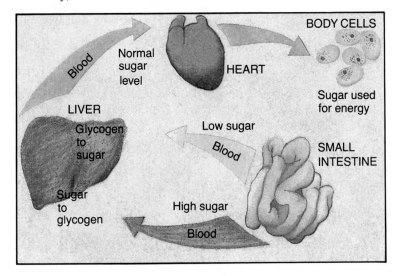

Fig. 16–12 This diagram shows how the liver regulates the amount of sugar in the blood.

Fig. 16–13 The liver works to protect the body from poisonous chemicals, such as those in some paints.

Keeping a normal blood sugar level is very important. Sugars are the main source of energy for all body cells. Without energy, our cells wouldn't be able to function. What happens when the liver's supply of glycogen runs out? The liver begins to change fats and amino acids into sugars. In this way, a normal blood sugar level is kept.

REMOVAL OF SUBSTANCES

Amino acids from protein foods contain nitrogen. When excess amino acids are changed into sugars, glycogen, or fats, the nitrogen in them is removed. The liver changes this nitrogen into *urea* and releases it into the blood. The blood then takes the urea to the kidneys, and it is eventually removed from the body.

The liver also changes poisonous substances into less harmful ones. These less harmful substances are released into the blood as waste. Many poisonous substances are in the foods you eat and the air you breathe. These include some food additives, pollutants in the food itself—such as mercury in fish or chemicals sprayed on foods, poisonous fumes from some paints, cleaning products, and chemicals in drugs. The liver is one of the body's most important defenses against these and other poisons. See Fig. 16–13.

SUMMARY

Most of the liver's work involves moving things into and out of the bloodstream. For example, it helps regulate the blood sugar level so that the body cells get the energy they need. The liver also releases a variety of wastes into the bloodstream.

QUESTIONS

Use complete sentences to write your answers.

1. How does the liver regulate the sugar level in the blood? What controls this function?
2. Why is keeping a normal blood sugar level important?
3. Which functions of the liver are part of excretion?
4. Why would eating too much protein produce more urea?

CAREERS IN SCIENCE

RESPIRATORY THERAPIST

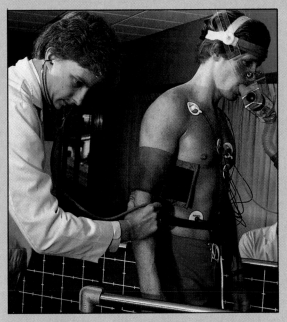

Whenever the breath that supports human life is threatened, respiratory therapists are brought in. They have patients in the newborn nursery (especially premature infants), in the surgical and medical wards, the emergency room, the outpatient clinic, and the intensive care units of a hospital. Respiratory therapists are beginning to work in the field of sports medicine, evaluating lung capacity and athletic performance. Also, colleges and private industry employ them as teachers, researchers, and advisers.

Preparation for this career includes a high-school diploma with a background in biology, chemistry, and mathematics plus attendance at a two-year AMA-approved training program. Registered respiratory therapists have to pass a qualifying examination given by the National Board for Respiratory Therapy.

For further information, write: American Association for Respiratory Therapists, Suite 101, 7411 Hines Place, Dallas, TX 75325.

DIALYSIS TECHNICIAN

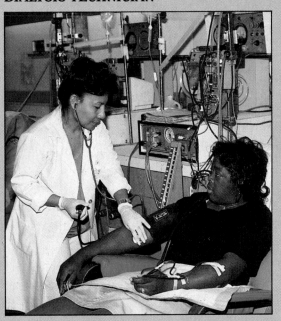

When the kidneys, the body's filtering system, no longer function, fluids build up in the body, and wastes collect in the bloodstream. Unless these fluids and wastes are removed, death will occur in about one week. In hemodialysis, blood is removed from a patient, carried by means of a tube to a kidney machine where it is purified and then returned to the patient. Monitoring both the patient and the machine is the job of the dialysis technician.

Technicians work mostly in hospitals and health centers. Almost all training is on the job, combining classroom and closely supervised clinical experience. Usually a high-school diploma and some college work in the biological sciences is expected. Trainee positions frequently go to those with prior medical experience, such as nurses or former army medics.

For further information, write: American Association of Nephrology Nurses & Technicians, 2 Talcott Road, Suite 8, Park Ridge, IL 60068.

CHAPTER REVIEW

VOCABULARY

On a separate piece of paper, match each term with the number of the statement that best explains it. Use each term only once.

liver   kidneys   alveoli   dermis
excretion   metabolism   trachea   nephrons
epidermis   diaphragm

1. The outermost layer of the skin is called the _____.
2. _____ is the sum of the chemical processes that take place in an organism.
3. The _____ is the muscle that forms the floor of the chest cavity and helps you breathe.
4. The _____ are the main excretory organs that remove wastes from the blood and help maintain water balance in the body.
5. The organs identified in sentence 4 are made up of tiny filtering structures called _____.
6. _____ is the process by which an organism gets rid of metabolic wastes.
7. Exchange of gases takes place between the _____ in the lungs and the capillaries.
8. The largest gland in the body, the _____, performs functions related to both digestion and excretion.
9. The _____ is the thick, active layer of skin.
10. Another word for the _____ is the windpipe.

QUESTIONS

Give brief but complete answers to each of the following questions. Unless otherwise indicated, use complete sentences to write your answers.

1. What is the purpose of any respiratory system?
2. Pretend that you are a molecule of oxygen and describe your travels from the time you enter the human nose or mouth until you are absorbed into the bloodstream. Be sure to describe the structures through which you pass. Also describe how you are affected by each structure.
3. How do the kidneys help maintain the body's water balance?
4. How do liquids enter and leave the urinary bladder?
5. How are wastes produced in the body?
6. What is urea and where does it come from?

7. In what form are wastes excreted by the skin? What is another function of this substance?

8. What are the metabolic wastes excreted by each of the following: lungs, kidneys, skin?

9. The blood from the small intestine travels directly to the liver. Why must the blood travel there before going on to the rest of the body?

10. Why is the skin considered an organ? What are the functions of this largest organ of the body?

11. Why does exercising cause a person to breathe faster?

APPLYING SCIENCE

1. Why are people told to breathe into a paper bag if they are hyperventilating (breathing too fast and getting too much oxygen as a result)?

2. The disease polio causes muscle paralysis. During polio epidemics, it was common to hear of people needing to be in "iron lungs" to survive. Why would a lung machine be necessary for someone severely stricken with polio?

3. How hard your cells are working can be measured by how fast they are using oxygen. You can determine how hard your cells are working at various times of the day and during or after different activities. To do this, measure your oxygen intake in breaths per minute. This is your breathing rate. You will need a clock with a second hand and a data sheet with three columns—one for identifying the activity, one for the time of day, and one for recording breathing rate. Check your breathing patterns all day long. Then make comparisons. What is your breathing rate first thing in the morning? When you are concentrating? When you are angry? What do these rates tell you about your oxygen intake?

BIBLIOGRAPHY

Baldwin, Dorothy, and Claire Lister. *You and Your Body: Your Heart and Lungs.* New York: The Bookwright Press, 1984.

Franklin Watts Science World. *Biology: Plants, Animals and Ecology.* New York: Franklin Watts, 1984.

National Geographic Society. *Your Wonderful Body.* Washington, DC: National Geographic Society, 1982.

Ward, Brian R. *The Human Body: The Lungs and Breathing.* New York: Franklin Watts, 1982.

A construction worker, working on a skyscraper high above the ground, must have "nerves of steel." His complex nervous system controls his senses, movement, and balance.

CONTROL SYSTEMS

CHAPTER GOALS

1. Describe the functions of three types of neurons.
2. List the organs of the nervous system and their functions.
3. Describe the structures and functions of the sense organs.
4. Describe the endocrine system and its functions.
5. Describe the physical and social effects of taking drugs.

17-1. Neurons

At the end of this section you will be able to:

- ▨ Identify the three functions of a nervous system.
- ▨ Describe the three types of *neurons*.
- ▨ Trace the path an *impulse* travels in a *reflex*.

In the classic horror thriller, Dr. Frankenstein used a great surge of electricity to "shock" his monster into life. Today, controlled electrical shocks help to bring heart attack victims back to life. Yet too much electricity can be deadly. What does electricity have to do with our bodies? Are the phrases "wired" and "tuned in" more accurate than we think they are?

THE NERVOUS SYSTEM

The ability to respond to a stimulus is common to all living things. In fact, it is a characteristic of protoplasm, the substance of living matter. Most animals have a nervous system. A nervous system allows one part of the organism to communicate with the other parts. The more complex an animal is, the more important this communication is to it. The functions of a nervous system are:

1. To detect changes (stimuli) inside or outside the body.
2. To transfer and interpret this information.
3. To respond to, or act upon, this information.

The human nervous system is made up of the brain, the spinal cord, and the nerves. The brain and spinal cord make up the central nervous system. The nerves, made up of many nerve cells, carry information to and from the brain and spinal cord. Look at the picture on page 382.

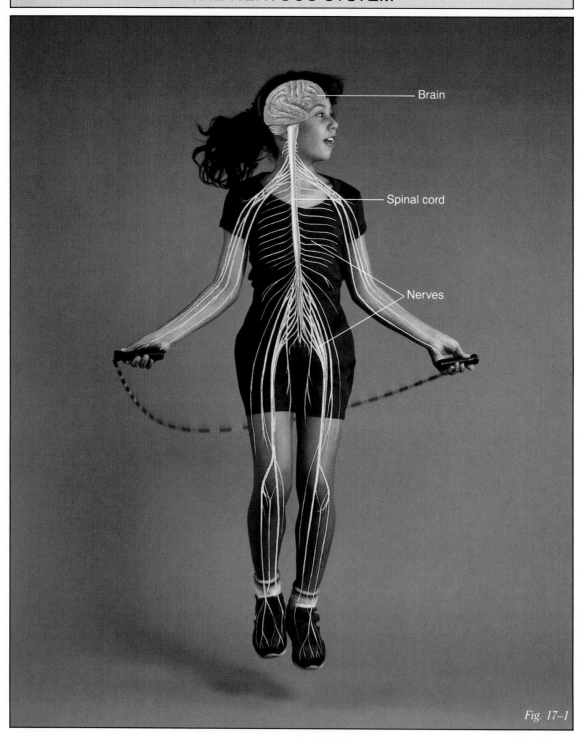

Brain

Spinal cord

Nerves

Fig. 17–1

TYPES OF NEURONS

The billions of nerve cells that make up the nervous system are called **neurons** (**nur**-onz). A typical neuron has an enlarged, though still microscopic, area called the cell body. See Fig. 17–2. All cellular functions take place here. A long strand, or thread, called an *axon* branches out of the cell body. Its job is to carry messages from the cell body. Some axons, like those that connect your fingertips to your spinal cord, may be as long as 90 cm. The cell body may also have many other fine threadlike branches called *dendrites*. Dendrites receive messages from axons and carry the messages into their cell body. How this works is described below.

Neuron A nerve cell that is the basic unit of the nervous system.

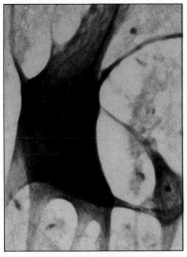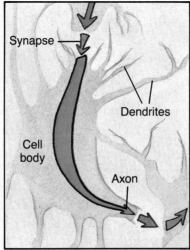

Fig. 17–2 *A photo of a nerve cell* (left) *and its parts* (right).

The information that travels from one nerve cell to another is called an **impulse.** An *impulse* is a tiny, electrical charge that moves along the axon somewhat like a burning fuse. The impulse can travel up to 90 m per second. Look at Fig. 17–2 again. When the impulse reaches the end of the axon, it must cross a gap to reach the next nerve cell. The gap between the axon carrying the message and the dendrites of the nerve cell body receiving the message is called a *synapse* (**sih**-naps). In order to cross a synapse, the electrical impulse must change to a chemical impulse. When it reaches the dendrites of the next nerve cell, it changes back to an electrical impulse.

Impulses travel in one direction along definite pathways. These pathways include three special types of neurons: sensory, motor, and association. **Sensory neurons** transfer

Impulse Information from a stimulus that travels along neurons.

Sensory neuron A neuron that carries impulses from sense organs to the spinal cord or brain.

Motor neuron A neuron that carries impulses from the brain or spinal cord to a muscle or gland.

Association neuron A neuron that carries impulses from sensory neurons to motor neurons.

Reflex An automatic response to a stimulus in which the brain is not directly involved.

impulses from sense organs to the spinal cord or brain. The nerve fibers from the fingertips to the spinal cord are *sensory neurons.* **Motor neurons** transfer impulses to muscles and glands. Impulses that travel along motor neurons cause the body to respond to the stimulus. However, *motor neurons* do not receive impulses from sensory neurons directly. It is the **association neurons** that connect the sensory neurons and motor neurons. *Association neurons* are located in the brain and spinal cord. They transfer impulses between sensory and motor neurons. This works somewhat like switching stations that transfer telephone calls.

REFLEXES

One of the simplest pathways along which an impulse travels is called a **reflex.** A *reflex* is an involuntary action that occurs in response to a stimulus. A reflex happens so quickly that you do not think about it until after it happens. Reflexes include sneezing, coughing, and blinking. Pulling away from a hot object, jumping when frightened, and recovering your balance when you slip are also reflexes.

Some reflexes occur that we are not even aware of. These include the release of digestive juices at the sight or smell of food and the movement of the *pupil* of the eye in response to light. Doctors test some of these reflexes to help diagnose disorders of the nervous system. See Fig. 17–3.

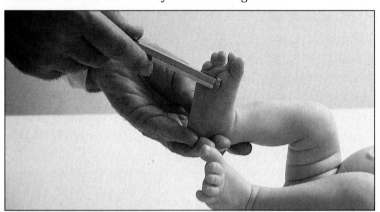

Fig. 17–3 A doctor tests the reflexes of a newborn by stroking the bottom of its foot.

The brain is not directly involved in reflexes. Reflexes can happen quickly because the impulse must travel only to and from the spinal cord for the body to react. The brain will receive a message after the reflex action has occurred. Thus, you realize that a reflex happened a bit later, after the brain receives the message.

Using Fig. 17–4, trace the path that an impulse travels in a reflex. Suppose you accidentally touch a hot iron. The sensory nerve endings in your fingers receive the stimulus. The impulse travels along the sensory neurons to the spinal cord. In the spinal cord, association neurons transfer the impulse to motor neurons. The impulse travels along motor neurons to the muscles in the arm. This causes the muscles in your arm to contract. Your hand pulls away from the hot iron. Part of the original impulse from the stimulus is still on its way to the brain. When the brain receives this impulse, you feel pain and inspect the damage.

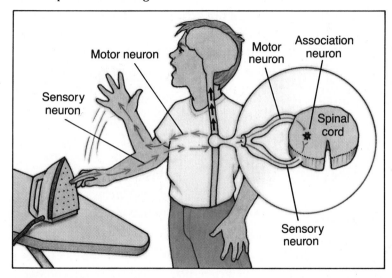

Fig. 17–4 A diagram of the pathway of a reflex impulse.

SUMMARY

The nervous system detects and responds to stimuli. The basic unit of the nervous system is the neuron. There are three types of neurons. Neurons may work together in an automatic response called a reflex.

QUESTIONS

Use complete sentences to write your answers.

1. What does a nervous system allow an animal to do?
2. What is a neuron? What are the three types of neurons?
3. What is a reflex? Why does it happen so quickly?

INVESTIGATION

REFLEXES

PURPOSE: To demonstrate and analyze two reflexes.

MATERIALS:

lightweight ball   clear plastic wrap

PROCEDURE:

A. You will need a partner for this investigation. One person should relax and sit with one leg crossed over the other. The other person will make one hand rigid. Using the edge of this hand, gently tap about 3 to 4 cm below the partner's kneecap. See Fig. 17–5. This area is soft in comparison to the kneecap or the bone below the knee. The tap must be done in the right place with the right amount of force. Otherwise, the reflex will not occur. You may have to try this a few times to get it right.

    1. What is the stimulus for the knee jerk reflex?

    2. What is the response?

B. Hold a sheet of plastic wrap in front of your face. Have your partner toss a little ball at the sheet of plastic.

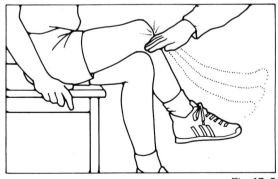

Fig. 17–5

    3. How did you respond?

C. Repeat step B. However, this time try not to blink.

    4. Were you able not to blink?

    5. What is the function of the blinking reflex?

D. Completely cover your eyes for one minute. When you remove your hands, have your partner immediately look at the pupils of your eyes. You may need to try this several times.

    6. How did your pupils respond?

    7. What is the function of this reflex?

E. Reverse roles and repeat the steps from A through D.

F. On a piece of paper, draw a sketch of the pathway for the knee jerk reflex. Label the stimulus, response, spinal cord, and types of neurons. Use arrows to show the direction the impulse travels.

    8. Which of the three types of neurons detects the stimulus?

    9. Where does this type of neuron carry the impulse?

    10. What type of neuron is the impulse transferred to next? What is the job of this neuron type?

    11. What type of neuron carries the impulse that causes a response?

    12. Where does this type of neuron carry the impulse?

CONCLUSION:

    1. What advantages do reflex reactions have over "thought reactions"?

17-2. The Central Nervous System

At the end of this section you will be able to:

- ☐ Identify the parts of the *central nervous system.*
- ☐ Describe the structures and functions of the human brain and spinal cord.

Look at Fig. 17–6. What do you see? Show it to your family and friends. Ask them what they see. You will probably get as many different answers as the number of people you ask. Although the picture remains the same, each person interprets it differently.

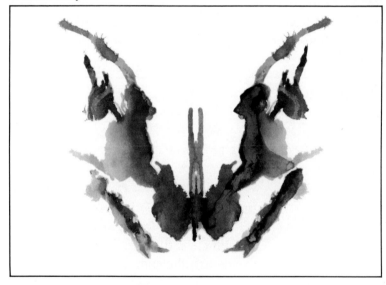

Fig. 17–6 An example of an inkblot test, which may be used by a psychologist.

THE BRAIN

Interpreting information from the senses is a function of your brain. So are remembering, learning, dreaming, imagining, coordinating, controlling, and decision making.

Your brain is very busy. Like the heart, it works 24 hours a day. The heart must send it a large supply of oxygen-rich blood. Although only about 2 or 3 percent of your body weight, the brain uses 25 percent of all the oxygen you take in. Brain cells are very sensitive to the amount of oxygen available. That is why you cannot hold your breath or pant indefinitely. Both too little and too much oxygen to your brain will cause you to faint.

Central nervous system Made up of the brain and spinal cord.

Cerebrum The largest region of the human brain. It is the center of human intelligence.

The brain and spinal cord are the two major centers of the human **central nervous system.** Both organs are very delicate. They are well protected by fluid and bone. The brain is also protected by three membranes.

There are about 10 billion (10,000,000,000) association neurons in the brain. These neurons are organized into three main areas: the cerebrum, the cerebellum, and the medulla. See Fig. 17–7.

The **cerebrum** (seh-**ree**-brum) is the largest area of the brain. It is divided into two halves. The surface of the *cerebrum* looks like the surface of a walnut because of all its bumps and folds. The cerebrum controls many thought processes, including memory and learning. It also controls many of our voluntary muscle movements. Information from the sense organs is interpreted by special areas of the cerebrum. See Fig. 17–7 again.

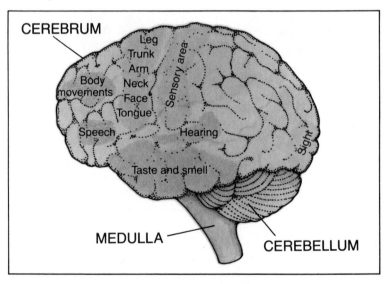

Fig. 17–7 Specific areas of the brain control certain functions.

Each half of the cerebrum controls the activities of the opposite half of the body. Thus, the left half of the cerebrum controls the right side of the body; the right half controls the left side of the body.

Cerebellum Area of the human brain between the cerebrum and the medulla. It coordinates muscular activities and body balance.

The **cerebellum** (ser-eh-**bel**-um) is about the size of a pear. It is found just below the back of the cerebrum. The *cerebellum* works with the cerebrum to control muscular activity. Without a cerebellum, you would not be able to walk, run, write, or play ball. These activities require you to repeat a pattern of movements. The cerebellum coordinates these movements. The cerebellum also helps you keep your balance. It does this by using information from your eyes and ears.

The nerve fibers from the cerebrum and cerebellum meet in the *brain stem*. The brain stem connects the spinal cord with the brain. The enlarged area of the stem at the base of the brain is the **medulla** (meh-**dul**-ah). The *medulla* controls the activities of the internal organs. These include breathing, heart action, and movements of the digestive system.

Medulla An enlarged area of the brain stem. It controls the functions of the internal organs.

THE SPINAL CORD

The brain sends and receives impulses by way of the **spinal cord.** The *spinal cord* is a thick bundle of nerves. It reaches from the medulla to near the base of your spine. Thirty-one pairs of spinal nerves branch off from the spinal cord. These nerves then branch out many more times. They form the tiny nerve fibers found throughout the body.

Your spinal cord is protected by the bones called vertebrae. See Fig. 17–8. All the vertebrae together form your spine. It is important to protect the neck and back from injuries. If the spinal cord were cut, impulses would not reach below that point. The ability to feel and move would be lost.

Spinal cord A thick bundle of nerves that reaches from the medulla to near the base of the spine.

SUMMARY

The central nervous system is made up of the brain and spinal cord. The brain directs muscular activity, controls thoughts, and interprets information from the sense organs. The brain sends and receives messages by way of the spinal cord. Nerves branch out from the spinal cord and reach all areas of the body.

QUESTIONS

Use complete sentences to write your answers.

1. What are the parts of the central nervous system?
2. List the three main parts of the brain and explain briefly the functions of each.
3. If the left side of the cerebrum were injured, where on the body would activities be affected?
4. Why can a broken neck make a person paralyzed?

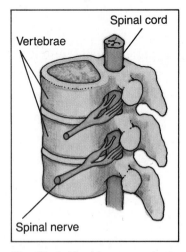

Fig. 17–8 Hard bones protect the delicate spinal cord.

SKILL-BUILDING ACTIVITY

PROBLEM SOLVING

PURPOSE: To practice solving problems.
How you approach a problem is based upon what you already know. Your brain uses this information to come up with a solution. Your past experiences play an important role in problem solving. You will see this to be true as you solve the following problems.

MATERIALS:
none

PROCEDURE:

A. One half of you is probably more coordinated than the other half. You probably write with one hand instead of the other. You may also use this hand to eat and to throw a ball. This is because one side of your brain is dominant over the other. Most people are right-handed and favor the right side. This means that the left side of the brain is dominant. The people who favor the left side have a dominant right side of the brain. Some don't favor either side. You can check dominance patterns with these tests.

B. To test your eyes: Hold a pencil at arm's length in front of your eyes. With both eyes open, sight along the top of the pencil to a distant object. See Fig. 17–9. Keeping your pencil in place, close one eye at a time.

1. With which eye does the pencil appear to jump? Which eye sees the pencil centered over the object? This is your dominant eye.

C. On a separate piece of paper, create a data table to record the results of each test. Include the following categories: the part of the body tested, the way it was tested, and which side was dominant.

D. You also have a dominate leg, ear, side of the face, and side of the tongue. Design some tests to determine dominance for each of these. Add your observations to your data table.

CONCLUSIONS:

1. Is one side of your body dominant? Which one?

2. How did you solve this problem?

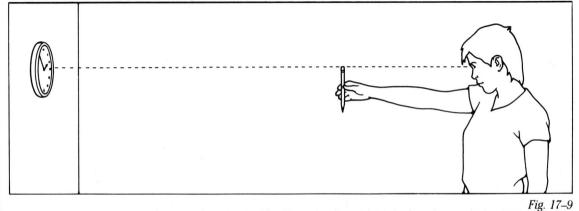

Fig. 17–9

17-3. The Senses

At the end of this section you will be able to:

- ☐ Identify the major sense organs of the body.
- ☐ Describe the sensations associated with each sense organ.
- ☐ Explain how each sense organ functions.

Imagine being in a windowless, soundproof room. There is no telephone, TV, or even a radio. You could not know what was happening outside the room. This is the situation in which your brain exists. It is enclosed in the skull. It, too is cut off from the world. However, your eyes, ears, nose, mouth, and skin send messages to the brain. They inform the brain about changes inside and outside your body.

The eyes, ears, nose, mouth, and skin are **sense organs.** These *sense organs* allow us to feel, smell, taste, see, and hear. Besides these five sense organs, we also have organs that detect blood pressure, body temperature, muscle contractions, and the direction of gravity.

A sense organ contains one or many *receptor cells*. The function of receptor cells is to receive stimuli.

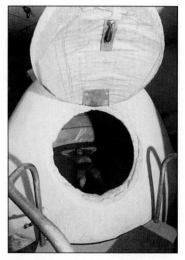

Fig. 17–10 Scientists designed a sealed tank to find out what it is like to have no stimuli from the environment.

Sense organs Structures of an organism, such as eyes and ears, that are sensitive to certain specific stimuli.

SENSES IN THE SKIN

There are several types of receptor cells in the skin. Each has a specialized function. See Fig. 17–11. Receptor cells can

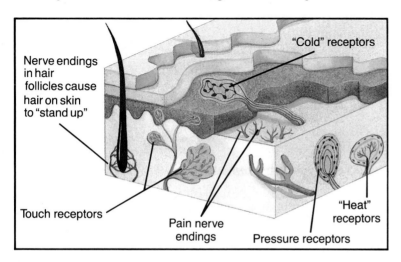

Nerve endings in hair follicles cause hair on skin to "stand up"

"Cold" receptors

Touch receptors

Pain nerve endings

Pressure receptors

"Heat" receptors

Fig. 17–11 The location of different touch receptors in a section of the skin.

receive the stimuli of touch, pressure, pain, or heat and cold. The receptors for touch are close to the skin's surface. Some areas of the skin are more sensitive than other areas. This is because they have more touch receptors than the others. Pressure receptors lie deeper in the skin. They require a stronger stimulus than the touch receptors. Pain and temperature receptor cells help protect the body. They warn the brain of either real or possible injury.

TASTE AND SMELL

The senses of taste and smell are very much alike. The receptor cells of each are sensitive to certain chemicals. The receptor cells for taste are located in structures called **taste buds.** *Taste buds* are found on the tiny bumps all over your tongue. It seems that there are only four basic tastes: sweet, sour, bitter, and salty. The buds for these are grouped on different parts of the tongue. See Fig. 17–12. Why do you think the ice cream in an ice-cream cone tastes sweeter when you lick it with the tip of your tongue?

The receptor cells for smell are much more sensitive than those for taste. They can detect up to 10,000 different chemicals. These receptor cells are located on a postage-stamp-sized area in the upper part of each nostril, just behind the bridge of your nose. See Fig. 17–12. Sniffing helps the sense of smell. It moves the air into these cells in the upper part of the nose. Just a few scent molecules cause a chemical reaction to take place. This reaction sends a signal along sensory neurons to the brain. The brain decodes the signal as a particular smell.

The flavors of foods are mostly smells, not tastes. When you have a cold it is hard to "taste" food. The nose passages become lined with thick mucus. When this happens, the smell receptors cannot work.

SIGHT

Your eyes are like a camera. See Fig. 17–13. Both eyes and a camera have a lens that focuses light onto a special surface.

Out of its protective socket, the eyeball looks something like a ping-pong ball. It is a thin shell, filled with a jellylike material that enables it to keep its shape. Light first passes

Taste buds Special receptor cells on the tongue that are stimulated by flavors.

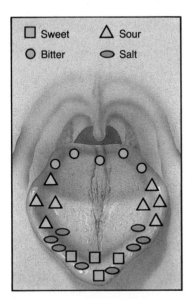

Fig. 17–12 The organs of taste on the surface of the tongue.

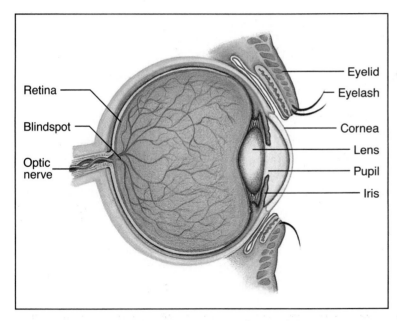

Fig. 17–13 A cross section and side view of the eye.

Labels on figure: Retina, Blindspot, Optic nerve, Eyelid, Eyelash, Cornea, Lens, Pupil, Iris

through the *cornea.* The cornea is the front of the tough, outer layer known as the "white of the eye." It is curved and clear. If the surface of the cornea is irregular, things you see will be blurred. This condition is known as astigmatism.

Also at the front of the eye is a ring of colored tissue called the *iris.* This iris gives us the color of our eyes. In the center of the iris is an opening called the *pupil.* The pupil is a hole, not a structure. It is covered by the clear cornea. The iris contains muscles that change the size of the pupil. The pupil automatically gets larger or smaller. The pupil's size controls the amount of light coming into the eye. Behind the pupil is a flexible *lens.* Muscles in the eye adjust the lens to focus the light. To do this, your lens changes shape about 100,000 times a day.

The lens focuses light on the innermost layer of the eye, called the **retina** (**reht**-in-ah). The *retina* contains receptor cells and nerve fibers. *Rod cells* and *cone cells* are two types of receptor cells in the retina. Rod cells are very sensitive. They can respond to very dim light. Cone cells react to very bright light. They also sense colors.

The rods and cones send this information to sensory neurons. These neurons join to form the *optic nerve,* which leads to the brain. The brain interprets the nerve impulses and forms a picture or image.

Retina The innermost layer of the eye made up of receptors that are sensitive to light.

HEARING, BALANCE, AND MOTION

Sound is made when molecules move back and forth in a pattern. Our ears are sensory organs that can change sound energy into nerve impulses. See Fig. 17–14. The *outer ear* collects sound waves and directs them along a narrow canal to the *eardrum.* The eardrum is a membrane that separates the outer ear and the *middle ear.*

The middle ear contains three little bones that make the sound louder. It also contains a tube that keeps the pressure equal between the outer ear and inner ear. You are sometimes aware of this tube when you swallow and when your ears "pop."

From the middle ear, the sound travels to the *inner ear.* The inner ear has many fluid-filled chambers and canals. A chamber called the **cochlea** (**kok**-lee-ah) is the actual hearing organ. The *cochlea* contains hairlike receptor cells. The sound causes the fluid in the cochlea to move. This movement causes the tiny hairs to bend. The bending sends impulses along the *auditory nerve* to the brain. Very loud noises, including loud music, can cause nerve damage. Hearing is a sensitive system of chain reactions. If any link is broken, hearing is impaired.

The ear also contains the structure that controls balance and motion. There are three *semicircular canals* in the inner ear. Look at Fig. 17–14 again. These canals also contain fluid and hairlike receptor cells. Whenever the head moves or rotates, the fluid also moves. This, in turn, causes the hairs to

Cochlea The spiral canal of the inner ear that contains the receptors for hearing.

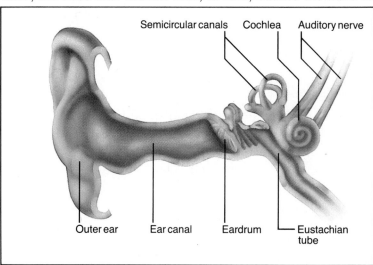

Semicircular canals   Cochlea   Auditory nerve

Outer ear   Ear canal   Eardrum   Eustachian tube

Fig. 17–14 The structures of the outer, middle, and inner ear.

move. Impulses are sent to the brain. The brain uses this information to determine the direction and speed that your body is moving in. The brain also uses impulses from your eyes for balance. If you stop suddenly after spinning, you get dizzy. This is because the fluid in the semicircular canals keeps moving. You can reduce the movement of the fluid and the amount of dizziness by looking at only one object while you are spinning. This technique is used by skaters and dancers as they turn.

Fig. 17–15 Because gravity is not pulling down on the fluid in the ear's semicircular canals, these space shuttle astronauts have a disturbed sense of balance.

SUMMARY

The human senses are touch, taste, smell, sight, and hearing. Balance and motion are controlled by structures in the inner ear. Each of these sense organs has specialized cells that detect stimuli and change them into nerve impulses. The nerve impulses are carried to specific areas of the brain, where they are interpreted.

QUESTIONS

Use complete sentences to write your answers.

1. What are the five sensations of the skin, and how are they different?
2. How is it that we can "taste" more than the basic four flavors?
3. How is the eye like a camera?
4. How do we hear?
5. What other senses are performed in the ear?

PRESSURE RECEPTORS

PURPOSE: To measure the distance between pressure receptors, and to use the distance measured to determine sensitivity of the body parts tested.

MATERIALS:
blindfold   large pencil eraser
ruler   or piece of cork
2 straight pins

PROCEDURE:

Part I

A. Move the nonwriting end of a pen or pencil lightly over the skin of your arm.

    1. How does this feel?

B. Repeat step A, but now press a little harder into the same area. Do not press too hard. You want to stimulate pressure receptors, not pain receptors.

    2. Describe how this movement feels compared to the first movement in step A.

Part II
You will need a partner for this part.

A. On a separate piece of paper, make three copies of Data Table 1. Label the first copy "Palm," the second copy "Outside of Forearm," and third copy "Forehead." Also, make one copy of Data Table 2.

B. Place two pins into the eraser or cork so they are 3 cm apart. Record this distance in Data Table 1.

Part of Body Tested: _____		
Trial	Distance Between Pins (cm)	Number of Pins Felt
1.		
2.		
3.		

Data Table 1

C. Blindfold one person. Gently press both pinheads into the blindfolded person's palm. See Fig. 17–16. Ask how many pins

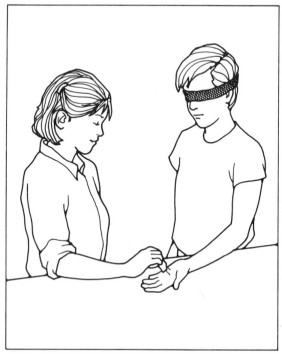

Fig. 17–16

are felt. If one pinhead is felt, only one pressure receptor was stimulated. If two pinheads are felt, two pressure receptors were stimulated. Record the number of pins felt under Trial 1 of your Data Table 1 labeled "Palm."

D. If only one pin was felt, move the pins farther apart. If two pins were felt, move the pins closer together. Use the ruler to measure the distance between the pins in cm. See Fig. 17–17. Record this distance under Trial 2 of Data Table 1.

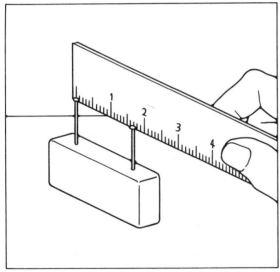

Fig. 17–17

E. Press the pinheads into the person's palm again. Ask how many pins are felt. Record this number on the data sheet.

F. Repeat these procedures on different areas of the palm. When you find the

shortest pin distance at which the blind-folded person can still feel two pins, stop. Circle the number indicating this distance on the chart.

G. Repeat procedures B through F for the outside of the forearm and the forehead. Record your observations in the Data Tables labeled "Forearm" and "Outside of Forehead."

H. Fill in Data Table 2.

Part of Body Tested	Shortest Distance at Which Two Points Can Be Felt
Palm	
Outside of Forearm	
Forehead	

Data Table 2

CONCLUSIONS:

1. On which area(s) of the body are the pressure receptors closest together?

2. On which area(s) of the body are the pressure receptors farthest apart?

3. Of the body areas tested, which is the most sensitive?

4. Which is the least sensitive?

5. How sensitive would you say the back is compared with the other body areas you tested?

6. Explain what you could do to test the answer you gave to question 5.

¡COMPUTE!

SCIENCE INPUT

Measuring a person's abilities can be difficult. Both physical and social scientists spend a great deal of time and effort devising ways to judge differences in abilities. The time it takes to accomplish something is one (and only one) measure of your ability. In this exercise, you will study the time it takes to use different senses to identify or locate an object.

COMPUTER INPUT

The computer has the capacity to be used as a counting and timing tool. This capacity is built into the hardware. In other words, every computer can send a message through its circuits in a specific time. That time will remain the same. Once you know how long it takes your computer to run the circuit, you can design a program where the computer is used as a timer or clock. In Program Timer, you will use the computer in an experiment to find the difference between your ability to locate an object through sight, hearing, and touch.

WHAT TO DO

Enter Program Timer and save it on disc or tape. Make a copy of the data chart.

There will be three conditions to the experiment. In condition (1), the object will be hidden in the classroom. The locater will run the program, starting to look for the object when the screen reads "READY" and pressing the return key to end the program when the object has been found. In condition (2), the object, which should be capable of producing a sound, will be in plain view; the locater will be blindfolded. When the screen prints "READY," the object

should be made to produce its sound and the locater will search for the object. When it is found, the locater should return to the computer without the blindfold to end the program.

In condition (3), the object is in view, the locater is blindfolded, and the object does not produce a sound. Using the sense of touch, the locater must try to find the object.

When the object is found, the locater returns to the computer and presses the appropriate key to end the program. Record the data for each condition on your chart. (In both blindfolded conditions, the floor should be cleared of safety hazards. In the touch conditions, objects should be within reach. Climbing should not be necessary to reach the object.)

Data Chart

Student	Locating time		
	Sight	Hearing	Touch
1.			
2.			
3.			
4.			
5.			

Total: _____
Average: _____

$$\text{Average time of all student tries} = \frac{\text{Sum of Trials}}{\text{Number of Students}}$$

GLOSSARY

FUNCTION KEY   a key on your computer's keyboard that performs a task other than printing a letter or number. A key may be programmed to function in a particular way, or it may be part of that computer's hardware design. In this program, pressing the "control" and "reset" keys will stop the program, and no more time will be recorded.

PROGRAM

```
100   REM TIMER
110   FOR P = 1 TO 24: PRINT : NEXT P:R$ =
      "READY"
120   PRINT : PRINT " FIRST, READ ALL DIRECTIONS
      CAREFULLY!"
130   PRINT : PRINT : PRINT : PRINT "    YOU ARE
      ABOUT TO FIND ONE MEASURE OF YOUR":
      PRINT "ABILITY TO LOCATE OBJECTS USING
      DIFFERENT SENSES."
140   PRINT : PRINT "       TO START, PRESS THE
      <RETURN/ENTER>": PRINT "KEY/THE TIMER
      WILL START WHEN 'READY' ": PRINT
      "APPEARS ON THE MONITOR/SCREEN."
143   PRINT : PRINT "       TO STOP THE TIMER,
      QUICKLY PRESS": PRINT "THE <RESET/
      CONTROL> OR ANOTHER FUNCTION": PRINT
      "KEY ON YOUR COMPUTER, THAT WILL END":
      PRINT "THE PROGRAM."
145   PRINT : PRINT "        AFTER RECORDING
      YOUR LOCATION TIME,": PRINT "TYPE 'RUN'
      THEN PRESS <RETURN/ENTER>": PRINT "TO
      RESTART THE PROGRAM."
147   INPUT "";Z$
150   S = 0:T = 0: FOR D = 1 TO 1000: NEXT D
155   HOME : REM USE YOUR COMPUTER'S
      FUNCTION TO 'CLEAR' SCREEN
160   PRINT : PRINT : PRINT ,"    "R$: PRINT :
      PRINT
170   PRINT : PRINT "    ","SECONDS"
175   VTAB 12: PRINT : PRINT : PRINT ,"    "S"."T:
```

```
      REM USE A VERTICAL-TAB FUNCTION &
      VALUE FOR MID SCREEN
180   FOR D = 1 TO 40: NEXT D: REM TIMER FOR 1/
      10TH SEC. - ADJUST W/YOUR COMPTR. AS
      NEEDED
190   T = T + 1: IF T = 10 THEN S = S + 1
200   IF T = 10 THEN T = 0: IF S = 60 THEN GOTO
      230
220   GOTO 175
230   PRINT : PRINT : PRINT : PRINT "    YOU
      NEEDED MORE TIME TO LOCATE THE
      OBJECT!": PRINT : PRINT "     RECORD THIS
      TIME ON YOUR CHART."
240   PRINT : PRINT : PRINT : PRINT "
      DIRECTIONS WILL APPEAR
      MOMENTARILY."
250   FOR D = 1 TO 4500: NEXT D: FOR P = 1 TO
      24: PRINT : NEXT P: GOTO 120
260   END
```

PROGRAM NOTES

Consult the manual of your computer to find out how long the looping mechanism takes. Set the timer appropriately in statement 180. The program counts and prints out tenths of a second. You can coordinate the speed of your computer with the program by changing the $D = 1$ to 40 (if necessary).

BITS OF INFORMATION

"Debugging" a program takes a lot of patience and ingenuity. Finding the exact cause of a problem is not always easy. Sometimes the problem is with the software itself, which may be poorly designed or simply have had incorrect input entered. Or, the problem may be a physical one. Computer legend has it that the word "debug" derives from such a physical problem. In the 1940's, a dead moth was found to be causing problems in a computer at Harvard. Ever since then, so goes the story, fixing these problems has been called "debugging."

17-4. Chemical Control

At the end of this section you will be able to:

- Describe the structure and functions of the human *endocrine system*.
- Explain the roles of *hormones*.
- Describe the sequence of events called *feedback*.

It's almost your turn. Your heart begins to pound. The palms of your hands begin to sweat, and your mouth becomes dry. The "butterflies" in your stomach won't settle down. As a result of your anxiety, your body is gearing you up for *flight or fight*— an important mechanism for survival.

Fig. 17–18 Some of the endocrine glands are different in males and females.

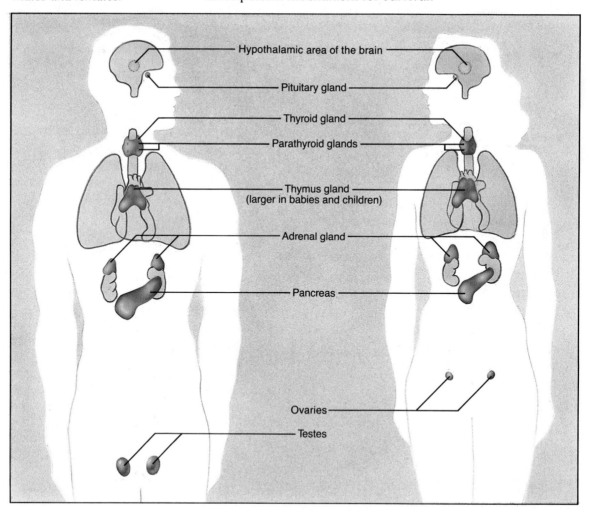

Hypothalamic area of the brain

Pituitary gland

Thyroid gland

Parathyroid glands

Thymus gland
(larger in babies and children)

Adrenal gland

Pancreas

Ovaries

Testes

THE ENDOCRINE SYSTEM

In the scene just described, the body reactions were the result of the cooperation between two main systems in the body. One, which you learned about earlier, is the nervous system. The other is a chemical-controlled system called the *endocrine* (**en**-doe-krin) *system*. Not all organisms have nervous systems. However, all organisms do have chemical-control systems.

The chemicals made by these control systems are called **hormones**. *Hormones* act as chemical messengers. In humans, hormones are made by special glands. These glands are called **endocrine glands**. See Fig. 17–18. *Endocrine glands* are different from digestive glands. They do not have a duct (tube) through which the hormone is released. Instead, hormones are released directly into the bloodstream. These chemical messengers are carried to where they must work by the circulatory system. In this way, hormones can affect cells that are far from where the hormones are made.

Hormone A chemical messenger that regulates and balances body functions. Hormones are carried in the blood.

Endocrine glands Ductless glands that release hormones into the bloodstream.

FUNCTION OF HORMONES

Hormones are also called "chemical regulators." As regulators, hormones act in three ways. First, some hormones may speed up body processes. For example, there is a hormone that speeds up your heartbeat. Second, others are able to hold back or slow down certain processes. This is true for the hormones that control the distribution of food, salts, and water in the body. Third, some hormones influence growth by producing changes in body structure.

Hormones are very powerful. Only tiny amounts of them are needed to control the body functions. However, the body must have the exact amounts it needs. For example, the *pituitary* (pih-**too**-ih-tare-ee) hormone regulates the growth of the skeleton. If too much of this hormone is produced, a giant will result. Too little causes a midget. Table 17–1, on page 402, lists some other hormones and their functions.

For the most part, the body does not store hormones. The glands produce hormones in small amounts as they are needed. If not enough hormone is produced by an endocrine gland, the hormone must be supplied in the form of medication. For example, many of the people who have diabetes must take some form of the hormone insulin on a regular basis.

Once some hormones have done their jobs, they are carried to the liver. There, they are either broken down and removed from the body, or they are used to make new hormones.

FEEDBACK

Feedback An automatic "turn on" or "turn off" process that occurs when a certain level of a hormone or other substance is reached.

The amounts of hormones produced in the body are regulated by a mechanism called **feedback.** *Feedback* works like a thermostat in a building. The thermostat is set at the desired temperature level. When the temperature of the building drops below this level, a signal turns the furnace on. When the building temperature reaches the set temperature, the thermostat signals the furnace to turn off. See Fig. 17–19.

The signals that turn different glands on or off vary. Sometimes the signal is the amount of the substance the hormone regulates. For example, the *parathyroid* (**par**-ah-**thie**-roid) gland releases a hormone that controls the amount of calcium in the blood. When the calcium level is low, the parathyroid gland releases its hormone. This hormone causes certain bone cells to release calcium. It also slows down the kidney's excretion of calcium. When there is enough calcium in the blood, the parathyroid gland stops releasing the hormone.

THE ENDOCRINE SYSTEM		
Gland	Hormone	Function
Thyroid	Thyroxine	Regulates release of energy in body. Iodine is needed to make thyroxine.
Parathyroid	Parathyroid hormone	Controls the body's use of calcium.
Pituitary	Growth hormone, other pituitary hormones	Regulates growth of skeleton; controls the release of hormones from other glands; controls kidney function; regulates blood pressure.
Adrenals	Adrenalin (epinephrine)	Increases heartbeat, blood flow, and the amount of sugar in the blood; activates the nervous system; regulates water and mineral balance in body tissues.
Pancreas	Insulin	Allows liver to store sugar; regulates body's use of sugar.
Ovaries (female)	Female sex hormone (estrogen)	Controls development of reproductive organs and female characteristics.
Testes (males)	Male sex hormone (testosterone)	Controls development of reproductive organs and male characteristics.

Table 17–1

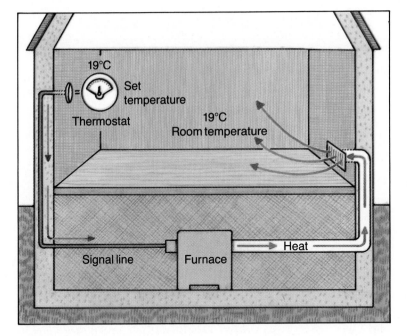

Fig. 17–19 This thermostat is set at 19°C. If the room temperature falls below 19°C, the thermostat sends a message to the furnace to provide more heat. When the room reaches 19°C, the thermostat sends a message to the furnace to switch off.

The nervous system also helps turn glands on or off. Have you ever heard of someone having superhuman strength during times of danger? By releasing more of the hormone adrenalin, the adrenal gland helped provide the sudden strength. In this way, the emotions of fear or anger stimulated the endocrine system.

SUMMARY

The functions of an organism are partly controlled by chemical messengers called hormones. In humans, hormones are produced by the endocrine system. The level of hormones produced is regulated by the feedback mechanism.

QUESTIONS

Use complete sentences to write your answers.

1. How are endocrine glands different from other glands?
2. In what ways do hormones act?
3. What are some specific functions of hormones?
4. What is feedback? How is feedback used in the endocrine system?

17-5. Drugs and Your Health

At the end of this section you will be able to:

☐ List the different groups of addictive drugs and give examples for each group.

☐ Describe the effects that drugs, alcohol, and tobacco have on the body.

☐ Describe the social problems associated with alcohol, tobacco, and drugs.

This morning Olive, an office worker, overslept. She didn't have time for breakfast and missed her morning coffee. At work, she was so busy that she missed her coffee break. By noon, Olive had a splitting headache. She noticed that any day she didn't have her coffee, she would get a headache. With her lunch, she drank a cup of coffee and took two aspirin. Did Olive take drugs on her lunch hour? Is Olive *addicted* to coffee?

Fig. 17–20 *Many people use drugs, such as these, every day.*

Drug Any substance, other than food, that changes the body or mind.

Stimulant A drug that speeds up body functions, especially the central nervous system.

DRUG USE AND ABUSE

What is a **drug?** A *drug* is any substance other than food that alters the mind or the body. Based on this definition, Olive took two kinds of drugs. One, the aspirin, is a nonprescription pain reliever. The other one, found in coffee, is called caffeine. Caffeine is a mild **stimulant.** A *stimulant* speeds up body functions, especially those of the central nervous system. This means that the person feels more active and nervous. Drugs such as stimulants are classified based on their effects on the body. Other drug classes are identified in Table 17–2.

ADDICTIVE DRUGS			
Type	Effects	Examples	Dangers and/or Aftereffects
Stimulants	Act on central nervous system. Increase alertness and body functions, such as pulse rate and blood pressure.	Cocaine, amphetamines.	Severe depression, irritability, nervousness, long periods of sleep, and possible personality change.
Depressants	Slow down nervous system. Slow down actions of the heart, nerves, skeletal muscles, and breathing. Lower blood pressure.	Tranquilizers, sleeping pills, bromides, barbiturates, and alcohol.	Prevent people from facing and solving problems; anxiety, inability to sleep.
Hallucinogens	Affect functions of sense organs. Person sees things that do not exist and has a false idea of what is going on around him/her. Change person's thinking, emotions, self-awareness, sense of time, and sense of distance.	LSD, mescaline, PCP or angel dust.	Flashbacks—unexpected and frightening return of symptoms, even though person has not recently used the drug. Abuser may have difficulty in thinking and concentrating, may become overly concerned with death.
Cannabis	Quieting, and effects similar to those of hallucinogens.	Marijuana and hashish.	Produces steady changes in personality; lowers resistance to disease; may affect hormones produced by pituitary gland, causing problems with development and reproduction.
Narcotics	Produce sleep, relieve pain. Medical use carefully controlled.	Opium, morphine, codeine, heroin (not allowed to be used medically), methadone.	Sick feeling with severe headache and muscle cramps. Addict is overcome with desire to take more drugs to avoid feeling sick.
Inhalants (chemical vapors that are inhaled into the body)	Loss of muscle coordination, dizziness, slurred speech, double vision, staggering.	Model airplane glue, household cleaners, paint thinner, varnish, shellac, and aerosol propellants.	Damage lungs, liver, blood cells, and brain tissue. May cause unconsciousness and sudden death.

Table 17–2

Fig. 17–21 Discussing drugs with a trained counselor is one way to prevent drug abuse.

Addiction A physical and/or psychological dependence on a drug.

Withdrawal The addict's physical reaction to stopping the use of a drug.

Many widely used drugs are physically and socially harmful. So why would anyone use them? Some people take drugs to escape problems or cope with stress. Others take drugs to be sociable or to be accepted by friends. In general, people who take drugs do so to change the way they feel. But, this can also be dangerous, because it can lead to addiction. There are two types of **addiction.** One type is emotional dependence on the drug. This type is called *psychological addiction.* The other type is *physical addiction.* In this type, the body needs a constant supply of the drug. The addict's body gradually becomes dependent on the drug. With some drugs, both types of addiction can occur. Usually, an addict cannot stop taking the drug without treatment. If the drug is no longer taken, the body goes through a period of adjustment. This is called **withdrawal.** *Withdrawal* sickness may be serious and even lead to death.

Now let's return to Olive's story. Is Olive addicted to coffee? Her symptoms and behavior may lead you to believe that this is possible.

ALCOHOL

Alcohol is a drug, too. People can also become addicted to alcohol. This addiction is called *alcoholism.* Unfortunately, many people view alcohol as merely a "beverage." But alcohol can upset the delicate balance within the body.

Alcohol enters the bloodstream within two minutes after being swallowed. Then it is circulated throughout the body. In the blood, alcohol combines with oxygen and forms a poisonous substance. The brain is especially sensitive to this poison. However, every system in the body is affected. Heavy alcohol use can damage the heart and kidneys. Perhaps the organ most seriously affected is the liver. Alcohol causes the liver tissues to harden. This prevents the liver from performing its many functions. Eventually, death may result.

The healthy liver can break down the poison formed from alcohol, but only in very small amounts; and the body gets rid of it very slowly. For this reason, drinking large amounts of alcohol in a short period of time is especially dangerous. Sudden death from alcohol poisoning can result when the level of alcohol in the blood builds up very rapidly.

Alcohol is a **depressant.** *Depressants* slow down the activities of the central nervous system. Nerve impulses do not travel as quickly. As a result, the body processes slow down. After only one or two drinks, alcohol seems to "loosen up" the drinker. With more drinks, speech becomes slurred. Muscles lose some coordination. The brain's thinking processes slow down. Eventually, all the drinker wants to do is sleep.

Because of its effects on the nervous system, alcohol makes it difficult and dangerous for people to drive. Only one or two drinks affects a person's judgment and reaction time. Many innocent people are killed each year by people who drink and then drive. See Fig. 17–22.

DRINKING AND DRIVING CAN KILL A FRIENDSHIP.

Fig. 17–22 Alcohol is involved in 50 percent of highway accidents each year.

Depressant A drug that slows down the activities of the central nervous system.

TOBACCO

Almost any use of tobacco is harmful. Tobacco contains a drug called *nicotine.* Nicotine is a stimulant. Just one cigarette makes blood vessels become smaller and speeds up your heartbeat. This upsets the flow of blood and air into and out of your lungs. Cigarette smoke also paralyzes the tiny hairs that line your trachea and bronchi. These tiny hairs are needed to keep your lungs free of disease organisms. Studies show that smokers have damaged tissue and tar deposits in their lungs. Tar is a black substance in cigarette smoke. Many smokers' lungs show abnormal cells. These may be the beginning of cancers.

WHY START A LIFE UNDER A CLOUD?

Smoking is harmful to your baby's health. Quit for both of you. For help call your American Cancer Society.

AMERICAN CANCER SOCIETY

Fig. 17–23 Stillbirths and low birthweight infants are more common among smoking mothers.

The chances of early death from lung cancer are about ten times greater for smokers than for nonsmokers. Smokers also increase their chances of dying early from heart and respiratory diseases. These risks begin to decline as soon as the smoker stops smoking.

Although many people know these facts, they continue to smoke. Part of the problem is that nicotine is an addictive drug. Smokers develop a physical desire for the *nicotine* in tobacco smoke. Smokers get used to the body changes it creates. When they stop smoking, they feel uncomfortable.

EFFECTS DURING PREGNANCY

All of these substances—drugs, alcohol, and tobacco—are especially harmful to unborn children. Until 1961, scientists believed that harmful substances could not be passed from the mother to her unborn child. In that year, over 6,000 European babies were born with birth defects caused by their mothers' taking a supposedly safe drug. The drug, called thalidomide, caused the growth of flipperlike limbs on the babies. Since then, we have learned that many drugs, and other things such as X-rays, can and do harm the developing baby. For this reason, pregnant women should avoid these substances. Every child has the right to start life as healthy as possible. It is up to each person to ensure this right. See Fig. 17–23.

SUMMARY

Substances taken into the body, such as alcohol, tobacco, or drugs, can upset its delicate balance. Misuse of some of these substances may lead to physical, psychological, and social problems.

QUESTIONS

Use complete sentences to write your answers.

1. What type of drug is caffeine? Why do many people start their day with coffee or tea?
2. How does drinking alcohol make it difficult to drive safely?
3. What are some effects smoking has on the body?
4. Why do drugs cause social problems?

NEUROTRANSMITTERS: THE BODY'S MESSENGER SERVICE

Why do we act the way we do? What causes us to feel certain emotions or remember something that happened a long time ago? Research on the nervous system and on the biochemistry of the human body suggests that it may be a matter of different chemicals being released inside our brains.

It has been known for many years that chemicals are released in order to carry messages from one nerve cell to

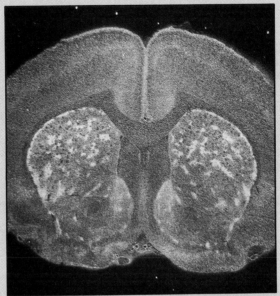

another. These chemicals help the message travel across the gap that lies between the end of one nerve cell and the beginning of another. Known as neurotransmitters, these chemicals are found at the ends of long nerve cells. They are also found where nerve cells stimulate a gland or muscle into action. A number of neurotransmitters, such as acetylcholine (ah-see-tul-koh-leen), are produced in body tissues. Others are produced in glands. Adrenalin, for instance, is produced in the adrenal glands, which are located above the kidneys.

You may have experienced the effects of adrenalin when you have been frightened or in a dangerous situation. You feel more alert, your heart beats faster, and you perspire. You are able to respond more energetically. This is because adrenalin is being pumped into your system as a result of a message sent from your brain. You will experience adrenalin's effects for as long as it is being released. When release stops, the communication between nerve cells slows, your heart rate decreases, and you stop perspiring.

Acetylcholine is a neurotransmitter that seems to affect the ability to learn and remember. In a recent experiment, a scientist worked with aged laboratory mice that could no longer run along a bar in their cage. They were given a transplant of brain tissue from young, healthy mice. Because this tissue produced high levels of acetylcholine, the aged mice were once again able to run along the bar. Medical researchers are investigating the possibility that certain diseases that cause humans to forget may be the result of loss of acetylcholine.

Endorphins (en-dor-finz) are another group of neurotransmitters. Certain endorphins stimulate the appetite. Other endorphins are able to block messages of pain going to the brain. In fact, these kinds of endorphins are our built-in pain relievers.

Where do these neurotransmitters come from? Our bodies make them from the foods we eat. Recently, scientists fed laboratory rats foods rich in several amino acids. As a result, the rats produced more neurotransmitters.

This information might cause us to ask questions such as: "Are there really 'smart' foods?" "Do some foods improve a person's memory and learning ability?" These questions are yet to be answered.

CHAPTER REVIEW

VOCABULARY

On a separate piece of paper, unscramble each term in capital letters so that it is correctly defined by the phrase that follows it.

1. RUNEON: A complete nerve cell.
2. ATERNI: The innermost layer of the eye, made up of receptors that are sensitive to light.
3. LEEFRX: An automatic response to a stimulus not directly involving the brain.
4. BEERMUCR: The center of human intelligence.
5. SPILUME: Information from a stimulus that travels along neurons.
6. DEEKFACB: An automatic regulating process.
7. HOCLACE: The spiral canal in the inner ear that contains the receptors for sound.
8. TIIDODCAN: A physical and/or psychological dependence on a drug.
9. NEHOORM: A chemical messenger that is carried in the blood and helps regulate body functions.
10. ALMULED: Part of the brain that controls the functioning of internal organs.
11. LIICCRRAESUM LAASNC: Three curved passages in the inner ear associated with maintaining balance.
12. DEENNRIOC DLASNG: Ductless structures that release chemical messengers directly into the blood.

QUESTIONS

Give brief but complete answers to each of the following questions. Unless otherwise indicated, use complete sentences to write your answers.

1. Describe what happens after a stimulus is detected by your body.
2. List and describe the parts of a typical neuron.
3. How does an impulse travel from neuron to neuron when the neurons are separated by gaps called synapses?
4. What neuron causes the contraction of muscles?
5. Choose a reflex and describe its nerve pathway, including a description of the three types of neurons. How do some reflexes help protect the body?
6. What is the central nervous system?
7. Describe the location of the spinal cord. What branches off of the spinal cord? To what system do these belong?
8. What are the major parts of the brain? List them. Next to each part, describe its major function(s).

9. How do sounds stimulate the receptors in the inner ear?
10. How do your ears help you to keep your balance?
11. Compare and contrast the senses of taste and smell. Include the receptor cells in your answer.
12. How does light travel in the eye? What are the two types of receptor cells in the eye? Where are they located? How do they differ?
13. What sensations can the receptors in the skin detect?
14. What is a hormone? In what ways do hormones affect the body? Give one example of a hormone and describe its function.
15. What is the difference between an endocrine gland and a digestive gland? In your answer, include how the means of secreting from these two glands differ.
16. How is a thermostat similar to the endocrine system?
17. Describe the effects of the following on the body: tobacco, alcohol, and drugs. Why shouldn't these substances be used during pregnancy?
18. Is nicotine a drug? Explain your answer.

APPLYING SCIENCE
1. Why is it dangerous for a person to drink alcohol and drive a car?
2. If smoking is such a harmful habit why do people continue to smoke?
3. What factors should be considered before taking a nonprescription drug?
4. Caffeine is a mild stimulant. Some foods, beverages, and nonprescription drugs contain this drug. Most products containing chocolate have caffeine. At your local supermarket and pharmacy, find as many products as you can that contain caffeine. Organize and group these products by their type. Are there patterns you can identify? Next to each caffeine-containing product, you might list a similar product that is caffeine free.

BIBLIOGRAPHY

Baldwin, Dorothy, and Claire Lister. *You and Your Body: Your Brain and Nervous System.* New York: Franklin Watts, 1984.

Baldwin, Dorothy, and Claire Lister. *You and Your Body: Your Senses.* New York: Franklin Watts, 1984.

Rahn, Joan Elma. *Keeping Warm, Keeping Cool.* New York: Atheneum, 1983.

Ward, Brian R. *The Brain and Nervous System.* New York: Franklin Watts, 1981.

Ward, Brian R. *The Ear and Hearing.* New York: Franklin Watts, 1981.

CHAPTER

18

This photograph shows a set of quad-ruplet babies, all four born to the same parents. Two are boys, and two are girls. In what other ways might they be different from each other?

DEVELOPMENT AND HEREDITY

1. Compare and contrast the male and female human reproductive systems.
2. Describe the development of a human from the fertilization of an egg to old age.
3. Describe the experiments of Gregor Mendel and relate his work to what is now known about heredity.
4. Describe genetic inheritance in humans.

18-1. Human Development

At the end of this section you will be able to:

- ☐ Explain the purpose of meiosis.
- ☐ Compare and contrast human eggs and sperm.
- ☐ Describe the human reproductive systems.
- ☐ Describe human development from egg to old age.

Do you remember your tenth birthday? It comes at one of the fastest growing times of your life. At ten, people begin a journey that will take them from childhood to adulthood. They not only grow taller after age ten, they also mature both physically and emotionally.

This part of their lives is called adolescence. During this time, humans reach *puberty.* Puberty is the time when the body matures and becomes able to reproduce, or create offspring. Many changes take place in the body. These changes are directed by the endocrine system. In males, body and facial hair grow, the voice deepens, and muscles further develop. In females, the breasts develop, the hips widen, and some body hair appears.

MEIOSIS

All organisms must reproduce to continue their own kind. Vertebrates, such as birds and mammals, reproduce by sexual reproduction. In sexual reproduction, a new individual begins with the joining of two sex cells, a male and a female.

You may recall that the process by which new cells are

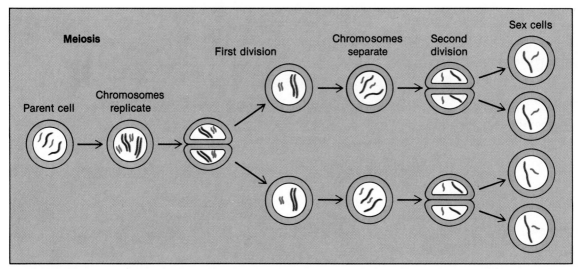

Fig. 18–1 The end result of meiosis are sperm and egg cells. Each sperm or egg has half the usual number of chromosomes.

Meiosis Cell division that reduces the number of chromosomes by half to produce sex cells.

formed is called *mitosis.* Mitosis ensures that the daughter cells each end up with the same number of chromosomes as the parent cell. When sperm and egg cells are formed, a different kind of cell division takes place. The sperm and the egg each have only half the required number of chromosomes needed for an organism. This is the result of a process called **meiosis**. *Meiosis* is a cell division process that reduces by half the number of chromosomes. This takes place during a second cell division after mitosis. When the two sex cells, one from each parent, join during fertilization, the number of chromosomes doubles. In this way, the correct total number is restored. Thus, the new organism receives half of its chromosomes from its mother and the other half of its chromosomes from its father. Use Fig. 18–1 to trace the two divisions that take place during meiosis.

REPRODUCTIVE SYSTEMS

Human egg cells are larger than sperm cells, because they contain a stored food supply. Egg cells are formed by female reproductive organs called **ovaries** (**oh**-vah-reez). The *ovaries* also produce the female sex hormones. The female has two ovaries located inside her abdomen, one on each side.

At birth, the ovaries contain the beginnings of all of the eggs a female will ever have. Usually, however, only one egg is released by one ovary at a time. The process by which one of the ovaries releases an egg is called **ovulation** (aw-vyoo-**lay**-

Ovaries Female reproductive organs that produce eggs and a female sex hormone.

Ovulation The process in which a mature egg is released from an ovary.

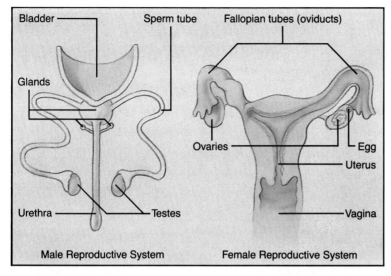

Fig. 18–2 *The organs of the male and female reproductive systems.*

shun). Once the egg is released, it is drawn into the open end of the *oviduct* (**oh**-vih-dukt). The oviduct is also called the *Fallopian tube* (fah-**loe**-pea-un). The Fallopian tubes are not directly connected to the ovaries. If fertilization takes place, it will occur here. See Fig. 18–2.

After *ovulation,* the ovary releases a sex hormone. This hormone causes the lining of the *uterus* (**yoot**-eh-rus) to become thicker and spongier with blood-rich tissue. The uterus is a pear-shaped organ. It has thick walls and is very muscular. If the egg is fertilized, it travels from the Fallopian tube into the uterus and attaches to the lining of the uterus. Here, the egg grows and eventually develops into a baby.

If the egg is not fertilized, the ovary stops releasing its sex hormone. This causes the uterus to shed its blood-rich lining. This lining leaves the woman's body by way of the *vagina,* or *birth canal.* The flow of the lining out of the woman's body is called **menstruation.** The entire process then begins all over again. This cycle, which averages 29.5 days, is summarized in Fig. 18–3.

The *menstrual cycle* continues until a woman reaches *menopause.* Menopause marks the time in a woman's life when the reproductive process ends. Eggs will no longer be released from her ovaries. This occurs most often between the ages of 47 and 50. It is sometimes called "the change of life."

Males are usually able to reproduce from the time they reach puberty until they are very old. Sperm cells are produced in

Menstruation The stage of the menstrual cycle in which the uterine lining and an unfertilized egg are released.

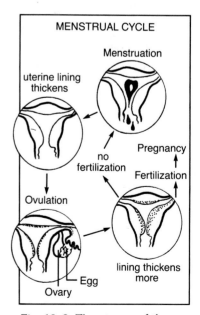

Fig. 18–3 *The stages of the female menstrual cycle.*

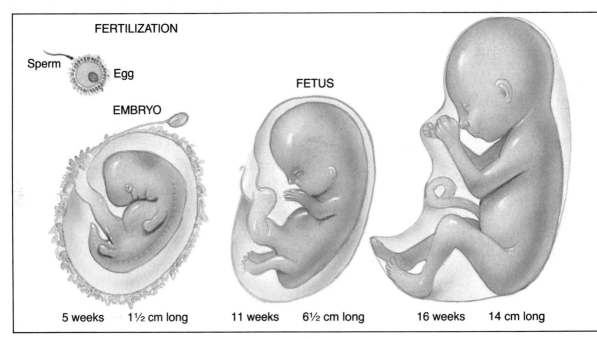

FERTILIZATION		
Sperm Egg	FETUS	
EMBRYO		
5 weeks 1½ cm long	11 weeks 6½ cm long	16 weeks 14 cm long

Testes Male reproductive organs that produce sperm and male sex hormones.

the male reproductive organs called **testes** (**tes**-teez). See Fig. 18–2. The testes also produce the male sex hormones.

The sperm is made up of a head and a tail, or flagellum. The head contains the nucleus. The tail helps the sperm swim to the egg.

Sperm production requires a slightly lower temperature than that of the body. For this reason, the testes are located outside the male's abdomen in a saclike structure called the scrotum. This keeps the testes cooler than the rest of the body.

After the sperm are formed, they travel to a chamber, where they mature and are stored. From the storage chamber, the sperm travel through a duct that is connected to glands. The glands secrete a fluid in which the sperm can swim. The sperm and fluid are released through the *penis*.

FERTILIZATION TO BIRTH

Sperm swim from the woman's birth canal, through the uterus, to the Fallopian tubes. Fertilization takes place here. Only one sperm can fertilize the egg. The fertilized egg will then take about seven days to reach the uterus. See Fig. 18–4.

Once an egg has been fertilized, it begins to divide. A ball of many cells forms. These cells keep dividing to form an embryo (**em**-bree-oe). The embryo attaches to the wall of the

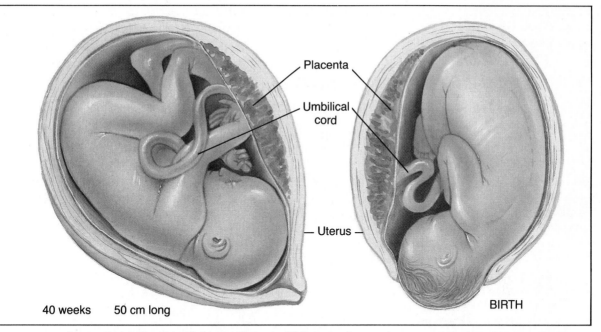

Placenta

Umbilical cord

Uterus

40 weeks   50 cm long   BIRTH

Fig. 18–4 Development of a human, from fertilization to birth, takes about 40 weeks.

uterus by means of a special membrane. This membrane is called the **placenta** (plah-**sen**-tah). The *placenta* is rich with blood vessels. These blood vessels come both from the mother and from the embryo. The vessels from mother and embryo are close together but do not join. Thus, the blood of the mother does not mix with the blood of the embryo. Food and wastes are exchanged between the mother's blood and the embryo's blood across the walls of the blood vessels in the placenta. When a pregnant woman smokes, drinks alcohol, or takes drugs, the baby is also affected. The embryo also develops a membrane that forms a sac filled with fluid. The fluid acts as a cushion. It protects the embryo from injury.

As the embryo develops, it grows an **umbilical cord** (um-**bil**-ih-kul). The *umbilical cord* contains large blood vessels. It connects the embryo to the placenta. Your navel, or belly-button, marks the spot where your umbilical cord was attached. When the baby is ready to be born, the smooth muscles of the mother's uterus begin to contract. These contractions widen the opening of the uterus and push the baby down and out through the birth canal. After birth, the uterus contracts again to push the placenta out of the mother's body.

Up until birth, all of the baby's needs are provided by its mother. During pregnancy, the mother's breasts prepare for

Placenta A structure that attaches the embryo to the wall of the uterus.

Umbilical cord A structure that connects the embryo to the placenta.

Fig. 18–5 Young and old people can learn from each other.

the production of milk. After the baby is born, breast milk provides the proper nourishment for the baby. It contains all the nutrients needed for almost the first year of a baby's life.

GROWING OLDER

With nourishment and love, a baby grows steadily. Once babyhood is over and childhood begins, this growth continues at a relatively steady pace. However, very rapid growth occurs during adolescence, and during this time the body will usually grow to resemble what its adult form will be. Somewhere between the years of 18 and 21, nearly all the body systems are fully developed and physical growth stops. Adulthood has then begun. At a certain point later in your adult life, your body systems will begin to weaken. Your joints may become stiff, and it may become harder for you to see or hear. This aging process also causes our bodies to lose some of the ability to fight disease and repair injuries.

Aging is nothing to be afraid of, though. Today, many senior citizens lead active lives and keep fit by exercising and eating nutritious foods. They have an understanding of themselves and the world based on years of experience. Some consider the later years to be the best time for living. See Fig. 18–5.

SUMMARY

Sexual and physical maturity are reached during adolescence. The male and female reproductive systems begin to produce special reproductive cells. The fertilized egg develops inside the mother's body until the birth of a baby. After birth, the baby continues to grow and develops into an adult.

QUESTIONS

Use complete sentences to write your answers.
1. What changes take place during adolescence?
2. How are eggs and sperm different?
3. What is ovulation?
4. Where are sperm produced?
5. How is the developing baby nourished and protected in the mother's uterus?
6. What happens to the body systems as the body ages?

SKILL-BUILDING ACTIVITY

SEQUENCING

PURPOSE: Sequencing includes describing something in the proper order. You will trace the process of meiosis to practice this skill.

MATERIALS:

scissors   construction paper
marker or pen   4 blue sheets,
paper clips   1 red sheet, and
                  1 green sheet.

PROCEDURES:

A. Cut out eight blue, 20 cm semicircles. These will represent cells.

B. Cut four red strips 1.5 cm by 6 cm. To represent one chromosome, mark two strips with an X and fasten them with a paper clip. For the other chromosome, fasten the unmarked strips.

C. Cut four green strips 1.5 cm by 4 cm, then mark and fasten as in step B to make two more chromosomes.

D. Place two semicircles together so that they form a circle. This will be a "cell."

E. Inside this cell, place two pairs of chromosomes. Line them up as in Fig. 18–6.

F. To make the "cell" divide, pull the semicircles apart in opposite directions. Each half should have one of each kind of chromosome. Look again at Fig. 18–6. To complete the division, position matching empty semicircles to form two new, complete "cells." Each cell now has one chromosome from each pair.

G. The cells will divide another time. Pull the semicircles apart again. Now remove the paper clips and separate the chromosomes. Put one red strip and one green strip into the new "cells." Each cell now has one half of each chromosome. Complete the new cells by adding empty semicircles to the four "half-cells." Each new cell now has half the original number of chromosomes, as in Fig. 18–6.

CONCLUSION:

1. How do the first division and the second division in meiosis differ?

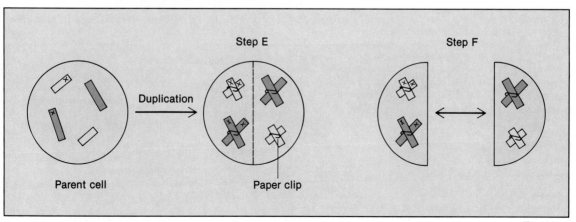

Fig. 18–6

18-2. Heredity

At the end of this section you will be able to:

- ☐ Describe the experiments of Gregor Mendel and explain their importance.
- ☐ Compare Mendel's conclusions with what we know today about *heredity.*

Unless you are an identical twin, you are a one-of-a-kind human. The chance of your parents' producing another you hardly exists at all. What makes you so unique?

The answer involves the traits you inherited from your parents. This is called **heredity.** Little was known about *heredity* until 1900. In that year, the work of Gregor Mendel, an Austrian monk, was found by three scientists. Mendel had done research on heredity in garden pea plants. He had published his results in 1865, but scientists did not pay attention to them. When his report was rediscovered, it formed the basis for a new branch of science called **genetics** (jeh-**net**-iks).

Heredity The passing of traits from parent to offspring.

Genetics The science that studies the laws of heredity.

Pure Refers to an organism or cell in which the pair of genes required to produce a trait is identical.

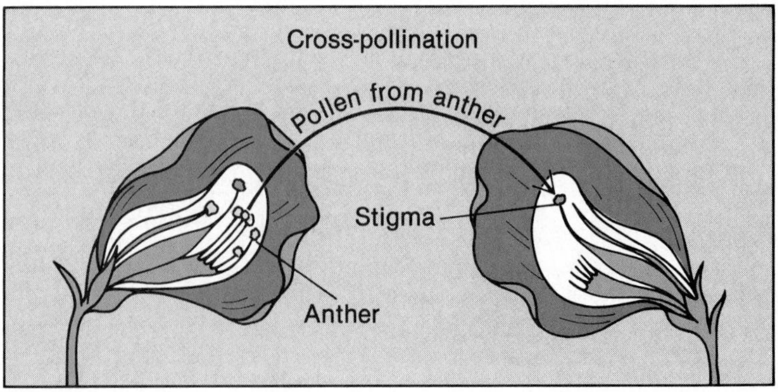

MENDEL'S DISCOVERIES

Mendel chose garden pea plants because they have different varieties and reproduce by self-pollination. This meant that the pea plants usually produced seeds that grew into plants similar to the parent plant. Tall plants produced seeds that grew into tall plants. These plants were called **pure** tall plants.

Mendel noted that his pea plants had some traits that were easy to tell apart. For example, seed color was either yellow

Fig. 18–7 Pea flowers usually pollinate themselves. The flower does not even open. Mendel was able to hand-pollinate one flower with another to make his crosses.

Cross-pollination

Pollen from anther

Stigma

Anther

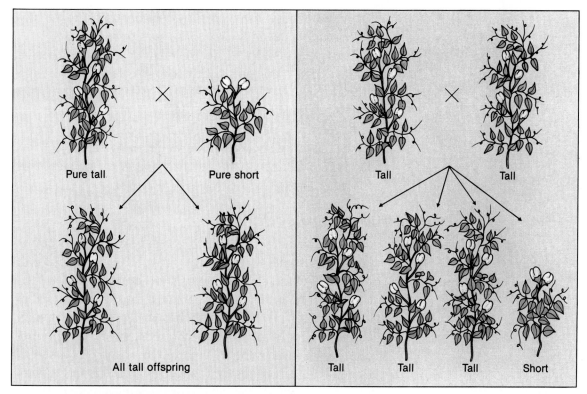

or green, and the stem length was either tall or short. Mendel studied a total of seven characteristics, each having two very different forms.

For each characteristic, Mendel would cross two different traits. He did this by cross-pollination. See Fig. 18–7. In one experiment, he fertilized *pure* short plants with pollen from pure tall plants. He planted the seeds that were produced. Then he counted the types of offspring that grew. He was surprised to find that all the offspring were tall. He also got the same results when he fertilized pure tall plants with pollen from pure short plants. Any cross between a pure tall plant and a pure short plant produced only tall plants. See Fig. 18–8.

Mendel found similar results when he crossed plants that were pure for each of the seven pairs of traits. All offspring from these crosses showed only one of the pairs of traits. It appeared that for each pair one was stronger than the other. Mendel called the stronger one the **dominant** trait and the hidden one the **recessive** trait.

Mendel's next step was to use the offspring from these pure crosses. He allowed the plants to self-pollinate. Three fourths

Fig. 18–8 (left) *The offspring from a cross between pure tall and pure short pea plants.* (right) *A cross of these offspring produced short and tall plants.*

Dominant Refers to the stronger of a pair of traits.

Recessive Refers to the weaker of a pair of traits.

of the next generation showed the *dominant* trait. The remaining one fourth showed the *recessive* trait. See Fig. 18–8.

From these experiments, Mendel made several conclusions:

1. Each pea plant has two hereditary "factors" for each characteristic.

2. When a sex cell is formed, the two factors separate.

3. Each factor passes into separate sperm or egg cells. Thus, each sex cell contains one factor for each trait.

4. Therefore, a new plant receives one factor for each trait from each parent.

GENES

Genes Locations on chromosomes that determine heredity traits.

Today these factors are called **genes.** *Genes* are locations on a chromosome. They can be thought of as beads on a string. Think of the chromosome as the string.

Each cell in an organism contains a certain number of chromosomes. The number depends on the species. Human cells, for example, contain 46 chromosomes. The chromosomes are in pairs in the nucleus. Each member of a pair of chromosomes has genes for the same characteristics. Therefore, as Mendel predicted, an organism usually has two genes (factors) for each trait. These two genes may or may not be identical. Cells with identical genes for the same trait are *pure* for that trait. Cells whose genes are different for the same trait are **hybrid** (**hie**-brid) for that trait. The organism itself is often described by these terms. The offspring that result from crossing a pure tall and a pure short plant are all *hybrid* tall plants.

Hybrid Refers to an organism or cell in which the genes for a trait are different.

Mendel's second conclusion agrees with what happens during meiosis. In this process, the chromosome pairs separate.

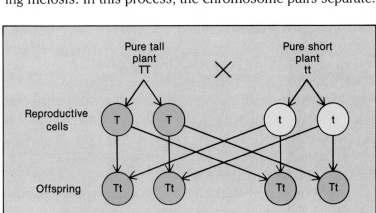

Fig. 18–9 A cross between a pure tall and a pure short plant, showing possible gene combinations.

The egg and sperm receive one of each pair of chromosomes. Refer back to Fig. 18–1 on page 414 to review meiosis.

Let's look again at Mendel's cross between pure tall and pure short pea plants. Tallness is dominant and shortness is recessive. A pure tall pea plant has two genes for tallness. This is shown by the symbol TT. When it produces reproductive cells by meiosis, each one will have a gene T for tallness. A pure short pea plant has two genes for shortness. This is shown by the symbol tt. When it produces reproductive cells, each will have a gene t for shortness. Look at the cross shown in Fig. 18–9. A tallness gene T from the tall plant (TT) will combine with a shortness gene t from the short plant (tt). This is the only combination possible. The result is offspring that are hybrid for tallness (Tt).

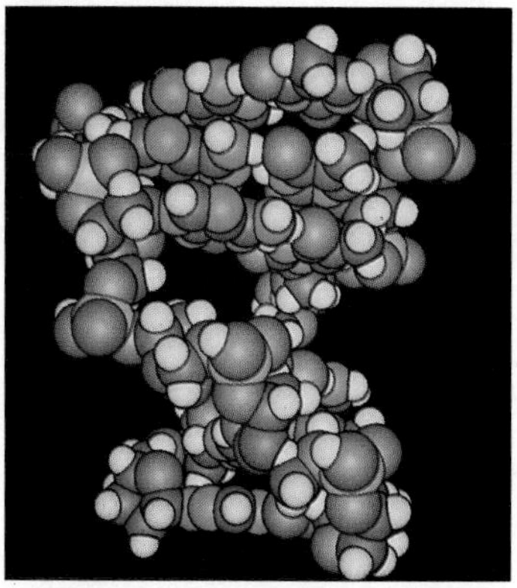

Fig. 18–10 A computer-drawn model of a DNA molecule. Each colored shape represents a different type of atom.

DNA

By 1953, it was clear to biologists that chromosomes were made up of large molecules of **DNA.** At that time, James Watson and Francis Crick proposed a model for the *DNA* molecule. DNA contains hereditary information. It also controls activities in the cell, such as making protein.

Parts of the genetic code (DNA molecules) are the same in all of us. However, no two people, except identical twins, have exactly the same DNA molecules. Therefore, except for identical twins, all people have different inherited traits.

DNA Deoxyribonucleic acid —a large molecule that contains the hereditary information of the cell and controls the cell's activities.

SUMMARY

Little was known about heredity until Gregor Mendel's experiments with pea plants were rediscovered in 1900. He showed that there is a pattern to the way some traits are inherited. Mendel's theories were proved to be correct as scientists learned more about *genetics.*

QUESTIONS

Use complete sentences to write your answers.

1. Why did Mendel choose pea plants for his experiments?
2. What were Mendel's conclusions? How do they compare with our knowledge of genetics today?
3. How are genes and DNA molecules related?

SHOWING POSSIBLE GENE COMBINATIONS

PURPOSE: To determine the combinations of genes that would result from various crosses.

A Punnett square is a special grid that is used to show gene combinations. The possible genes for a certain trait in the female egg cells are written down the left side. The possible genes for the corresponding trait from the male sperm cells are written across the top. Fig. 18–11 (A) shows how the square would look if you crossed a pure tall (TT) pea plant with a pure short (tt) pea plant. Each box shows a possible combination of genes in a fertilized egg.

MATERIALS:
none

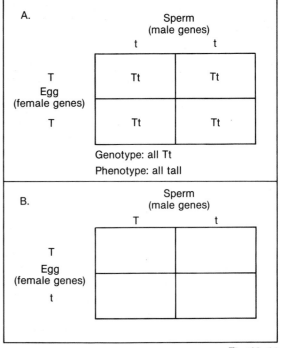

Fig. 18–11

PROCEDURE:

A. On a separate piece of paper, copy the square in Fig. 18–11 (B). In each row, copy the gene symbols for the female cells all across the row. Next to them in each box, write the gene symbol from the male above. Be sure to use a letter from a female and a letter from a male in each box. Make sure you carry the letters across and down correctly.

1. Of the four possibilities, what fraction will have the genes TT?

2. Of the four possibilities, what fraction will have the genes Tt?

3. What fraction will have the genes tt?

4. How will each of these plants look?

B. Mendel also studied seed color. He found that yellow was dominant over green. Using the symbols Y for yellow and y for green, complete a Punnett square for the crosses YY × yy; YY × Yy; yy × Yy. Next to the square, tell what the color of each of the four plants would be.

CONCLUSIONS:

1. Why can't you always tell the genes of an organism from the way it looks?

2. How does the Punnett square show what traits could appear in addition to showing the possible gene combinations of those traits?

18-3. Human Genetics

At the end of this section you will be able to:

- ☐ Apply Mendel's conclusions to human heredity.
- ☐ Describe the appearance and functions of *sex chromosomes.*
- ☐ Identify some ways the environment influences inherited traits and explain why identical twins are useful for this purpose.
- ☐ Explain the importance of *sex-linked* genes in human heredity.

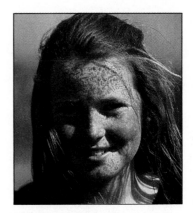

Fig. 18–12 Are freckles an inherited trait?

Do you have freckles? Do you know anyone who does? What are freckles? Are they inherited? Why do freckles seem to "pop out" in the sunlight? Freckles are small spots of skin coloring. The tendency to have them seems to be inherited through a dominant gene. They are an example of a human trait that can be affected by something in the environment. Sunlight can cause more freckles to appear.

MENDEL AND HUMAN HEREDITY

Although Mendel experimented with pea plants, his conclusions apply to the way almost all organisms, including humans, inherit traits. His conclusions have been supported and confirmed by later theories of genetics. Expressed in modern terms his work showed that:

1. The genes for some traits are dominant over others. The hidden traits are called recessive.

2. The inheritance of most traits involves pairs of genes, one from each parent.

3. If a dominant gene and a recessive gene are present, only the trait of the dominant gene appears.

4. The offspring from a cross between two hybrids may have different traits than either parent. The recessive trait appears in one fourth of the offspring when hybrids are crossed.

Many human traits are inherited in the same way as the traits Mendel studied in pea plants. These include the shape of the hairline, dimples, and long eyelashes.

ENVIRONMENTAL INFLUENCE ON GENES

Many human traits are not inherited as simply as Mendel described. Some traits, such as height, weight, facial shape, and body build, are also influenced by the environment. A person's diet and general health can influence these traits. Even other genes can affect a trait. This is true for eye color. Other genes can affect the way brown-eyed and blue-eyed genes are expressed. As a result, parents with blue eyes (a recessive trait) sometimes have a brown-eyed (dominant trait) child. Your genes provide the information for what you can become. However, the kind of individual you do become depends upon the influence of both heredity and environment.

Scientists study identical twins to learn more about the effect of environment on the development of human traits. Since identical twins have the same genes, differences between them may be caused by differences in their environment.

SEX DETERMINATION

What determines whether a person is male or female? Remember that chromosomes occur in pairs. The two chromosomes in a pair are much alike in size and shape. In many organisms, however, there is one pair that is not alike. Compare the chromosomes of a male and female human. See Fig. 18–14. There is one such pair. Can you see it?

The chromosomes in this pair are called the **sex chromosomes.** They determine the sex of an organism. One of the *sex chromosomes* has many genes and looks like all the other chromosomes. This one is called the *X chromosome.* The other chromosome has only a few genes and is a different shape. This one is called the *Y chromosome.* Females have two X chromosomes. Males have one X chromosome and one Y chromosome.

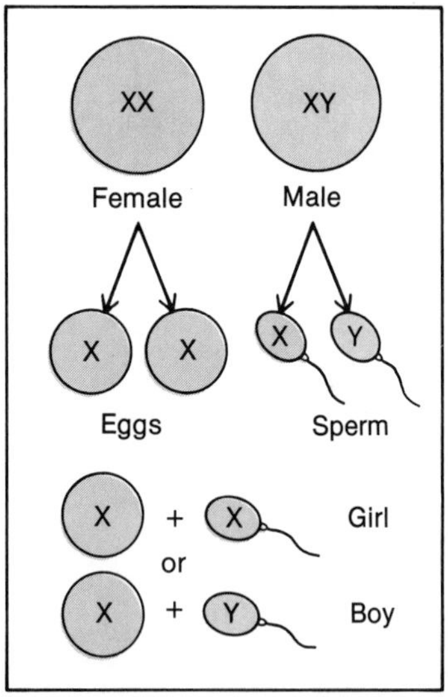

When sex cells are produced, all the cells receive one of each pair of chromosomes. Since the sex chromosomes of a female are both X, each egg cell receives an X chromosome. Males have both an X and a Y chromosome. Therefore, when a male produces sperm, half of the sperm cells receive an X chromosome and half receive a Y chromosome. See Fig. 18–13.

Sex chromosomes A pair of chromosomes that determines the sex of an individual.

Fig. 18–13 The distribution of X and Y chromosomes when eggs and sperm are produced.

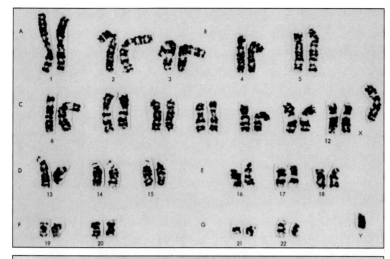

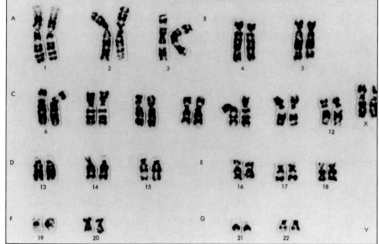

Fig. 18–14 The chromosomes of a male (top) *and female* (bottom). *Look at the right edge of each photo. The sex chromosomes are labeled X and Y.*

The sex of a baby depends upon which type of sperm fertilizes the egg. When a sperm with an X chromosome fertilizes an egg, the fertilized egg is XX. It will develop into a female. When an egg is fertilized by a sperm with a Y chromosome, the fertilized egg is XY. It will develop into a male. This means that a male gets an X chromosome from his mother and a Y chromosome from his father. A female gets an X chromosome from both her mother and her father.

SEX-LINKED GENES

In humans, the Y chromosome is smaller than the X. Thus, most genes on the X chromosome do not have matching genes

Sex-linked gene A gene that is carried only on the X sex chromosome.

on the Y chromosome. Genes located on the X chromosome but not on the Y chromosome are called **sex-linked genes.**

Sex-linked genes play an important role in human heredity. When a male receives a sex-linked gene that is recessive, he inherits the trait. Do you see why this should be? Males do not have a matching gene on the Y chromosome. The recessive trait appears, since there is no dominant gene to hide it. Can a female show a sex-linked trait that is recessive? She does *only* if she inherits two recessive genes, one from her mother and the other from her father. If a female is hybrid for the sex-linked trait, she will not show the trait.

One sex-linked recessive gene is the gene for red-green colorblindness. People who have this trait see shades of gray instead of the colors red and green. Fig. 18–15 shows a test for colorblindness that is often given by the school nurse. How can a male inherit this trait? His mother must have one or two genes for this trait. If she has two, she is colorblind. If she has one, she is known as a *carrier* of the trait. See Fig. 18–16. How can a female inherit the trait? If the father is colorblind and the mother is not a carrier, will their children be colorblind? See Fig. 18–16.

Another sex-linked recessive gene is the gene for *hemophilia* (hee-moh-**fill**-ee-ah). This is an inherited disease in which a person's (usually male's) blood does not clot easily. There are about 25,000 hemophiliacs in the U.S. who can die from bruises or cuts if the bleeding cannot be controlled.

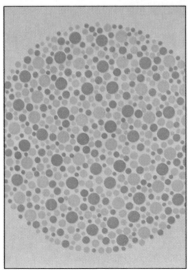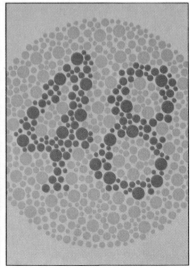

Fig. 18–15 Colorblind persons have trouble seeing the numbers in these paintings.

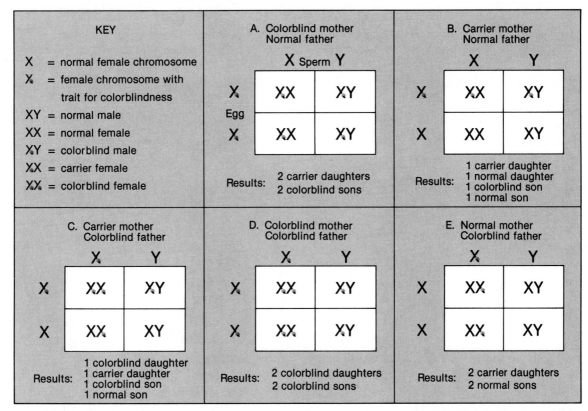

KEY		
X	=	normal female chromosome
X̷	=	female chromosome with trait for colorblindness
XY	=	normal male
XX	=	normal female
X̷Y	=	colorblind male
X̷X	=	carrier female
X̷X̷	=	colorblind female

A. Colorblind mother Normal father

X Sperm Y

Egg	XX	XY
X̷	X̷X	X̷Y
X̷	X̷X	X̷Y

Results: 2 carrier daughters
2 colorblind sons

B. Carrier mother Normal father

X Y

X̷	X̷X	X̷Y
X	XX	XY

Results: 1 carrier daughter
1 normal daughter
1 colorblind son
1 normal son

C. Carrier mother Colorblind father

X̷ Y

X	X̷X̷	X̷Y
X	XX̷	XY

Results: 1 colorblind daughter
1 carrier daughter
1 colorblind son
1 normal son

D. Colorblind mother Colorblind father

X̷ Y

X̷	X̷X̷	X̷Y
X̷	X̷X̷	X̷Y

Results: 2 colorblind daughters
2 colorblind sons

E. Normal mother Colorblind father

X̷ Y

X	XX̷	XY
X	XX̷	XY

Results: 2 carrier daughters
2 normal sons

Fig. 18–16 Inheritance combinations for the sex-linked trait of colorblindness.

SUMMARY

The experiments of Gregor Mendel help explain how some human traits are inherited. Traits such as height and weight are also influenced by a person's environment. Identical twins are often studied to determine the effects of environment on inherited characteristics. Certain traits are sex-linked. Hemophilia is an example of a sex-linked recessive trait.

QUESTIONS

Use complete sentences to write your answers.

1. In humans, brown eyes are dominant over blue eyes. If two brown-eyed parents have a blue-eyed child, what gene combinations may the parents have?
2. How are the sex chromosomes of a male and female different?
3. What factors may influence inherited traits?
4. Why are more males red-green colorblind than females?

INVESTIGATION

DOMINANT AND RECESSIVE TRAITS

PURPOSE: To determine the frequency of some traits in the classroom population.

MATERIALS:
graph paper

PROCEDURE:
A. On a separate sheet of paper, copy Table 18–1.
B. Identify the traits you have by placing a mark (|) next to them.
C. Survey your classmates for each of the traits listed. Again place marks next to the traits on your chart. Group marks by five (|||||) to make them easier to count.
D. For each trait, determine the percentage of the class that has that trait. To determine percentage, divide the number of people that have the trait by the total number of people in the class. Then multiply this number by 100.

$$\frac{\text{number of people with trait}}{\text{total number of classmates}} \times 100$$

For example: If 6 people have a cleft chin and there are 30 people in the class, the percentage would be:

$$6 \div 30 = 0.2 \qquad 0.2 \times 100 = 20\%$$

E. Graph the percentages you obtained for each of the traits you surveyed.

CONCLUSIONS:
1. For each trait, is the dominant characteristic usually shown by a majority of the class?
2. Does dominance seem to be determined by the percentage of people who have the trait? Give examples to support your answer.

INHERITANCE OF SOME HUMAN TRAITS				
Dominant	Number of Students	Recessive	Number of Students	Percent with the Trait
1. Tongue rolling U		Cannot roll tongue into U		
2. Cleft chin		No cleft in chin		
3. Dark hair		Light or red hair		
4. Free ear lobes		Attached ear lobes		
5. Curly hair		Straight hair		
6. Widow's peak		Straight hairline		
7. Pigmented skin color		Albino		
8. Brown, green, or hazel eyes		Blue or gray eyes		
9. Dimples		No dimples		

Table 18–1

CAREERS IN SCIENCE

GENETICIST

Genetics is the science of heredity. The work geneticists do ranges from answering theoretical questions to applying genetic principles to practical, economically significant subjects. Geneticists study animals, plants, and microorganisms to learn how a gene codes traits from generation to generation.

Geneticists have made great practical contributions to human life. A huge increase in the production of foods such as rice, has come about through research in the genetics of crops. The hybrid corn industry owes its existence to geneticists' study of the corn plant.

Most geneticists do undergraduate work in biology and have a strong background in physics, chemistry, and mathematics. An advanced degree is usually necessary for good jobs in the field.

For further information, contact: American Society of Human Genetics, Medical College of Virginia, Box 33, Richmond, VA 23298.

CHILD DEVELOPMENT SPECIALIST

Child development specialists study the way children grow physically, psychologically, emotionally, mentally, and socially from birth until about the age of twelve. They are interested in the way children learn, the way they behave, and how one kind of development relates to others.

Most child development specialists work directly with children in schools, nurseries, day-care centers, and hospitals. Some work as teachers or therapists with children who have physical or mental impairments. Others are counselors to emotionally disturbed children. Many are certified teachers.

Most jobs in this field require a minimum of a bachelor's degree. Teachers in colleges or universities, as well as most counselors and therapists, have advanced degrees.

For further information, contact: American Psychological Association, Educational Affairs, 1200 17th Street NW, Washington, DC 20036.

VOCABULARY

On a separate piece of paper, write the word TRUE next to the number of each statement that is true. Next to the number of each false statement, write FALSE; then make it true by writing the correct term in place of the underlined incorrect term.

1. Hybrid is the passing of traits from parent to offspring.
2. Ovaries are the female reproductive organs.
3. A person who has a gene for a trait but does not have the trait is called a sex-linked recessive.
4. Locations on chromosomes that determine hereditary traits are called genes.
5. The process in which an ovary releases a mature egg is called ovulation.
6. The developing embryo is attached to the wall of the uterus and nourished through a structure called the placenta.
7. The embryo is connected to the placenta by the oviduct.
8. The male reproductive organs are the testes.
9. The stronger of a pair of traits is referred to as the recessive trait.
10. A pair of DNA molecules determines the sex of an individual.
11. Meiosis is a process of cell division in which the number of chromosomes is reduced by half.

QUESTIONS

Give brief but complete answers to each of the following questions. Unless otherwise indicated, use complete sentences to write your answers.

1. In what way does an offspring formed by sexual reproduction receive one of each pair of chromosomes from each parent?
2. How does the number of sperm cells produced compare with the number of egg cells produced? Why is this difference important?
3. What is ovulation?
4. In what part of the male reproductive system are sperm produced? Where are these structures located? How does this location help sperm production?
5. Briefly compare and contrast meiosis and mitosis.
6. How did Mendel discover the basic rules of heredity?
7. Black hair (B) is dominant over blond hair (b). A mother with black hair and a father with blond hair have four children. Two children have black hair and two children have blond hair.
   (a) What combination of genes does the mother have?
   (b) What combination of genes does the father have?

(c) How many children inherited the dominant gene?

(d) How many children inherited the recessive gene?

(e) What combination of genes do the blond-haired children have?

(f) What combination of genes do the black-haired children have?

(g) In theory, could these parents have produced only dark-haired children?

8. How did the later discoveries of cell division and meiosis support Mendel's conclusions?

9. What is a sex-linked trait? Who determines whether a man will have a sex-linked trait—the mother or the father? Explain your answer.

10. Colorblindness is a sex-linked recessive trait. What gene combinations in the parents would result in a colorblind female offspring?

11. What combination of sex chromosomes produces a female? A male?

APPLYING SCIENCE

1. Why is it important to the health of a developing baby that its mother eat a healthy diet and not smoke, drink alcohol, or take drugs while the infant is in her uterus?

2. A pedigree chart shows a marriage and the offspring from that marriage. A circle means a female and a box means a male. Draw a pedigree chart for your family and fill it in with information you get based on the following trait:
The ability to curl one's tongue into a U shape is determined by a dominant gene. Survey your family members and ask them to try to curl their tongues into a U shape. Record your results using your pedigree chart. Shade the box or the circle of each family member that is able to curl his or her tongue. Try to determine for those members who can roll their tongue if their gene combination is pure or hybrid.

BIBLIOGRAPHY

Baldwin, Dorothy, and Claire Lister. *You and Your Body: How You Grow and Change.* New York: Franklin Watts, 1984.

Gonick, Larry, and Mark Wheelis. *The Cartoon Guide to Genetics.* New York: Barnes and Noble, 1983.

National Geographic Society. *Your Wonderful Body.* Washington, DC.: National Geographic Society, 1982.

Weiss, Ellen. "Busy Bodies: Growth." *3–2–1 Contact,* March 1982.

A sea otter treads water as it gets ready to eat a prickly sea urchin. The otter, urchin, and the floating kelp around them are all parts of the ocean ecosystem along the coast of California.

ECOSYSTEMS

CHAPTER GOALS

1. List and describe the parts of an ecosystem.
2. Trace the cycles of key substances through an ecosystem.
3. Diagram the transfer of energy within an ecosystem.
4. Identify factors that affect populations.
5. Describe how and why communities change.

19-1. Environments and Communities

At the end of this section you will be able to:

- ☐ Describe the living and nonliving parts of an *ecosystem.*
- ☐ Define the term *community.*
- ☐ Identify an organism's *habitat* and *ecological niche.*

By the end of this century, the United States is planning to have a permanent space station in orbit. Everything needed for human survival—food, water, and oxygen—will be on board. Most of these materials will be reused again and again. Solar batteries will change sunlight into electricity to provide energy. A space station such as the one just described will be an artificial environment.

ECOSYSTEMS

A lake is an example of a natural environment. It is made up of both living and nonliving parts that affect one another. The living things in any environment are called the **biotic** (bye-**ot**-ik) parts. The nonliving things, such as water, soil, light, and temperature, are called the **abiotic** (**ay**-bye-ot-ik) parts.

A typical lake environment is shown in Fig. 19–1 on page 438. The *biotic* parts include the raccoon, sunfish, frog, cattail, and other plants. There are also insects, protozoa, and plants too small to be seen. All these living parts depend on each other for food. The *abiotic* parts include all the nonliving things. Light, water, and the mud on the bottom are some of the abiotic parts of a lake.

Biotic Refers to the living organisms in an ecosystem.

Abiotic Refers to the nonliving materials and energy in an ecosystem.

Fig. 19–1 What are the biotic parts of this ecosystem? What are the abiotic parts?

Ecosystem A group of organisms and their physical environment.

A group of organisms and their nonliving environment are called an **ecosystem.** An *ecosystem* may be as small as a plant and some soil in a pot. It may be a vacant lot or pond. It may be a small, rocky beach, a marsh, a large lake, or an entire forest. A city park very often contains several different ecosystems.

Materials are always moving into and out of an ecosystem. Look at Fig. 19–1 again. What might move into and out of a lake? The raccoon leaves the lake's shore to sleep all day in the woods. Birds may stop at the lake. They might carry in seeds and small organisms from other ecosystems. Before the birds leave, they may feed on organisms from this lake.

Water and air are two abiotic parts that move freely between ecosystems. They carry many living and nonliving materials into and out of the ecosystem. Rain, snow, or streams flowing in add water to the lake. Evaporation and streams flowing out carry water away. Sand, silt, carbon dioxide, and oxygen are a few of the abiotic materials moved by water and wind. Seeds and insects are some of the biotic parts wind and water may move into or out of the lake.

Energy also enters the ecosystem. Most of it enters in the form of sunlight. Organisms with chlorophyll store this energy during photosynthesis. However, energy is needed by all organisms. The energy is passed on to animals that eat these organisms. Still other animals eat these animals and get *their* energy. Both plants and animals must use some energy to carry on their life activities. The energy that they use in this way can no longer be passed on to other organisms. Some energy is changed to heat and is released into the environment. Ecosystems need a constant supply of energy coming in to replace what is lost in this way.

THE LIVING COMMUNITY

A **community** is made up of the plants and animals living in a certain area. In other words, it is the biotic part of an ecosystem. The name of a *community* often comes from where it is located. Thus, all the plants and animals living in a forest are called a forest community. A stream that runs through the forest may lead to a pond community some distance away. A seashore may have a sandy beach, a rock breakwater, a salt marsh, or a tidepool. Each of these is a separate community of living things. A city park might contain pond, open-field, and flower-bed communities. Even a single tree can be a community, because it is home to insects, birds, and squirrels.

Conditions such as temperature and the supply of water change with the seasons. These changes cause some organisms to leave the community. Other organisms become inactive. See Fig. 19–2. Many communities change drastically over long periods of time.

Community All the plants and animals living in a particular area.

Fig. 19–2 How do conditions change for this polar bear in winter (left) *and summer* (right)*?*

HABITATS AND NICHES

Each organism in a community has a certain kind of place in which it lives. This place is called its **habitat** (**hab**-ih-tat). Think of this as the organism's *habitat* must supply all the conditions the organism needs to survive. See Fig. 19–3.

Within the forest community are many different habitats. The soil is the habitat of earthworms and many small insects. Salamanders may find homes in rotten logs. Deer live among trees and shrubs. Earthworms, insects, salamanders, and deer have different habitats. Yet they all live together in the same community.

Organisms may share the same habitat. The green plants of the forest grow in the same soil in which earthworms live. Insects share the rotten logs with salamanders. Many birds live among the trees and shrubs with the deer.

Every organism plays a special role in its community. In a grassy field, the plants change energy from the sun into food. Insects, rabbits, and cattle eat the food the plants produce. The role each organism plays in its community is called its **ecological niche** (nich). Think of this as the organism's occupation. A field community may have many other *niches*. Some animals and birds may eat the animals that feed on plants.

Fig. 19–3 A sandy shoreline contains several different habitats.

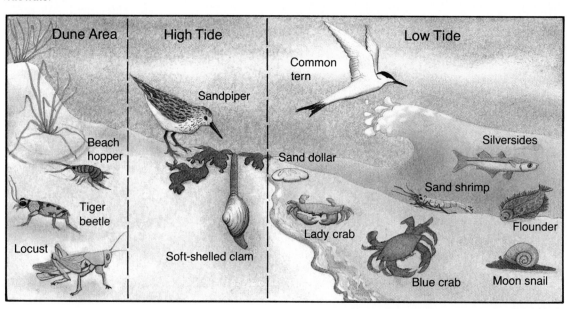

Dune Area — High Tide — Low Tide

Common tern

Sandpiper

Beach hopper

Silversides

Sand dollar

Sand shrimp

Tiger beetle

Flounder

Locust

Lady crab

Soft-shelled clam

Blue crab

Moon snail

Two different kinds of organisms may try to occupy the same niche in the habitat. When this happens, the organisms are said to be in competition with each other. The one that is better adapted to the habitat will crowd the other out. For example, sheep and cattle both eat grass. However, sheep bite the grass close to the ground. As a result, they make the grassland unfit for cattle, which cannot feed on such short grass. See Fig. 19–4.

Each ecosystem has a different combination of biotic and abiotic parts. However, each must provide everything that the organisms living there need to survive. Each ecosystem is affected by what happens in its neighboring ecosystems. All the individual ecosystems of the earth make up one large ecosystem. We call it the biosphere.

Fig. 19–4 Two different species cannot occupy the exact same niche. Here, sheep have made it impossible for cattle to graze.

SUMMARY

An ecosystem consists of biotic communities and an abiotic environment. The sun supplies most of the energy to ecosystems. Each organism lives in one or more habitats and has a niche within its community.

QUESTIONS

Use complete sentences to write your answers.

1. List the biotic and abiotic parts of an ecosystem near your home.
2. What is a community?
3. What habitats may be present in a desert ecosystem?
4. How do the abiotic parts of freshwater and ocean ecosystems differ?

INVESTIGATION

STUDYING AN ECOSYSTEM

PURPOSE: To describe the abiotic factors and biotic community of a typical ecosystem.

MATERIALS:
none

PROCEDURE:
Answer the following questions about the flow of energy and materials in the ecosystem shown in Fig. 19–5. Your teacher may assign other ecosystems for study, such as a pond, open field, or wooded area.

1. What would you name the community pictured here?
2. What members of this community can you identify?
3. What other biotic parts might be present even though they are too small to be seen?
4. What are some of the abiotic parts of this ecosystem?
5. What materials may be entering this ecosystem from other ecosystems? How?
6. What materials may be leaving this ecosystem? By what means do they leave the system?
7. What is the major source of energy in the ecosystem? How does energy enter and pass through this ecosystem?
8. List several habitats that are available in this ecosystem.
9. What organisms take in the sun's energy in this ecosystem?

CONCLUSIONS:
1. What characteristics do all ecosystems have in common?
2. How are abiotic and biotic factors related in an ecosystem?
3. What might happen if cattle were introduced to this ecosystem?

Fig. 19–5

19-2. Recycling Matter

At the end of this section you will be able to:

- ☐ Diagram and explain the *water cycle.*
- ☐ Describe the *carbon dioxide–oxygen cycle.*
- ☐ Explain the role of living organisms in the *nitrogen cycle.*

When humans finally do travel to other planets, they will do so in artificial ecosystems. Their spacecraft will contain everything needed to support human life. But it will be impossible to carry enough water and fresh air for a long trip. What is carried will be recycled and reused many times. In a sense, earth, too, is a spaceship. It has a limited amount of the materials needed for life. Luckily, these materials are constantly being recycled and used again.

THE WATER CYCLE

Earth's water is recycled by the *water cycle.* See Fig. 19–6. The sun's heat causes water to change from a liquid to a gas. This process is called *evaporation.* Water evaporates and enters the atmosphere as a gas. When the temperature of the air

Fig. 19–6 The water cycle.

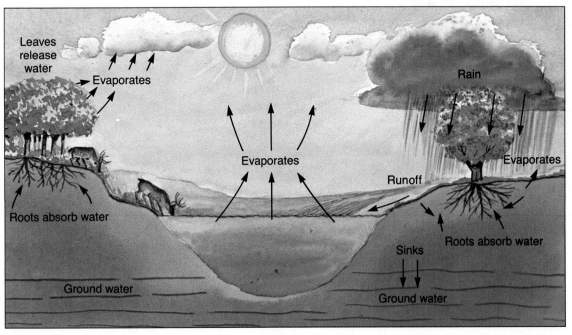

Leaves release water

Evaporates

Roots absorb water

Evaporates

Rain

Runoff

Evaporates

Roots absorb water

Sinks

Ground water

Ground water

falls, the gas changes back into a liquid. If the temperature is cold enough, the gas may form a solid—snow or sleet. When the drops of water or snow crystals are large enough, they will fall to the ground. This water may evaporate again, or it may run along the surface of the land. The surface water collects in streams, lakes, and the ocean. There it evaporates again. Water may also sink right into the ground. Here it becomes part of the supply of ground water stored in the soil.

This soil water is the source of water for land plants. Water is taken up by the roots of the plants. It is then carried up to the leaves, where it is used in foodmaking. While making food, plants produce more water. This waste water is lost from the plants' leaves by evaporation.

Animals also need water. They may drink it or get it as part of their food. Much of this water is excreted as waste. It also returns to the *water cycle.*

THE CARBON DIOXIDE-OXYGEN CYCLE

Without algae and plants, there wouldn't be enough oxygen on earth for animal life. All of these organisms are part of the *carbon dioxide–oxygen cycle.* See Fig. 19–7. During photosynthesis, algae and green plants take in carbon dioxide from the

Fig. 19–7 The carbon dioxide–oxygen cycle.

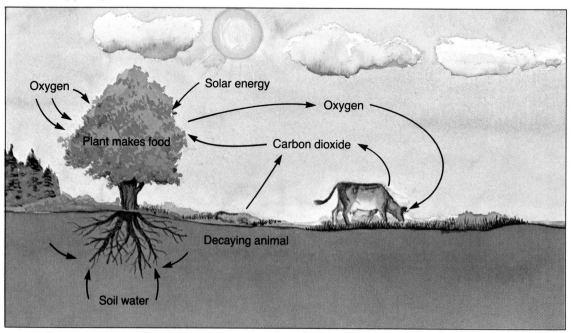

air. They use the carbon dioxide, along with water and the sun's energy, to make their food. In the process of making food, oxygen is released as waste.

Plants and animals need oxygen to release energy from food. In this process, oxygen is taken in and carbon dioxide given off. Animals do this when they breathe. Many of the compounds that make up living organisms contain carbon. Carbon is released as CO_2 when the organism dies and decays.

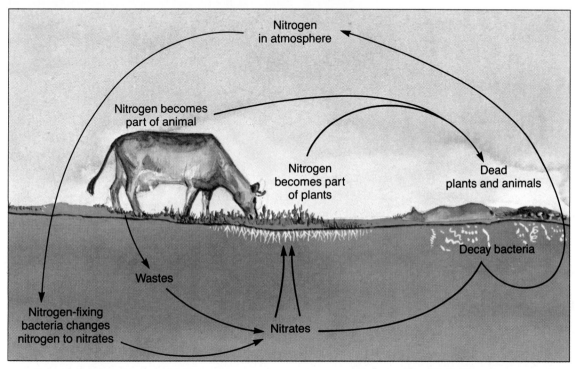

Fig. 19–8 The nitrogen cycle.

THE NITROGEN CYCLE

The *nitrogen cycle* also involves living things. See Fig. 19–8. Nitrogen makes up some 80 percent of the atmosphere. Living cells need this nitrogen to build protoplasm. Animals obtain it from plants. However, most plants cannot use nitrogen directly from the air. There are special kinds of bacteria that can combine nitrogen with other elements to make nitrogen compounds. These compounds are called *nitrates*. The bacteria are called *nitrogen-fixing bacteria*. See Fig. 19–9 on page 446. The nitrates they make are released into the soil.

These nitrates in the soil are used by plants. This is the way plants get the nitrogen they need. The nitrates are taken in

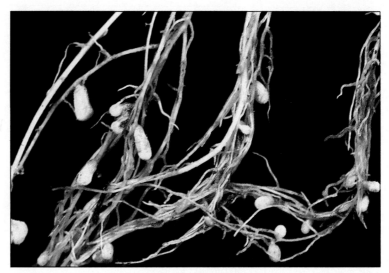

Fig. 19–9 These white clover roots contain nitrogen-fixing bacteria.

along with water by plants' roots. The nitrogen becomes part of plants' protoplasm. The plants are eaten by animals, and the nitrogen then becomes part of the animal protoplasm. Wastes, and dead plants and animals are broken down into nitrogen compounds by another special kind of bacteria called *decay bacteria.* These nitrogen compounds reenter the soil and can be used again by plants.

SUMMARY

Many compounds needed by living things are recycled naturally through the environment. If such recycling did not occur, the available supplies of these materials would eventually be used up.

QUESTIONS

Use complete sentences to write your answers.

1. Explain why the sun is important to the water cycle.
2. How do plants and animals get the nitrogen they need for making protoplasm?
3. For long space trips, scientists are considering sending simple green plants or algae with the astronauts. Explain how this arrangement would help duplicate the carbon dioxide-oxygen cycle.
4. Why are these cycles necessary to the survival of life on earth?

SKILL-BUILDING ACTIVITY

SEQUENCING

PURPOSE: To determine the flow of materials through natural cycles.

MATERIALS:

water cycle cards   nitrogen cycle cards
CO_2–O_2 cycle cards   string

PROCEDURE:

A. Obtain the envelope labeled "The Water Cycle." In it you will find a card for each of the major parts of the water cycle. You will use the string to connect each of the parts of the cycle.

B. Reread the description of the water cycle. Using the cards and string, make a diagram that shows this cycle. Have your teacher check your diagram. Now answer the following questions.

    1. How is the sun involved in the water cycle?

    2. What happens to water that is lost from a plant's leaves?

    3. Why does the water level in an aquarium become lower over several days?

    4. How do organisms that live in water fit into the water cycle?

C. Copy the diagram onto a separate sheet of paper.

D. Obtain the envelope marked "CO_2–O_2 Cycle." Each of the ten cards represents a major step in this cycle. As before, reread the section on this cycle. Use the cards and string to construct a diagram. Have your teacher check the diagram. Then answer the following questions.

5. Would the CO_2–O_2 cycle operate without sunlight?

6. What three things must green plants obtain before they can make their food?

7. Besides food, what else is produced in the food-making process?

8. Would this cycle operate in an aquarium? In a terrarium?

E. Copy the diagram onto a separate sheet of paper.

F. Obtain the envelope marked "Nitrogen Cycle." Reread the section on this cycle. Use the ten cards to construct a diagram. Have your teacher check the diagram. Answer the questions below.

    9. What is the difference in the roles decay bacteria and nitrogen-fixing bacteria play?

    10. Why are the products of nitrogen-fixing bacteria called compounds, not elements?

    11. What two things may happen to the nitrogen that is released from decaying organisms?

    12. Do you think there are any nitrogen-fixing bacteria in a terrarium? How did they get there?

    13. Would the nitrogen cycle be working in an aquarium? Explain how.

G. Copy the diagram onto a separate sheet of paper.

CONCLUSION:

    1. Why is recycling of these compounds necessary to living things?

19-3. Energy Pathways

At the end of this section you will be able to:

☐ Describe *food chains* and *food webs.*

☐ Contrast *herbivores, carnivores,* and *omnivores.*

☐ Explain food pyramids of numbers, mass, and energy.

A grasshopper sits on a tall pondweed chewing on a leaf. Suddenly, it is seized by a hungry bullfrog. The frog fails to see a snake moving closer. The snake strikes, and the frog becomes a victim. The snake's movement catches the eye of a hawk. There is a sudden swoop and a flurry of wings. The hawk flies off with the snake dangling from its claws.

FOOD CHAINS

The chain of events just described, and shown in Fig. 19–10, is called a **food chain.** All the organisms in this chain need energy to carry on their activities. The sun is the source of energy for all life. Only organisms with chlorophyll can capture this energy. The energy is stored in the food they make. Because they can produce their own food, these organisms are

Food chain The passing of energy in the form of food from one organism to another.

Fig. 19–10 A food chain diagrams the transfer of energy from one organism to another.

called **producers.** Algae and green plants are producers. Animals that eat them to get energy are called **consumers.** A *consumer* may also eat other animals. Because the food chain is the way in which energy is passed up through the chain to various consumers, food chains are also called energy chains.

In the *food chain* in Fig. 19–10, the pondweeds are the *producers.* These plants pass energy along to a consumer, the grasshopper. Because the grasshopper feeds on plants, it is called a **herbivore** (**hur**-bih-vore), or plant eater. The grasshopper was eaten by a higher-level consumer, the frog. The frog, snake, and hawk are another type of consumer, called a **carnivore.** *Carnivores* are animals that eat other animals. Their name means "meat eater." Some animals, such as mice, bears, and humans, eat both plants and animals. Such animals are called **omnivores,** or "everything eaters."

Each food chain leads to an animal that is not eaten by other animals. But this is not the end of the chain. The wastes and remains of these organisms are food for **decomposers.** Decay bacteria and fungi are examples of *decomposers.* They break down wastes and dead organisms all along the food chain. These materials are returned to the soil. Here they are recycled and used again. See Fig. 19–11.

Producer A green plant able to make its own food by photosynthesis.

Consumer An organism that depends on other organisms for its food needs.

Herbivore A consumer that eats only plant tissue.

Carnivore A consumer that eats only animal tissue.

Omnivore A consumer that eats both plant and animal tissue.

Decomposer An organism that obtains its food from wastes and dead organisms.

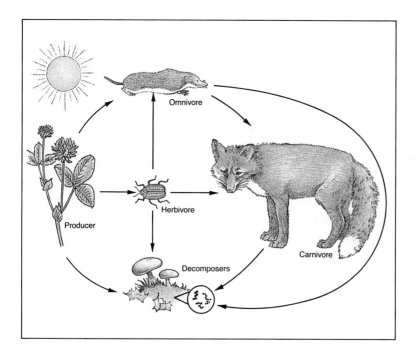

Omnivore
Producer
Herbivore
Carnivore
Decomposers

Fig. 19–11 Which organisms are the producers, consumers, or decomposers?

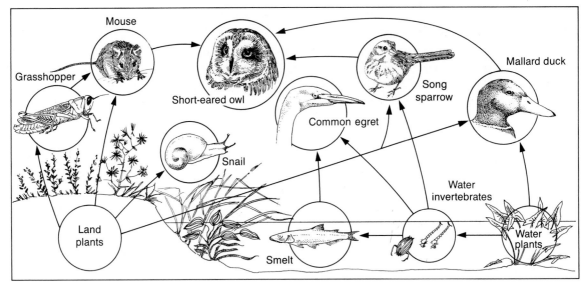

Fig. 19–12 A food web of some organisms that live in a salt marsh.

Food web All the feeding patterns in an ecosystem.

FOOD WEBS

An ecosystem may have many food chains. Most animals in a food chain eat a variety of foods. An animal in one chain often eats animals from other chains. As a result, food chains overlap. This network of overlapping food chains is called a **food web.** A simplified *food web* of a salt marsh is diagrammed in Fig. 19–12.

Food webs exist in every ecosystem. Most animals have more than one source of food. Those with only one source are less likely to survive if something should happen to that single food source.

FOOD PYRAMIDS

Food chains are often shown as *food pyramids.* The shape of the pyramid shows a decrease in some factor from the bottom to the top. The base is made up of the producers. The last level of consumer is at the top. There are usually more producers than other organisms in a food chain. As you move to the next level, the number of organisms decreases. A grass field chain shows this decrease. The grasses feed thousands of plant-eating insects. These insects feed a few hundred sparrows. Finally, the sparrows feed only a few hawks. This example is called a *pyramid of numbers.* The numbers decrease as you move up the chain. See Fig. 19–13 (left).

Another way to study food chains is to compare the total

mass of all the organisms at each level. This is shown in a *pyramid of mass.* See Fig. 19–13 (right). Each level contains only about 10 percent of the mass of the previous level. What happens to the rest of the mass? At each level, many organisms are not eaten. Only parts of others are eaten. Much of the mass is lost this way. More mass is lost as it is changed into energy at each level.

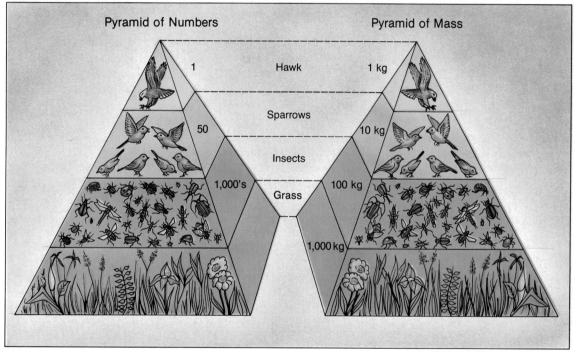

The transfer of energy is shown in a *pyramid of energy.* See Fig. 19–14 on page 452. Only about one percent of the sun's energy that strikes the biosphere is used by algae and green plants. The rest is either reflected or converted to heat in the atmosphere. Most of the energy a plant gets from the sun is used to carry on the plant's life activities. Only a small part of the energy is left to be stored in the plants. This stored energy is passed on as food to the next level. At each level, most of the energy is used to carry out the organisms' life processes. In these life processes, much of this energy is changed to heat and lost to the environment. Only the energy that remains is passed on to the next level when the organism is eaten. As a result, each level has less energy available to it than the previous level. This helps to explain why there are fewer

Fig. 19–13 A pyramid of numbers and a pyramid of mass.

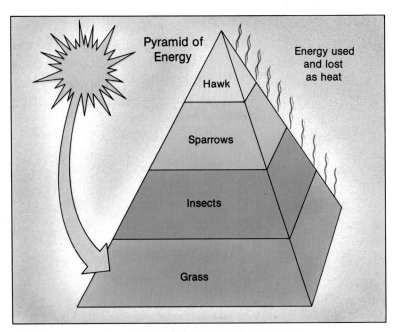

Fig. 19–14 *Why must every food chain have a constant source of energy?*

organisms at the top of the pyramid. There is less energy available for life. All food pyramids show the same thing. Matter and energy in a food chain decrease as the levels increase. Each pyramid shows this in a different way.

SUMMARY

The relationships between the organisms in a community can be described with food chains and food webs. In order to do this, one must know what each animal eats. Pyramids organize the community into levels, which decrease in size from the base upward.

QUESTIONS

Use complete sentences to write your answers.

1. Construct a simple food web for the following organisms: mountain lions, grasses, humans, wolves, deer, sheep, rabbits, shrubs.
2. Identify each of the above organisms as a producer, herbivore, carnivore, or omnivore.
3. Why are food chains often called energy chains?
4. How are food chains different from food webs?
5. What do the pyramid shapes of food chains show?

19-4. Changes in Populations

At the end of this section you will be able to:

☐ Explain how limiting factors affect a population.

☐ Describe some predator/prey relationships.

Isle Royale National Park is a large island in Lake Superior. See Fig. 19–15. Around 1900, a few moose either swam to the island or crossed on the winter ice. They found plenty of food. They had no natural enemies there. As a result, the number of moose increased rapidly. Within 20 years, over 2,000 moose were living on the island.

Fig. 19–15 The location of Isle Royale.

POPULATIONS AND LIMITING FACTORS

All the organisms of the same species within a community are called a **population.** The moose of Isle Royale are such a *population.* The island also has populations of beavers, fir trees, birch trees, and foxes. See Fig. 19–16.

Scientists keep track of what is happening to populations in isolated areas like Isle Royale. It helps them understand what normally goes on in a community. They try to find out what populations are present. They also attempt to determine how many individuals are in each population. They check the age, range, *birthrate,* and *death rate* of each population as well.

By the 1920's, the moose population had increased so that there were four moose for each square km of land on Isle

Population A group of one kind of organism in a particular community.

Fig. 19–16 A moose on Isle Royale. What other populations are present?

Density The number of organisms found in a certain area at a given time.

Royale. This is the **density** of the population. The population *density* was so great that it caused a food shortage. The moose began to compete with each other for food. In the next few years, the death rate increased. All but a few hundred moose died. Most of them died of starvation. Such a drastic decrease in numbers is called a population *crash.*

In the summer of 1936, a forest fire burned over a quarter of the island. Over the next several years, the burned area began to fill in with lichen, mosses, and young trees. The herd thrived on these excellent sources of food. More young moose survived long enough to reproduce. The birthrate increased and many offspring were born. The death rate decreased because of plentiful food supplies. When the birthrate is higher than the death rate, a population will increase. As the number of moose steadily increased again, competition for food increased too. By the 1940's, park rangers again began to find dead moose. It looked like another crash would occur.

Limiting factor A condition in the environment that may cause a change in the size of a population.

The amount of food available to a population is an example of a **limiting factor.** Any factor that can cause the size of a population to change is a *limiting factor.* In both the 1920's and 1940's, a shortage of food caused the death rate of moose to increase. This, in turn, caused the population to decrease. In this example, the amount of food was a limiting factor. Other factors can also limit population density. Disease and **predators** may kill large numbers of organisms. Weakened and old organisms are easy **prey** for *predators.*

Predator An animal that hunts other animals for food.

Prey Animals that are hunted by other animals as food.

A combination of limiting factors determines the size of a population. Every ecosystem has a maximum number of organisms that it can support. As the population nears this maximum number, the limiting factors begin to affect it.

PREDATORS AND PREY

Fig. 19–17 shows how predators can be a limiting factor. The predator is the Canadian lynx. The *prey* is the Varying hare. As the number of hares increases, so does the number of lynxes. The increased number of predators will cause a drop in the number of prey. Then the predators begin to starve to death. This decrease in predators allows the prey to increase again. This is sometimes called the *predator/prey cycle.*

What finally happened to the moose of Isle Royale? In the late 1940's, a pack of wolves crossed the winter ice to the

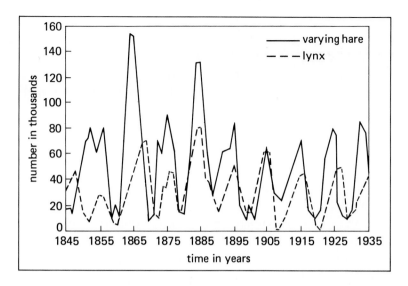

Fig. 19–17 This graph shows changes in the Varying hare and lynx populations over a 90-year period.

island. Many people feared that the wolves would increase and then destroy the moose herd. But this did not happen. The wolves killed mainly the old, sick, or injured moose. Their actions reduced but did not destroy the herd. The herd size slowly began to increase again.

In the 1970's, the size of the moose herd peaked. With many old and weak moose available for food, the number of wolves doubled. Soon after, the moose population declined rapidly, or crashed. In response, in the early 1980's, the wolf population also crashed. The numbers of both groups were back to the low points of the late 1940's. But not for long. The moose population is making a rapid comeback. The wolf population is beginning to increase also.

SUMMARY

The population of any species in a particular area is affected by several limiting factors. Predator/prey cycles are a natural part of all populations.

QUESTIONS

Fig. 19–18 Wolves surround a single moose on Isle Royale.

Use complete sentences to write your answers.

1. What limiting factors might affect the size of an insect population in a garden?
2. How would an increased death rate affect a population?
3. Give two examples of predator and prey in a pond.

INVESTIGATION

SAMPLING A POPULATION

PURPOSE: To determine the size of a population without counting each individual.

MATERIALS:

4 meter sticks or a wire coat hanger

marker (such as a square of cardboard)

PROCEDURE:

A. You have been asked to determine how many dandelions and plantains there are on a lawn. You do not wish to count each one. You will use a population census method instead.

B. Determine how many square meters (m^2) of grass are to be surveyed. If the area is not a rectangle, ask your teacher for help.

C. Construct your sampler. For large areas, connect four meter sticks together at their ends. This is a one square meter area. For smaller areas, bend a wire coat hanger into a square 20 cm on a side. This is 0.04 of a square meter.

D. Stand with your back to the area and toss a marker over your shoulder. Place the sampler so the marker is in the center.

E. Count the dandelions and plantains that are found inside the sampler. On a separate piece of paper, copy and record your data in a table like the one shown in Table 19–1.

F. Repeat the procedure ten times in other parts of the area.

G. Determine the average number of dandelions per square meter by dividing the total number of dandelions counted by 10. Repeat for the plantains. If you used the small sampler, multiply the total by 25 before dividing.

H. To determine the total population, multiply the average per square meter by the number of square meters in the entire survey area (determined in step B).

CONCLUSIONS:

1. What is the average population of dandelions per square meter?

2. What is the total population of dandelions in the survey area?

3. Repeat questions 1 and 2 for the plantain population.

4. List several reasons why these answers may not be 100 percent accurate. (This is called experimental error.)

5. What procedure would you use to sample the number of bass in a lake?

POPULATION CENSUS													
Area Sampled: _____										*Size (m²):* _____			
	SAMPLE NUMBER											Average per	Total
ORGANISM	1	2	3	4	5	6	7	8	9	10	Total	square meter	population
Dandelion													
Plantain													

Table 19-1

19-5. Changes in Communities

At the end of this section you will be able to:

- ☐ Define and give examples of *succession*.
- ☐ Contrast primary and secondary succession.
- ☐ Describe a *climax community*.

Mt. St. Helens erupted in 1980. The blast destroyed 550 square kilometers (km^2) of forest. Hillsides were stripped clean. Then they were buried under several meters of volcanic ash. Almost nothing remained alive in the blast area.

Fig. 19–19 *Plants have begun to grow again in the ash fields surrounding Mt. St. Helens.*

PRIMARY SUCCESSION

Areas such as Mt. St. Helens allow scientists to study the sequence of events that occur as a new community develops. This series of changes is called **succession.**

If *succession* occurs in an area where there was no previous life, it is called *primary succession*. Plant growth on a solid rock lava flow is an example of this type.

Succession A series of changes occurring in the plant and animal community of an ecosystem.

In primary succession, barren rock must be broken down into a soil. This is done by the action of weather and organisms such as lichens and bacteria. These first organisms to appear are called "pioneers." You may have seen a crust of lichens growing on rocks. Lichens give off acids that break down the rock to form a simple soil. The soil is then a suitable habitat for plants such as mosses.

Mosses and lichens trap windblown dirt. They also add organic material to the crumbling rock when they die. This will make the developing soil richer. The area eventually becomes favorable for the growth of other types of plants.

SECONDARY SUCCESSION

Secondary succession occurs where soil has already developed. The ash deposits on Mt. St. Helens or a forest fire area are examples of places where secondary succession might take place. Volcanic ash is a low-grade soil rich in minerals. Other places where secondary succession may occur are vacant lots or an old field. When a farmer abandons a field, succession begins very quickly. The type of community that develops depends on the soil and climate of the area. For instance, a field in Florida will develop a different community from a field in New England. We will use a New England field as an example of secondary succession.

Imagine that a farmer plowed a field and left the soil bare. See Fig. 19–20. Within a year or two, the field is covered with grasses. These grasses are called *pioneer* species. These plants thrive in direct sunlight. Grasses can withstand extremes of temperature and moisture. They also produce many seeds that sprout very quickly.

Fig. 19–20 Succession in an old New England field.

Soon woody shrubs sprout and grow among the grasses. A short time later, white pine and birch trees appear. The white pine seedlings thrive in the open sunlight. As the years pass, the white pines dominate the field. This is the beginning of a forest. The pines grow quickly. Within 20 to 30 years, the ground beneath the pines is so shaded that their own seedlings cannot grow. However, oak and maple seedlings are able to grow in the shade. Thus, they thrive beneath the pines. Both the shade and the soil are now suitable for other species. Beeches and hemlocks also appear beneath the pines. Woodland mosses, ferns, and flowers begin to grow.

What was once open farmland has changed back to forest. Soon the forest will reach the final stage of succession. Only the seedlings of the hemlocks and beeches will survive. They will grow and replace their own kind. This stage of balance is called a **climax community.** The climax community will remain unchanged for many years.

ANIMAL SUCCESSION

As the plants change, so do the animals. As soon as there are plant remains for them to feed on, worms arrive from the surrounding areas. Insects appear as the plants reach the stage where the insects can survive.

Larger animals move into the field as their food supply becomes abundant. These may include field mice, snakes, and weasels. As the grasses appear, deer may browse. As the area changes from shrubs to forest, different animals come and go. Animals are part of the *climax community,* too. See Fig. 19–21.

Succession takes a very long time to be completed. It takes about 25 years before the pine forest stage will replace an abandoned field. Close to 100 years might pass before the oak-maple forest will take over. The hardwood-hemlock climax would take even longer. Scientists studying succession around Mt. St. Helens have many years of study ahead of them.

SUMMARY

Succession can begin on bare rock or where there is some soil. Pioneer organisms begin to grow. Eventually, the area becomes suitable for other organisms. The original community is replaced by a new one. This process continues until a climax community forms.

QUESTIONS

Use complete sentences to write your answers.

1. What comes first, succession or a climax community? Explain what each term means.

2. Describe what might happen to a vacant city lot over a long period of time.

3. Why are the plants in each stage of succession different from those in the last stage?

4. In 1984, volcanoes in Hawaii erupted, covering a large area of land with lava. Will a primary or secondary succession develop here? Why?

Climax community The final stage of a community at the end of a succession.

Fig. 19–21 These Pileated woodpeckers require large trees for their nesting holes. They are members of the climax forest.

SUCCESSION

PURPOSE: To identify the different stages of succession in decaying logs.

MATERIALS:

thermometer   trowel   magnifier

Optional: soil thermometer, light meter, hygrometer, hatchet, field guides.

PROCEDURE:

A. Do this study in woods that have not been logged or cleared for several years.

B. Within the study area, locate examples of the following stages of succession (all stages should be trees of the same species):

   a. trees that are dead but still standing

   b. trees that have fallen fairly recently

   c. fallen trees with firm exteriors but rotten centers

   d. fallen trees that are completely rotten

C. On a separate piece of paper, construct a data table. Record and compare your observations.

D. Describe the following abiotic factors about areas close to an example of each stage. Use instruments to make accurate measurements when possible.

   1. What is the air temperature?

   2. What is the soil temperature?

   3. What is the relative humidity?

   4. How much light does the area get?

   5. Is the soil moist or dry?

   6. Is the soil loose or packed?

   7. What is the soil color?

E. Examine the standing dead tree. CAUTION: Dead trees are not safe. Your teacher will decide if a tree is appropriate for use. Do not climb or knock over such a tree.

   8. Does the tree show evidence of vertebrate activity (nests, woodpecker holes, etc.)?

   9. Are there fungi or plants growing on the tree? If so, describe them.

   10. Is the bark dry or moist?

   11. Remove some bark. Are there invertebrates beneath it? If so, describe the type and number.

F. Quietly approach the newly fallen log. Look around and under it.

   12. Do you see any mice, snakes, or salamanders? Is there any evidence of their presence such as tunnels, nests, or food supplies?

   13. Repeat questions 9–12 for the newly fallen log.

G. Examine the partially rotten log. Look around and under it.

   14. Repeat questions 9–12 for this log.

H. Use the trowel to break apart the rotten insides. Describe the color, moisture, and odor of this area.

I. Examine the completely rotten log.

   15. Repeat questions 9–12 for this log.

CONCLUSIONS:

   1. Why did each stage have different plant and animal communities?

   2. Which ecological niche is most important in a fallen log habitat?

TROPICAL BIOLOGIST

Life scientists usually specialize in one section of biology. Tropical biologists concentrate their research on the plant or animal life existing in tropical climates, such as the Amazon jungle in South America or the Congo basin in Africa. But since all plant and animal life is related through ecosystems, tropical biologists' work also applies to cooler climates. For example, to build a highway through the Amazon jungle, many trees and plants were cut down. But the plants in the Amazon produce a large percentage of the entire earth's oxygen. Continual research is needed to understand the relationship between plant growth in the tropics and the worldwide oxygen supply.

A tropical biologist has a Bachelor of Science in biology as well as an advanced degree in some specialized field of biology.

For further information, contact: American Institute of Biological Sciences, 1401 Wilson Boulevard, Arlington, VA 22209.

ENVIRONMENTALIST

People of many backgrounds enter the field of environmental action. They may be scientists with strong ecological interests, or concerned and knowledgeable citizens who decide to commit themselves to environmental activism. Environmentalists are, in many ways, lobbyists. That is, they support a cause, encourage others to do so, and act as an information source for this cause. Usually, they form groups. Some of these groups are well known and have existed for some time—like the Sierra Club, an organization that works to preserve the wildlife and natural land resources of the United States. Other groups are formed because of specific issues, such as the dumping of toxic wastes or the need for clean-air legislation.

Many volunteers work in this area, but paid work is available at all levels as well.

For further information, write: Environmental Protection Agency, Dept. of Public Information, 401 M Street, Washington, DC 20460.

VOCABULARY

On a separate piece of paper, match each term with the number of the statement that best explains it. Use each term only once.

consumer	ecosystem	abiotic	predator
herbivore	food web	community	climax
succession	producers	population	carnivore

1. All the possible food pathways in an ecosystem.
2. Green plants that carry on photosynthesis.
3. Changes in a community over a period of time.
4. A group of organisms and their physical surroundings.
5. An organism that depends on other organisms for its food needs.
6. An animal that hunts other animals for food.
7. An animal that eats only meat.
8. All the different organisms in an area.
9. The nonliving parts of an ecosystem.
10. All of the organisms of the same species found in a particular area.
11. The final makeup of a community at the end of a succession.
12. An animal that eats only plants.

QUESTIONS

Give brief but complete answers to each of the following questions. Unless otherwise indicated, use complete sentences to write your answers.
1. How might the abiotic factors of temperature, water, air, and soil be different from one ecosystem to another?
2. How might the parts and activities of a farm ecosystem affect a nearby pond ecosystem?
3. Diagram a simple food web using the following organisms: starlings, beetles, weeds, grubs, robins, grass, cats, worms.
4. Why are humans called omnivores? Draw two food chains that will help explain your answer.

5. Why does the available energy in a food chain decrease as the levels increase?
6. What happens to an area when succession takes place? How might the activities of humans interfere with the natural succession of an area?
7. Describe some of the effects a high population density might have on a population of rabbits.
8. How is it possible that organisms can occupy more than one niche in a community? Give an example.
9. What is a climax community? Why are they different in different locations?
10. Why are green plants at the bottom of any food chain or web?

APPLYING SCIENCE
1. The food web of a large city may extend far beyond the city limits. It may even be worldwide. Explain why.
2. A geranium plant and a mouse are placed in separate sealed containers with a supply of water. What do you think will eventually happen to each? Explain your answers. What will happen if they are both placed in the same sealed container?
3. Describe some of the effects a high population density might have on a human population.
4. List all the different foods served in your school cafeteria this week. Work backward to construct food chains to the sources of the food. Make a diagram to illustrate your answer.
5. Construct a balanced aquarium and terrarium. Use biology reference books for directions. Describe how the cycles of water, carbon dioxide and oxygen, and nitrogen operate in these ecosystems.

BIBLIOGRAPHY
Amos, William. "Life on a Rock Ledge." *National Geographic,* October 1980.
Johnson, Cecil. "The Wild World of Compost." *National Geographic,* August 1980.
Patent, Dorothy. *Plants and Insects Together.* New York: Holiday House, 1976.
Russell, Franklin. *At the Pond, Vol. 1 and Vol. 2.* New York: Four Winds, 1972.
Willman, Alice. *Africa's Animals: Creatures of a Struggling Land.* New York: Putnam's, 1974.

NATURAL SELECTION

CHAPTER GOALS

1. Explain how fossils provide evidence that the earth has changed.
2. Describe the parts of the theory of natural selection.
3. Explain the role of mutations in the adaptation of a species and the creation of new species.

20-1. The Fossil Record

At the end of this section you will be able to:

- ☐ Describe the forms in which *fossils* are found.
- ☐ Describe how fossils are dated and what scientists learn from them.
- ☐ Identify other evidence suggesting that organisms have changed over time.

Do you know who your great-great-grandparents were? What were they like? When were they born? How could you find out the answers to these questions? It may not be easy, but at least you know that records of that time were kept.

What was life on earth like in prehistoric times? *Prehistoric* refers to a time before events and ideas could be written down. Without a written history of what happened, how can scientists answer this question?

TYPES OF FOSSILS

Like detectives, scientists gather clues about life long ago. Fossils are clues. A **fossil** is any trace or remains of an organism that lived in the past.

Most dead organisms never become *fossils* because they either are eaten or decayed. A dead organism must be quickly preserved, and stay that way for a long time, to become a fossil. There are several ways that this can happen.

One way is for the organism to be buried beneath mud or sand sediments. This type of fossil mainly forms at the bottom of seas, lakes, and swamps. As rivers carry water into lakes and the ocean, sediments travel with the water. In the ocean,

Fossil The remains or imprint of an organism that once lived.

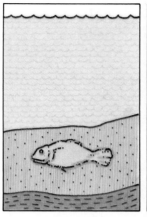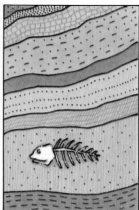

Fig. 20–1 Fossils may form when an organism is buried by layers of sediments.

the sediments settle to the bottom. In this way, dead organisms that are lying on the bottom are covered. See Fig. 20–1. These sediments protect the organism from decomposers breaking down its body structure. Gradually, through the years, more mud or sand is deposited on top. This puts great weight on the mud or sand below. After hundreds of thousands of years, this weight slowly causes the mud or sand to change to solid rock. If the organism eventually decays after this change, either its imprint or outline is left in the rock around it. This forms a *mold*. If the mold is gradually filled in with minerals, a fossil *cast* is formed. See Fig. 20–2 (a).

Imprints can also be made in mud or clay, just as you can make foot or hand prints in cement. If the mud or clay turns to stone, the prints are preserved. Look at Fig. 20–2 (b).

Sometimes a whole organism is preserved in mineral form. This happens when minerals filter in and slowly replace the tissues of the organism. The organism is said to be *petrified*. This means that the organism changed to rock. Petrified wood is a common example of this type of fossil. See Fig. 20–2 (c).

Not all fossils are formed in rock. Fossil skeletons of large animals, such as saber-toothed tigers, have been found in tar pits. These animals must have been trapped in the tar and died. The tar preserved their bones. Some small insects were preserved in the sticky sap of ancient trees. This sap hardened into a transparent solid called *amber*. Amber preserved the bodies of the insects so well that even the tiny hairs on their bodies can still be seen. See Fig. 20–2 (d).

Another substance that preserves organisms well is ice. Think of how well freezers today keep food. Woolly mam-

a

b

c

d

Fig. 20–2 Fossils are found in the form of (a) casts, (b) footprints, (c) petrified wood, and (d) trapped in amber.

moths, the ancestors of elephants, and woolly rhinoceroses have been found in deep ice cracks in Siberia.

THE AGES OF FOSSILS

By determining the ages of fossils, scientists can get an idea of how long ago the organisms were alive. The layer of rock in which a fossil is found can give clues to its age. In general, the deeper the layer of rock, the older the fossil is. However, this is only true if the rock layers have not been disturbed in some way, such as by earthquakes.

Radioactive dating gives a better estimate of age. In this method, radioactive forms of certain chemical elements are measured. These forms are unstable and give off nuclear radiation. As they do, they change into different elements. This always happens at a fixed, measurable rate. Scientists can

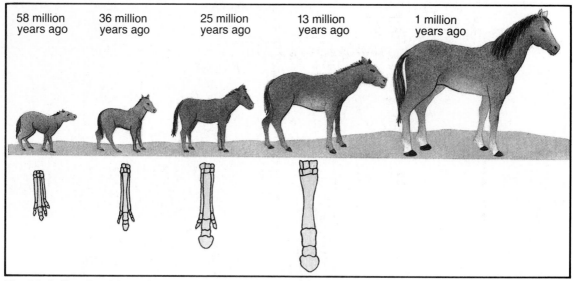

| 58 million years ago | 36 million years ago | 25 million years ago | 13 million years ago | 1 million years ago |

Fig. 20–3 Fossil evidence has shown how the horse has changed over time.

measure how much of a certain radioactive element has changed into a different element. By knowing the rate at which this change takes place, scientists can estimate the age of the fossil.

The estimated age of a fossil is often based on the age of the rock where it is found. The oldest rock ever found on earth is over four billion years old. Fossils indicate that life may have existed over three billion years ago.

Although many organisms may never have left fossils, the fossil record still shows some clear patterns. The oldest fossils found have all been of very simple organisms. These organisms include bacteria, algae, fungi, and protozoa. Fossils of more complex life forms date from much more recent times. *Complex* refers to a greater number of different body parts and systems.

The fossil record also shows that the numbers and kinds of organisms have changed over time. Some animals that exist today appear different than their ancestors might have looked. The horse is an example of this. Fossils of horses show that horses were once much smaller. The structure of their feet was also different than the horses of today. See Fig. 20–3. Other fossils do not look like any plant or animal living today. This suggests that some species have become **extinct.** They no longer exist today. For example, fossils show that dinosaurs once dominated the land. However, dinosaurs are now *extinct.*

Extinct Refers to an organism that no longer exists on earth.

OTHER EVIDENCE OF CHANGE

Sometimes, species seem to change rapidly. A species called the Peppered moth is one example. Up until 1850, Britain had very few factories. Almost all Peppered moths were light-colored. Most tree trunks in Britain were also light in color. These moths are active during the night and rest on tree trunks during the day. See Fig. 20–4. The light-colored moths blended with the color of the tree trunks. It was difficult for birds to find and eat them. There were some dark-colored Peppered moths, but they were rare. The dark-colored moths were quickly seen and eaten by birds.

Within a few years, many factories were built in Britain. The factories began pouring out polluting smoke, which covered the tree trunks with soot. Now, the dark-colored moths were hidden while resting on the soot-darkened tree bark. The light-colored moths were easily spotted and eaten by birds. In less than 75 years, most of the Peppered moth population was dark-colored.

We see evidence today of change as people breed plants and animals to meet human needs and tastes better. Modern varieties of fruits and most other crops do not look exactly the same as they did 50 years ago. Animals, such as the cow and pig, have been bred to give more milk or meat than their recent ancestors did.

Fig. 20–4 Two forms of the Peppered moth. How does their color affect their survival?

SUMMARY

Fossils can form in several different ways. Fossils show that many organisms have changed over long periods of time. There is also more recent evidence that organisms continue to change. The change in the population of Peppered moths in Britain and the breeding of animals are examples of this.

QUESTIONS

Use complete sentences to write your answers.

1. What are four ways that fossils are formed?
2. How do scientists estimate the age of a fossil?
3. What type of evidence do fossils provide?
4. How did the Peppered moths show that a species could change?

INVESTIGATION

MAKING A FOSSIL MODEL

PURPOSE: To demonstrate how one of the several types of fossils could have formed. In this investigation, you will make a fossil imprint.

MATERIALS:

plaster of Paris	leaves, shells, or
plastic container	small bones
water	tart pans or small
plastic spoons	pie or cake tins
petroleum jelly	masking tape

PROCEDURE:

A. In the plastic container, mix a small amount of water with the plaster of Paris until it looks like pancake batter. Stir this mixture vigorously with the spoon until it is smooth. Be sure to add only a little bit of water at a time.

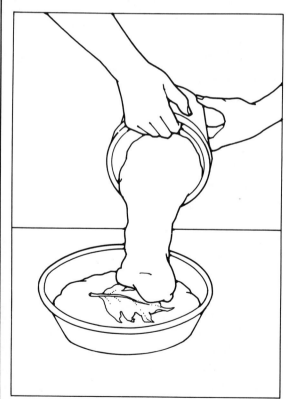

Fig. 20–5

B. Pour the plaster mixture you made into the tin, making a layer about 2 cm thick. Coat some leaves and/or other objects with petroleum jelly on all sides. Lay each one on the plaster you just poured. Cover these objects with another layer of plaster of Paris. See Fig. 20–5.

C. Write your name on a piece of masking tape. Use the tape to label your tin container. Let the plaster harden overnight.

D. The next day, remove the plaster from the tin pan. Gently crack it open.
   1. What do you find?
   2. What type of fossil can this be a model for?
   3. What would be different if this were an actual fossil?

E. Place your plaster cast on a table with those of your classmates. Place all of the objects that were cast on the table, too. Try to identify which object formed each cast.

CONCLUSIONS:
   1. What must be true for an organism to become a fossil?
   2. How does the fossil model made with plaster of Paris compare with the actual formation of fossils?

20-2. The Theory of Natural Selection

At the end of this section you will be able to:

- ☐ Discuss Lamarck's theory of acquired characteristics.
- ☐ Describe several influences on Darwin as he developed his theory of *evolution*.
- ☐ Describe the five parts of the theory of *natural selection*.

Has there ever been someone, something, or some event that caused you and your life to change? Perhaps it was a grand-parent, something you read, or a trip you took. All three of these were important influences on a man named Charles Darwin (1809–1882). See Fig. 20–6.

Darwin's grandfather, a doctor, wrote a book to try to explain the theory of diseases. In the same book he also tried to explain life by **evolution.** *Evolution* is a theory that attempts to ac-count for the slow, gradual development, or change, in living things over time.

Like many youngsters, Charles Darwin collected pebbles, insects, plants, and birds' eggs. Even when he got older, Dar-win continued making collections. This habit was to be an important part of his work.

LAMARCK'S THEORY

While he was at Cambridge University, Darwin read the work of Jean Baptiste Lamarck. Lamarck also had a theory of evo-lution. He, too, had studied fossil evidence. Lamarck sug-gested that during an organism's life it could *acquire* traits that would make the organism better able to live in its envi-ronment. He said that this organism could then pass these traits on to its offspring. He also believed that if certain traits were not used, they would eventually disappear. Lamarck's idea is often called the *theory of acquired characteristics.*

Lamarck's most famous example is the long neck of the giraffe. He said that as giraffes strained to reach the leaves on higher branches, their necks stretched and became longer. This trait of a longer neck was then passed on to their offspring. See Fig. 20–7 on page 472. Lamarck's theory is no longer accepted by most biologists. Today, we know that only

Fig. 20–6 Charles Darwin, in his later years (1881).

Evolution The process by which organisms change and become different from gener-ation to generation.

Fig. 20–7 Lamarck believed that traits gained during a lifetime could be passed on to offspring.

changes in genes, not changes in behavior, are passed on to offspring.

DARWIN FORMS HIS THEORY

In 1831, after graduation from the university, Darwin was invited to join a five-year voyage on a British ship called the HMS *Beagle*. During the voyage, Darwin studied many organisms, recorded his observations in notebooks, and kept a journal. His collections included many fossils. He kept at least one of every species of plants and animals he found.

When he returned home, Darwin began the long process of classifying all of the specimens he had collected. He also began to sort out his ideas on how organisms can change over time. Darwin knew that farmers selected and bred animals and plants with certain traits. However, he didn't know what could cause such selection in nature. Again, something Darwin read influenced him.

Darwin happened to read *An Essay on the Principle of Population* by Thomas Malthus. It describes how the number of people always gets larger when there is enough food and other necessities, and gets smaller when there is not. Malthus said that people would overrun the earth if starvation, disease, crime, and war did not keep their numbers down. Malthus called this the *struggle for existence*.

The struggle for existence was the key to Darwin's own

theory. Darwin realized that all organisms must compete for food, water, and other necessities in order to survive. Only the organisms that are better able to compete survive. Those that are not, die.

It was twenty years later before Darwin began to write out his entire theory. Before he finished, he received an essay in the mail from Alfred Wallace. Wallace was also a naturalist and worked in the jungles of Indonesia. He and Darwin had been writing to each other about their research and theories. In a few pages, Wallace stated the main points of the theory Darwin had worked on for so long. In 1858, friends of Darwin and Wallace presented the results of their work at a scientific meeting in London.

THE PARTS OF THE THEORY

The theory that Darwin and Wallace developed to explain evolution is called **natural selection.** It has five main parts. As you read, you will see differences between Darwin's and Wallace's theory and Lamarck's theory.

1. Each species produces many more offspring than can survive and reproduce. For example, female fishes lay enormous numbers of eggs. If all of these eggs hatched and the young survived, the ocean would quickly be overrun with fish. Most plants also produce large numbers of seeds. Even giraffes often have more young than can survive.

2. The overproduction of offspring leads to a struggle for existence. The members of any species must compete for food, water, and space in their environment. All of the giraffes must struggle to get enough food, water, and space to live. Some will not be successful and will die before they have young of their own.

3. All organisms of the same species are somewhat different from one another. Individuals are not exactly alike in all of their traits. Except for identical twins, we can easily see this in animals, including humans. For example, some giraffes may be stronger or run faster. Some may have longer necks than others. Humans also have varying traits.

4. Individuals with certain traits have a better chance of surviving and reproducing. Darwin called this *survival of the fittest.* Giraffes with longer necks were better able to survive.

Fig. 20–8 The Marine iguana was one of the animals that Darwin studied on the Galapagos Islands.

Natural selection The process by which only the organisms that are the best suited to their environment survive. As a result, they pass their traits on to their offspring.

Fig. 20–9 Most species produce many more offspring than will survive to adulthood. These are salamander eggs.

They could eat leaves from tall trees that other giraffes couldn't reach.

5. Those organisms that survive and reproduce pass their traits on to their offspring. These offspring also have a better chance of survival. The giraffes with longer necks survive and reproduce. They pass the trait for a longer neck on to their offspring. Their offspring are also better able to survive because they also can eat the higher leaves. Although it may take many, many years, species can change by the passing on of successful traits.

SUMMARY

In 1858, the theory of evolution by natural selection was presented by Charles Darwin and Alfred Wallace. This theory explains how species can change over time. Before their theory, a popular theory of evolution was Lamarck's theory of acquired characteristics.

QUESTIONS

Use complete sentences to write your answers.

1. What are the main points of Lamarck's theory?
2. Why was the voyage of HMS *Beagle* a great influence on Charles Darwin?
3. Briefly state the five parts of the theory of natural selection.
4. What differences would there be between Lamarck's and Darwin's explanation of how the giraffe's long neck evolved?

SKILL-BUILDING ACTIVITY

HYPOTHESIZING

PURPOSE: To practice hypothesizing.

MATERIALS:

blank paper	anything that could
clay	be fossilized, such
pencil	as bones, twigs,
petroleum jelly	and leaves

Background: Paleontologists study plant and animal fossils to know more about prehistoric forms of life. Most often, they do not find the fossil of an entire organism. Therefore, paleontologists must hypothesize what the entire organism looked like. In this activity, you will have an opportunity to hypothesize, or "make educated guesses," about fossils.

PROCEDURE:

Part I:

A. Fig. 20–10 shows four illustrations of fossils. Observe each of these closely. None of these shows the entire organism. Hypothesize what part of the organism the fossil represents. Then hypothesize what the whole organism might have looked like. Write your hypotheses on a separate piece of paper. For some of these, you may have more than one hypothesis.

B. For each fossil, draw a picture of what you think the entire organism looked like.

C. Your teacher will identify each of the fossils. Compare your hypothesis with the information given by your teacher.

Part II:

D. Draw or make a fossil of a part of an organism. You may press a bone, twig, or piece of a leaf into clay. Coat your object with petroleum jelly first. Remove the object and let the clay harden. You may also make a rubbing of leaves, bark, etc.

E. Give the fossil to the teacher. On the underside, label it with the name of the organism and the part it came from.

F. Look at several other fossils created by your classmates.

G. For each fossil, repeat steps A and B.

H. Your teacher will identify each fossil when you have finished. Compare your hypotheses with this information.

CONCLUSIONS:

1. What is a hypothesis?
2. What information is necessary in order to make a hypothesis?
3. Were your hypotheses true or false?
4. How were you acting like a paleontologist in this activity?

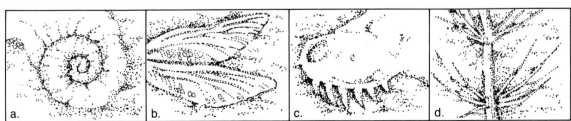

a.   b.   c.   d.

Fig. 20–10

SCIENCE INPUT

Evidence suggests that all forms of plants and animals on the earth today evolved from a few primitive forms beginning approximately three billion years ago. There are methods that scientists use for studying and dating fossil remains which give us a good idea of what animal and plant life existed millions of years ago. One way to compare and contrast different time periods is to draw a time scale, which shows how long or short these times were, relative to one another.

COMPUTER INPUT

The computer does not calculate using the numbers you ordinarily use to solve math problems. Just as it does with words, the computer translates the numbers into binary digits, called *bits*.

The computer is, after all, a machine. It cannot read as you do. It scans a machine language of electronic impulses. All it can actually do is tell if the electronic flow is present (1) or absent (0). The binary number system is one having two parts: 0 and 1.

Program Fossil Record takes advantage of the computer's ability to do quick mathematical calculations to produce the data that will allow you to make a scale model of geologic time periods.

WHAT TO DO

The geologic time scale chart will provide the data necessary for the program. Program Fossil Record will ask you to enter data on the beginning of different geologic periods and epochs. After you have entered all the necessary data,

the program will print out a scale to be used in drawing a geologic time map. The advantage of a scale is that it enables you to see relationships between objects without having to reproduce their actual size. You are probably familiar with the "legends" on maps that tell you what the scale is—for example, one kilometer may be represented by one centimeter. Therefore, if the map distance is 3 cm, the real distance would be 3 km.

Program Fossil Record will print out the measurements you should use for your map in meters, centimeters, and millimeters. Use the larger of the measurements to draw your time scale map.

Geologic Time Scale Chart			
	MYA		MYA
Cenozoic	63	Jurassic	180
Recent	?	Triassic	225
Pleistocene	1	Paleozoic	600
Pliocene	7	Permian	270
Miocene	26	Carbon	350
Oligocene	38	Devonian	400
Eocene	54	Silurian	440
Paleocene	65	Ordovician	500
Mesozoic	225	Cambrian	600
Cretaceous	135	Precambrian	5,000

■ Era   □ Epoch
□ Period   MYA = Millions of Years Ago

GLOSSARY

HIGH-LEVEL LANGUAGE
a computer language, such as BASIC, LOGO, or PASCAL, that uses everyday words to represent computer commands.

LOW-LEVEL LANGUAGE
machine language. It uses electronic data, not words, to execute the program. The binary number system is a low-level language.

PROGRAM

```
100   REM FOSSIL RECORD
110   FOR CS = 1 TO 24: PRINT : NEXT
      CS
120   DIM T$(16),R$(16),M(16),CM(16),
      MM(16)
130   PRINT : PRINT : PRINT "   AFTER
      VIEWING THE N.Y.S. GEOLOGICAL":
      PRINT " TIME SCALE CHART,
      ENTER THE DATA REQUESTED."
135   FOR TD = 1 TO 2000: NEXT TD
140   FOR X = 1 TO 16
150   READ T$(X): NEXT X
155   FOR K = 1 TO 16
160   FOR CS = 1 TO 30: PRINT : NEXT
      CS
170   PRINT K;". HOW MANY MILLIONS
      OF YEARS AGO DID": PRINT "THE
      ";T$(K);" BEGIN?"
175   PRINT : PRINT : INPUT "";R$(K)
180   IF VAL (R$(K)) < 0 OR VAL (R$(K)) >
      5000 THEN GOTO 170
185   IF VAL (R$(K)) = > 1000 THEN M(K)
      = VAL (R$(K)) / 1000
190   IF VAL (R$(K)) < 1000 OR VAL (R$(K))
      = > 100 THEN CM(K) = VAL (R$(K))
      / 10
195   IF VAL (R$(K)) < 100 THEN MM(K) =
      VAL (R$(K))
200   NEXT K
210   FOR CS = 1 TO 24: PRINT : NEXT
      CS
220   PRINT : PRINT : PRINT
      "AGE","    YEARS SCALE"
223   PRINT ,"M.    =CM. =     MM."
225   FOR K = 1 TO 16
230   PRINT : PRINT : PRINT K".
      ";T$(K),M(K)" "CM(K)"     "MM(K)
235   FOR TD = 1 TO 4000: NEXT TD
240   NEXT K
250   DATA     PLEISTOCENE,PLIOCENE,
      MIOCENE,OLIGOCENE,EOCENE,
      PALEOCENE,CRETACEOUS,
      JURASSIC
260   DATA     TRIASSIC,PERMIAN,
      CARBON.,DEVONIAN,SILURIAN,
      ORDOVICIAN,CAMBRIAN,
      P-CAMBRIAN,XXX
270   END
```

PROGRAM NOTES

(1) In data statement 260, CARBON. is short for Carboniferous, a period sometimes called the Pennsylvanian and Mississippian periods. (2) A number of computer terms and concepts have been presented in the COMPUTE! section. Can you apply them in Program Fossil Record? Which statements tell you how many numbers will be in your groups of data? What statement(s) have instructed the computer to print out the scale measurements on the screen or monitor? Which is the READ statement?

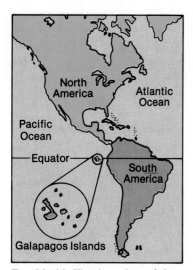

Fig. 20–11 The location of the Galapagos Islands.

20-3. Change Over Time

At the end of this section you will be able to:

☐ Define the term *adaptation*.

☐ Define *mutations* and relate them to differences in a population.

☐ Describe ways in which mutations and environmental changes can affect the formation of new species.

Imagine going on a long trip and arriving in a strange new place. Even the animal and plant life look strange. You find giant tortoises grazing like cattle. There are seafaring lizards and unusual, flightless birds. Yet some other forms of life look strikingly similar to those in other places you have been.

This was the situation Charles Darwin was in when he visited the Galapagos Islands. These more than 14 islands are located in the Pacific Ocean, on the Equator, 1,000 km off the northwestern coast of South America. See Fig. 20–11. Here, Darwin got many of his ideas that led to his important theory.

ADAPTATIONS

Darwin found 14 species of one bird group—the finches. He saw that in many ways these 14 species were all very similar. Yet, there were great differences in both the size and the shape of their beaks. See Fig. 20–12. To Darwin, these finches looked like they had all come from one or a few types of finch that had landed on the islands a long time ago. How could this have happened?

According to the theory of natural selection, there was a variety of beak shapes in the original finch population. These birds competed for the available food on the islands. Some beak shapes allowed birds to eat insects, while others allowed birds to eat nuts and seeds. Thus, the shape of the beak became an **adaptation** (ad-ap-**tay**-shun). An *adaptation* makes the organism more suited to its environment. The beak adaptations allowed some finches to find food at the shore and others to find food on hilltops. Eventually, the finches became so different that they were classified as separate species. This process took place over many years.

Adaptation A trait or behavior that enables an organism to survive in its environment better than similar organisms.

Fig. 20–12 Six of the fourteen species of Galapagos finches.

MUTATIONS

One of the important parts of Darwin's theory is that there are differences among members of the same species. Darwin could not explain the source of these differences. He knew nothing about genetics. Today, however, we know that genes can and do change. These changes often happen by chance. When one or more genes change, the trait that results is called a **mutation** (myoo-**tay**-shun). Traits such as colorblindness and hemophilia are thought to have been the result of *mutations* originally. The seedless orange and the fuzzless peach (nectarine) resulted from mutations. Hundreds of mutations have been found in many types of organisms. Mutations are the source of differences that Darwin could not explain.

If a mutation happens to a gene in an egg or sperm cell, the change can be passed on to the offspring. Whether or not a mutant gene continues to be passed from generation to generation depends upon the environment. If a mutation helps

Mutation A change in one or more genes that can result in a new trait.

Fig. 20–13 Two of these Bengal tigers have a mutation that causes albinism. Would this mutation be helpful or harmful to these animals in the wild?

the organism to survive and reproduce, the change will be passed along. Such mutations become adaptations. According to the theory of natural selection, the giraffe's long neck would be considered an adaptation.

How bacteria sometimes become resistant to antibiotics can also be explained by mutations and natural selection. In any population of bacteria, there may be a few that have mutations which allow them to survive the effects of the drug. These bacteria multiply quickly, since all the other bacteria have been killed. Soon, the disease-resistant bacteria become widespread.

In general, most mutations are harmful to the organism. They make it less able to survive. Organisms with a harmful mutation are less likely to live long enough to produce offspring. Thus, the mutant genes would probably not be passed on. See Fig. 20–13.

ENVIRONMENTAL EFFECTS ON EVOLUTION

Mutations are often helpful to the organism if its environment changes. There are two ways that the environment may change.

An organism may move to a new area by chance. For example, some of Darwin's finches may have been blown by storm winds from the mainland to the Galapagos Islands. Once there, they were *isolated* from their original populations and could not breed with them. Over many generations, new gene combinations and also mutations could cause this population to become a whole new species. In this case, the finches formed many new species.

An organism may move to a new area by choice. Many animals live in a certain area, which they defend against others of their own species. This area is called the animals' **territory.** Within their *territory,* they raise and feed their offspring. As populations increase, territories may get too crowded. Eventually, some individuals must move on to a new area. After several generations, many changes may result within the population of individuals that moved. They may not mate very often with individuals from the old area. Thus the two groups become more different over time. Darwin's finches are also an example of this type of environmental change. These finches are closely interrelated. However, they separated into types that feed and perch at higher or lower levels in the trees and bushes. This adaptation enabled them to move into new territories.

Territory The area in which an animal or group of animals lives and breeds. The area is also defended by the animal or group of animals.

The environment itself can change. An example is described in Section 20–1. In this case, the peppered moth population almost completely changed color in less than 75 years. This was a result of a change in the color of the tree trunks on which the moths rested.

SUMMARY

When Darwin reached the Galapagos Islands, he found how animal species varied on and between each island. He did not know that the source of these differences could be caused by mutations. Mutations and environmental changes can provide the beginning of the evolution of a new species.

QUESTIONS

Use complete sentences to write your answers.

1. What is a mutation?
2. What is an adaptation? How can a mutation become an adaptation? Give an example of each.
3. How can environmental changes cause new species to evolve?

INVESTIGATION

PEPPERED MOTH ADAPTATION

PURPOSE: To use a model showing selection and adaptation.

MATERIALS:

1 sheet black paper	150–200 black circles
(1 m square)	150–200 white circles

PROCEDURE:

A. Copy the data table below on a separate piece of paper. Do this five times—for five generations. Number these data tables 1 through 5.

B. Get a sheet of black paper at least one meter square. Place it on a cleared area of the floor. This will represent the bark of trees that are near factories.

C. You will need a partner. While one student is not looking, the other student will randomly scatter 30 white and 30 black circles on the paper. The white and the black circles will represent the first generation of light- and dark-colored moths. The student who scattered the moths will also act as timekeeper.

D. The other student will act as a bird. This student "bird" will have 15 seconds to gather ("eat") as many moths as possible.

E. At the end of 15 seconds, count the number of light- and dark-colored moths that were gathered. Record the results on your data table 1.

	Dark	Light
Generation (total) Number Eaten Number Survived		

Data Table

F. Subtract the number of moths that were caught from the number of moths in generation 1. Do this for each color. This will give you the number of survivors.

G. Let's assume that the surviving moths reproduced and doubled their numbers. On your data table 2, double the number of survivors from generation 1 to generation 2. On the black paper, the timekeeper will scatter circles of each color exactly equal to the number of survivors. This will double the population for the next generation because the survivors are already lying on the paper.

H. Repeat procedures C through G for at least five generations.

1. You began this activity with equal numbers of light- and dark-colored moths. What is the difference between the numbers of each color now?

2. Which color moth, if any, increased its numbers over the five generations?

3. Which color moth, if any, declined in numbers over the five generations?

4. Why was one color moth more likely to be eaten than the other color?

5. Was either color moth completely safe from the bird?

CONCLUSIONS:

1. Which color moth is better adapted to its environment?

2. How do organisms become adapted?

3. What would happen to the generations of moths if the tree trunks became lighter in color?

RECOMBINANT DNA: A SPLICE OF LIFE

Human genes control traits in humans. Bacterial genes control traits in bacteria. Right? Usually. But what if the bacterial DNA has just a tiny piece of human gene in it? Then, bacterial cells can be made to do things that human cells would do. In fact, this is actually being done—through genetic engineering—that is, by rearranging, or recombining, the DNA sequence on the bacterial gene. Recombinant DNA is DNA that has been changed by adding or removing a gene for a certain trait. In nature, such a change could occur only by mutation. But with recombinant techniques, genes can be modified in the controlled environment of the laboratory.

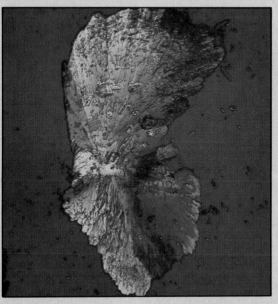

Using recombinant DNA is causing a revolution in the study of genes.

By being able to change the genes in an organism, scientists can redesign plants and animals to suit human needs. Plants could be bigger and more nutritious, and animals more productive, if certain genetic defects were erased. Also, certain microscopic organisms could be made to produce compounds they do not produce naturally but that are useful to humans. For example, yeast cells, which are fungi, can be made to produce human antibodies that fight against the disease hepatitis. The human body fights diseases by producing proteins, called antibodies, that attack specific bacteria or viruses.

Geneticists first isolated the antibody protein healthy people use to fight the hepatitis virus. They figured out the genetic sequence on the human chromosome that caused the production of this antibody protein. They took that sequence and inserted it onto the DNA of yeast cells. The yeast cells were grown in sterile laboratory conditions with plenty of growth medium. Because of the newly inserted gene, the yeast cells started to produce the antibody protein. Since yeast cells grow quickly, they could produce huge quantities of the protein in a short time. This protein is pure and can be used as a vaccine to prevent hepatitis. This was the first time recombinant DNA technique was used to produce a vaccine. However, gene splicing has been used for some time to help cure other human ills.

Why have yeasts and bacteria been used to receive the genes to produce materials that benefit humans? For one thing, their genetic structures are simple. They have fewer genes. The sequence of their genes can be mapped, so that scientists know where to splice the "new" gene. In addition, one-celled organisms reproduce quickly by binary fission—several times in an hour. Because of this, the amount of a product these cells can make "snowballs" rapidly.

There are, of course, many issues being hotly debated in this field. What guidelines should be developed for research and experimentation?

CHAPTER REVIEW

VOCABULARY

On a separate piece of paper, match each term with the number of the statement that best explains it. Use each term only once.

adaptation   natural selection   evolution
fossil   mutation   territory

1. Any trace or remains of an organism that lived in the past.
2. A change in one or more genes that results in a new trait.
3. A trait that allows an organism to survive better in an environment.
4. A definite area in which some animals feed and raise their young and defend the area from other animals of the same species.
5. The process by which organisms best suited to their environment survive and pass their traits on to their offspring.
6. The development or change in living things over time.

QUESTIONS

Give brief but complete answers to each of the following questions. Unless otherwise indicated, use complete sentences to write your answers.

1. What are four ways that organisms become fossils?
2. How could the layers of rock in which a fossil is found help date the fossil?
3. What is a more accurate way to date fossils?
4. How does the example of the Peppered moth provide evidence of natural selection? How did the environment affect the population of Peppered moths?
5. In 1915, nearly all the oysters in a bay of Prince Edward Island, Canada, were killed by a disease. In 1930, the disease was still present, but most of the oysters were resistant to it. By 1938, the oyster harvest was higher than it had been before 1915. Explain how this could happen. In your answer, include mutations and the five parts of the theory of natural selection.
6. What were some of the experiences that influenced Darwin as he developed his theory of evolution by natural selection?
7. How do mutations cause differences in organisms?
8. What must be true of a mutation for it to become an adaptation?
9. What was Lamarck's theory of evolution?
10. On the Galapagos Islands Darwin found a species of cormorant, a large, flightless bird with very small wings. (a) How would Lamarck explain the wing size of these birds? (b) How would Darwin explain the wing size of these birds?

APPLYING SCIENCE

1. Cattle become fatter faster when fed antibiotics. This means higher profits for the meat industry. However, many countries have outlawed the use of antibiotics in cattle feed. Hypothesize why.
2. Dinosaurs ruled the earth for 130 million years. Why might they have become extinct?
3. Why are some insecticides such as DDT no longer effective against some types of insects?
4. What are some adaptations that allow some organisms to live in the desert successfully? In the cold? In high altitudes? In the depths of the ocean?
5. Demonstrate the effect of crowding, or overpopulation, in a territory. Obtain any fighting animals such as crayfish, crabs, beetles, or fishes such as mollies, swordtails, or cichlids. Make a small container that will hold a "crowd" of animals within some larger container (for example, plastic container or jar in an aquarium, vial in a terrarium, box in a room). Be sure that the smaller container has an opening somewhere, such as at the top. Put enough animals in the inner container so that they are crowded. Soon, some animals will leave the container, either as a result of fighting or not. Count the number of animals remaining in the smaller container after 30 seconds. Continue to do this until the number remaining in the smaller container does not change.

BIBLIOGRAPHY

Baggett, James A. "Dinosaurs Ruled the Earth for 130,000,000 Years—Why Did They Die Out?" *Science World,* November 25, 1983.

Cobb, Vicki. *The Monsters Who Died: A Mystery About Dinosaurs.* New York: Coward-McCann, 1983.

The Diagram Group. *A Field Guide to Dinosaurs.* New York: Avon, 1983.

Freedman, Russell. *They Lived with the Dinosaurs.* New York: Holiday House, 1980.

Sattler, Helen Roney. *The Illustrated Dinosaur Dictionary.* New York: Lothrop, Lee & Shepard, 1983.

"Search for the Past." *National Geographic World,* May 1984.

Selkirk, Errol. "Disappearing Dinosaurs." *3-2-1-Contact,* February 1984.

Taylor, Ron. *The Story of Evolution.* New York: Warwick Press, 1981.

It is winter in an evergreen forest. The tree branches are weighed down by the snow. Animals, such as this lynx, prefer to walk under the trees, where the snow is less deep.

BIOMES OF THE WORLD

CHAPTER GOALS

1. List and explain the conditions that determine the *biomes* of the world.
2. Classify *biomes* according to their physical characteristics.
3. Identify the major plant and animal communities of each *biome*.
4. Explain how the communities in a *biome* interact with each other and their environment.
5. Describe the effect of humans on each *biome*.

21-1. Biomes

At the end of this section you will be able to:

- ☐ Describe the factors that determine climates.
- ☐ List the major *biomes* of the world.

Biogeography is a science that deals with where different plants and animals are found. The kinds of animals found in a given area depend on the types of plants found there. The kinds of plants that will grow differs with an area's climate and its type of soil.

CLIMATES OF THE WORLD

The climate of a place is determined mostly by how much sunlight falls on it. See Fig. 21–1 on page 488. Sunlight falling on the equator is direct sunlight. The sun is nearly overhead. Because of the earth's curved shape, sunlight strikes the poles at an angle. This makes the temperature at the poles much cooler than that at the equator.

A band of very warm air extends around the earth at the equator. This is called the *Tropical Zone*. It extends about 25° of latitude both north and south of the equator. Over each pole is a zone of very cold air. Each of these areas is called a *Polar Zone*. Between these areas is a third zone. It is called the *Temperate Zone* because it has a moderate, or temperate, temperature range.

Temperatures do not only get lower closer to the poles. They also get lower as height above sea level increases. The temperatures at the top of a mountain may be quite a bit lower than at its base.

The way the sun's rays strike the earth is not the only important factor that affects climate. Another factor is the amount of moisture. Different combinations of temperature and moisture have produced a variety of climates over large areas of the world.

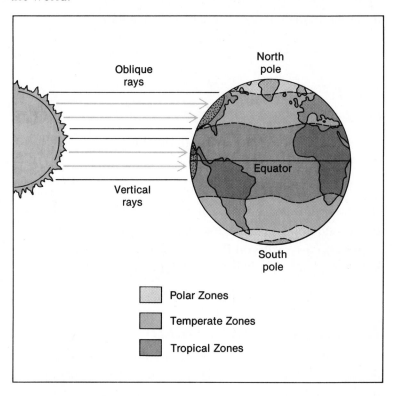

Fig. 21–1 Three major climate zones and how light rays from the sun strike them.

BIOMES

Large areas with similar climate and soil support certain kinds of plants and animals. These large areas are called **biomes.**

Biome A large area with similar climate that supports a certain community of plants and animals.

We will study six biomes of North America. These are: the tundra, grasslands, deserts, and three types of forests—coniferous forests, deciduous forests, and rain forests. Fig. 21–2 shows how these six biomes are distributed around the world. In reality, the boundaries between these biomes are not clear-cut. Scientists do not all agree on how many biomes there are.

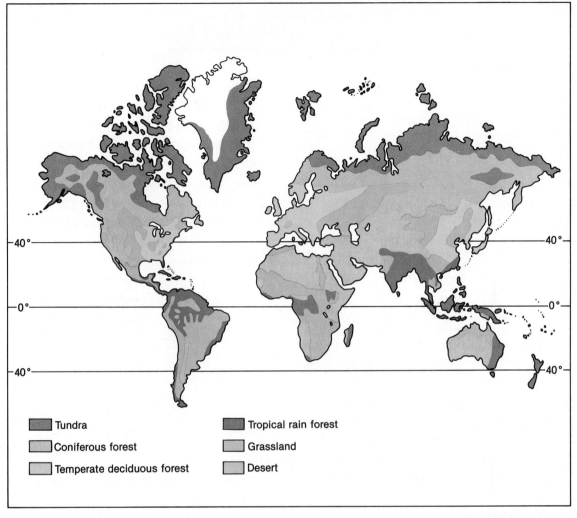

Fig. 21–2 The major biomes of the world.

SUMMARY

Both temperature and amount of moisture affect climate. Climate and soil type both have an effect on the kinds of plants and animals that can live in that climate. A particular climate with its related community is called a biome.

QUESTIONS

Use complete sentences to write your answers.

1. What are two factors that affect the climate of an area?
2. What is a biome?
3. List six major world biomes.

INVESTIGATION

DISTRIBUTING THE SUN'S ENERGY

PURPOSE: To determine the effect of the earth's shape on the distribution of solar energy.

MATERIALS:

1 basketball	1 light socket and
4 thermometers,	stand
each 50°C to	masking tape
− 20°C	200-watt, clear bulb

PROCEDURE:

A. Place a strip of masking tape around one of the seams on the basketball. Keep this seam parallel to the table top. Support the ball so that it will not roll. The masking tape strip will represent the equator.

B. Attach four thermometers to the ball as in Fig. 21–3. The bulb of each thermometer should be at the position shown. The stem will not lie flat. It should be held in place with tape. CAUTION: Do not cover the bulb of the thermometer.

C. Place the light 40 cm from the ball. Make sure it shines directly on position A.

D. With the light turned off, record the starting temperature of each thermometer.

E. Turn the light on. Record the temperature of each thermometer after 5 minutes. Record the temperatures again after 10 minutes. Then turn the light off.

1. Which position had the greatest increase in temperature after 5 minutes? After 10 minutes?

2. Which position had the smallest increase after 10 minutes?

3. Using the 10-minute data, list the positions in order, from highest to lowest temperature.

F. Make a graph of your data. Label the horizontal axis Time (min) and the vertical axis Temperature (°C). Use the readings from thermometer A to make a line graph. Repeat for thermometers B, C, and D.

CONCLUSION:

1. Make a general statement about distance from the equator and temperature. Use your graph and questions 1 through 3 to help you.

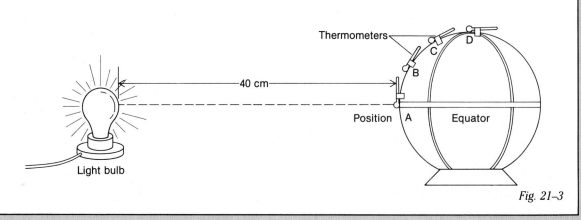

Fig. 21–3

21-2. Forest Biomes

At the end of this section you will be able to:

- ☐ Describe the physical conditions of three types of forest biomes.
- ☐ List examples of organisms from each forest biome.
- ☐ Discuss ecological problems of forest biomes.

There is a rain forest in Colombia, South America, that receives 10 m of rain each year. It is one of the wettest places on earth. Can you imagine a column of water 10 m high?

Forest biomes need plenty of rainfall. They are found all the way from the equator north to the Arctic Circle and south to the Antarctic Circle. The types of trees found in the hot, wet tropics are very different from those in the cold polar regions. Scientists name the three types of forest biomes the rain forest, the deciduous forest, and the coniferous forest.

RAIN FORESTS

Tropical rain forests are found where the days are hot and the rainfall is high. Their average rainfall is 212 cm per year. In some areas the rainfall is evenly spread throughout the year.

Fig. 21–4 Common langurs resting on tree branches in a tropical rain forest.

In other areas there are wet and dry seasons. The hottest rain forests have an average temperature of about 35°C. The coolest ones have a temperature of about 25°C. Since these conditions change little, the growing season lasts from 9 to 12 months.

The hot, wet climate of the rain forest produces the greatest variety of life anywhere on earth. Scientists think of rain forests as being divided into layers. Each layer has its own combination of light, temperature, and moisture conditions. This means that different species of plants and animals are found in different layers.

The top layer is formed by the tops of the tallest trees. This layer may be as high as 60 m above the ground. Here are found eagles, vultures, parrots, monkeys, and insects. Below this layer is the *canopy*. This layer has broad leaves, vines, and branches. The canopy is usually between 35 and 50 m above the ground. The foliage is so dense that as little as 0.1 percent of the sunlight reaches through it to the ground. Different species of birds, monkeys, insects, and snakes live here.

So little light reaches the floor of the rain forest that few plants can grow there. The typical "jungle," seen in the movies, is found only along river banks and in clearings where sunlight reaches the ground. In the lower layers are found wild pigs, armadillos, snakes, and jaguars.

DECIDUOUS FORESTS

Deciduous forests have a temperate climate. There are seasonal changes—the winters are cold and the summers are warm. In the northern parts of this biome, the winter low averages about −10°C. The summer high is about 21°C. The temperature in the southern parts ranges from 15°C to 27°C over the year. The annual rainfall of 75 to 125 cm in deciduous forests is evenly spread over the whole year. The growing season is about six months long. These conditions are good for trees that can store plenty of food during the long growing season. These trees lose their leaves and spend the winter in an inactive state. These are the deciduous trees.

These forests are also called the broadleaf forests. The broad, flat leaves of the trees found there make the most use of the available sunlight for photosynthesis. The trees are mainly beech, maple, ash, basswood, oak, hickory, and elm.

Like rain forests, deciduous forests are also layered. Here,

Fig. 21–5 Raccoons often live in the deciduous forests of North America.

enough light passes through the canopy so that some shorter trees can grow below. Beneath the trees, a layer of shrubs may reach a height of 2 to 5 m. Within a meter of the ground are ferns and wildflowers. This area is sometimes called the herb layer. Finally, there is the forest floor. Here, within a few centimeters of the ground, are mosses, lichens, and mushrooms.

Many different animal habitats may be found in a deciduous forest. Squirrels, for example, nest high in the trees. Shrews and moles burrow beneath the forest floor. Foxes, skunks, raccoons, turtles, snakes, chipmunks, grouse, and mice search the forest floor for food. See Fig. 21–5. Larger animals, such as deer, moose, bears, and lynxes, roam and feed in the shrub layer.

CONIFEROUS FORESTS

Coniferous forests are found across Canada and in the western United States. Conditions in this biome are just right for the cone-bearing trees to thrive. The average rainfall is from 25 to 125 cm per year. The average monthly temperature ranges from a winter low of $-30°C$ to a summer high of $20°C$. The growing season lasts only about two to five months.

The coniferous forest is full of spruces, firs, pines, hemlocks, and cedars. These are the needle-leaved conifers. The needles are small and have a thick, waxy coating. This keeps down the loss of moisture. The shapes of both trees and leaves let them shed heavy snow. Since the trees do not lose all their leaves, they can make food all year. See Fig. 21–6.

Ferns and mosses live in the shade beneath the conifers. Insects are plentiful during the summer. Insect-eating birds, such as chickadees, woodpeckers, and warblers, are common. Some seed-eating birds have beaks that can open the cones on the trees to eat the seeds. Owls, hawks, and eagles are also present. They hunt for small rodents such as shrews, lemmings, and mice.

Deer, moose, elk, caribous, hares, and squirrels find plenty to eat in these forests. They feed on seeds, plants, and needles. Wolves, cougars, black bears, lynxes, foxes, and weasels feed on the plant eaters. Some, like the bears, will also eat berries and other plant food.

Fig. 21–6 Elk in a coniferous forest in Yellowstone National Park.

PEOPLE'S USE OF THE FORESTS

Forests once covered large areas of the earth's surface. But people need the products of forests—such as lumber, firewood, paper, and chemicals. They also need the land for growing food and to live on. Large areas of both coniferous and deciduous forests have already been cut down.

Now the rain forests of the world are being cleared. Some scientists say that an area the size of New England is cleared every year. Rain forest soils are very poor. Attempts to farm the land cleared of rain forests have quickly met with failure. See Fig. 21–7.

Fig. 21–7 The rate at which tropical rain forests are cleared is a worldwide problem. Here, erosion is taking place along a road cut in Costa Rica.

As the forests are cleared, the habitats for forest animals disappear. Thousands of species are now found only in small parts of their former range. Hundreds have become extinct. Many more will follow. Humans are losing much more than they are gaining by cutting down the rain forests.

SUMMARY

Forest Type	Average Rainfall	Average Temperature	Length of Growing Season
Rain forest	212 cm	High: 35°C Low: 25°C	9–12 months
Deciduous forest	75–125 cm	High: 27°C Low: −10°C	6 months
Coniferous forest	25–125 cm	High: 20°C Low: −30°C	2–5 months

Table 21–1

QUESTIONS

Use complete sentences to write your answers.

1. How does climate determine the type of forest biome in an area?
2. Which forest biome has the least rainfall?
3. What tree types are found in each of these forest biomes?
4. How do humans affect forest biomes?

INVESTIGATION

THE BIOMES OF NORTH AMERICA

PURPOSE: To use climate data to determine the approximate boundaries of biomes.

MATERIALS:

pencils   colored pencils
tracing paper

PROCEDURE:

A. Use Fig. 21–8 to answer questions 1–4.

  1. Which parts of North America receive the most rain?

  2. Which parts receive the least?

  3. Which part is the coolest?

  4. Which is the hottest part?

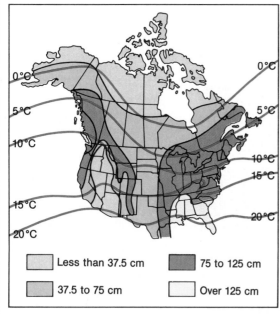

Less than 37.5 cm		75 to 125 cm
37.5 to 75 cm		Over 125 cm

Fig. 21–8

B. Place tracing paper over Fig. 21–8 and trace the outline of the continent.

C. Look at Table 21–2. It shows the average conditions of each biome. Based on the data, there are two areas of the continent dry enough to be tundra. Only one of these is cold enough. That is the northern part of Canada and Alaska.

D. Place the tracing paper in position over the map. Draw a line to show where the southern boundary of the tundra is probably located.

E. Using the data in Table 21–2, draw the boundaries of the desert, eastern deciduous forest, and the grassland biomes. Label and color each biome. The remaining parts of the continent are coniferous forests.

F. Look closely at the two southward extensions of the coniferous forests.

  5. What land features allow the cool temperatures that coniferous trees need to grow this far south?

Biome	Yearly Average Temperature	Total Yearly Rainfall
Tundra	Below 0°C	0–37.5 cm
Desert	10–22°C	0–37.5 cm
Eastern deciduous forest	5–25°C	75–150 cm
Grasslands	5–20°C	37.5–75 cm
Coniferous forest	0–5°C	25–125 cm

Table 21–2

CONCLUSIONS:

  1. Which factor appears to be more important in locating the boundaries of biomes, rainfall or temperature?

  2. In which biome do you live?

21-3. The Grasslands

At the end of this section you will be able to:

- ☐ Identify plants of the grasslands biome.
- ☐ Describe some animal habitats of the grasslands.
- ☐ Explain how the arrival of large numbers of humans changed the grasslands.

The covered wagons of the pioneers were sometimes described as "ships of the prairie, sailing on a sea of grass." The winds cause the grasses to sway and ripple like waves on the sea. This "ocean" once extended from the forests of the east to the Rocky Mountains and from the Gulf of Mexico to central Canada. Great herds of buffaloes once roamed these grasslands. Today, the buffaloes and the grasslands are almost gone. See Fig. 21–9.

Fig. 21–9 This tall-grass prairie in Kansas is an example of a grasslands biome.

THE GRASSLANDS

Look at the map of world biomes in Fig. 21–2 on page 489. Note that the grasslands in the United States are in the same latitude as the deciduous forest to the east and the desert to the west. All three have about the same average annual temperature range. Why, then, are the plant species found in these three biomes so different? The main reason is the amount of rainfall each biome receives.

The map in Fig. 21–8 on page 496 shows the differences in average rainfall. The southwestern deserts get less than 25 cm of rainfall per year. The southeastern coast receives close to 150 cm of rain per year.

In the grasslands, the annual rainfall averages between 25 and 75 cm. This is enough for grasses but not for trees. The western part of the biome gets the least rainfall because it is near the Rocky Mountains. They block rain clouds coming from the west. The eastern part averages much more rainfall. As a result, the types of grasses are different across the biome.

In the east, the growth of tall grasses is favored. They are about 1.5 to 2 m high. At the end of each growing season, the tall grasses die and decay. This helps to make a very rich soil. The tall-grass prairie makes excellent farmland.

Farther west, the tall grasses are replaced by midgrasses. These grasses are 60 to 120 cm in height. Still farther west are the short grasses. Short grasses are usually less than 40 cm high.

Winter on the grasslands can be very harsh. Temperatures can drop well below 0°C. The wind makes it even colder. Although the snow is not heavy, the winds blow it into deep drifts. Many animals move into the forests or mountains. Those that remain must dig through the snow to find food. Smaller mammals and reptiles spend the winter in their burrows.

THE BIOTIC COMMUNITY

There are many habitats in the grasslands. These include the different kinds of grasses, the forest edge, the soil, the humus layer, and the underground burrows. See Fig. 21–10.

Burrows provide protection for small mammals such as ground squirrels, gophers, prairie dogs, and some mice. The burrows are also used for nesting and napping during the winter. Prairie dogs build large communities, called towns, in the short-grass prairie. Empty prairie dog burrows provide homes for ferrets, badgers, snakes, and certain owls.

Most of the burrowing animals are first-level consumers. They eat mainly grass, but many also eat insects. Birds—such as prairie chickens, grouse, and meadowlarks—and jack rabbits, antelope, and buffaloes are also grassland herbivores.

Before the pioneers came, the carnivores of the grasslands

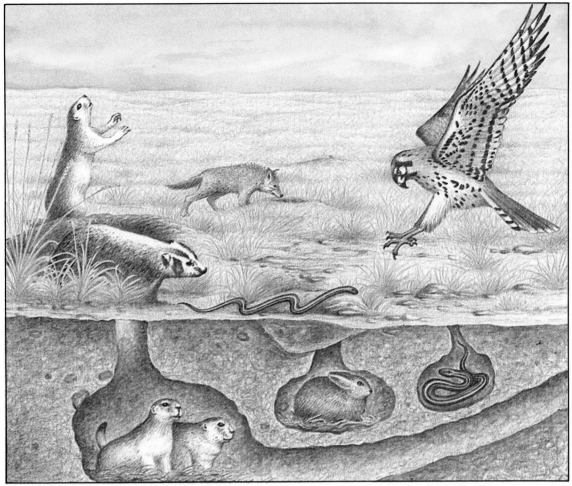

included the ferrets and badgers. They preyed upon the prairie dogs. Other predators, such as snakes, hawks, coyotes, and owls, fed on the smaller animals also. The largest predators were the wolves and an occasional bear. The wolves were the major predators of the buffalo herds. They attacked the weak or injured buffaloes that straggled behind the herds. Many of these species are rare in the grasslands now.

Fig. 21–10 A typical North American grasslands community.

THE CHANGING GRASSLANDS

The coming of the settlers and the railroads during the last century changed the grasslands. Buffalo herds were slaughtered for their meat and hides. Prairie dogs were poisoned to make room for farms and cattle. Coyotes and wolves were hunted. Cattle and sheep were brought in to graze on the grasses. Contained by fences, they overgrazed the land, ruining much of it. Most of the tall grasses were replaced by crops, such as wheat and corn.

Today, only small patches of true grasslands remain. Most of these are in state and national parks. In these areas, attempts are being made to reestablish the grasslands plant and animal communities. See Fig. 21–11. However, the great grasslands, or prairies, of the past are gone.

Fig. 21–11 Periodic fires is one way that natural grasslands maintain themselves. Wildlife specialists, at the Fermi Accelerator Laboratory in Illinois, set controlled fires to maintain a prairie nature sanctuary on the lab's grounds.

SUMMARY

The grasslands biome of the United States had three different types of grasses. The type that grew in an area depended on the rainfall. The grasslands provided food and shelter for many animals. Settlers found the land excellent for growing crops and raising cattle. Most of the grasslands have disappeared as people used them for these purposes.

QUESTIONS

Use complete sentences to write your answers.

1. What are the physical conditions of the grasslands biome?
2. What would happen to the grasslands biome if humans did not interfere and if the rainfall increased over a long period of time? If it decreased?
3. What wild organisms were replaced when settlers came to the grasslands?

INVESTIGATION

MOISTURE AND PLANT GROWTH

PURPOSE: To determine the effects of too little or too much water on plants.

MATERIALS:

potted plants, 3 each of 2 or more species	balance
	metric ruler
	bucket
3 trays, 10 cm or more deep	knife or scissors
	plastic bags and ties

PROCEDURE:

A. Obtain three potted plants of each of two different species. The plants of the same species should be equally developed.

B. Put one pot of each type plant in a tray. Add water to the tray to a depth of 5 cm. Keep the water at this level at all times. Label this tray number l.

C. Set up a second set of pots in a dry tray. Water these plants on a normal schedule. Label this tray number 2.

D. Set up a third set of pots in a third dry tray. Water these plants lightly *only* when they begin to wilt. Label this tray number 3.

E. Grow all the plants for two weeks in direct sunlight.

F. Then, carefully invert one pot, tap the bottom, and remove the plant.

G. Wash the soil from the roots by gently swirling the plant in a bucket of water. CAUTION: Do not wash the plants in a sink or empty the muddy water into the sink.

H. Measure the height of the upper growth from the soil line. On a separate piece of paper, enter the data in a table like the one below. Measure the root length from the soil line to the longest root tip. Enter the data in your table.

I. Repeat these procedures for each of the six plants.

J. Cut off the root material. Determine the total root weight. Enter this in your table.

K. Determine the weight of the upper growth. Enter the data in the table.

CONCLUSIONS:

Under which condition did:

1. the plants grow the tallest?
2. the longest roots develop?
3. the healthiest upper growth appear?
4. the greatest amount of plant matter form?

	Upper Growth							Lower Growth					
	Length			Weight				Length			Weight		
Species	#1	#2	#3	#1	#2	#3		#1	#2	#3	#1	#2	#3

Table 21–3

¡COMPUTE!

SCIENCE INPUT

"Shoreline" ecotones are environmental communities created by the overlapping of terrestrial (coniferous or deciduous forest, grassland, tundra, and/or desert) and oceanic biomes. As with all stable environmental communities, there is a balance that develops among all the consumers, producers, and decomposers. Consumers include carnivores, herbivores, and omnivores. Understanding the food pyramids in the community is important to predicting the future of a biome or ecotone.

COMPUTER INPUT

In Program Timer, you learned that there is a "loop" built into the electronic circuitry of a computer. A "loop" can also be created by program commands that instruct the computer to repeat certain steps. A program having "looping mechanisms" can use BASIC commands, such as FOR/NEXT, GOTO, or GOSUB, to create these repetitions. The looping, also called creating a subroutine, gives the computer user an opportunity to ENTER data again and again without actually writing the instructions over and over. In Program Pyramid, there are a number of "loops." Calculations will be repeated that will cause the computer to print out two different food pyramids for a shoreline ecotone.

WHAT TO DO

Enter Program Pyramid. Be especially careful. Remember that the longer the program, the more possible chances to make errors. Save the program on disk or tape and run it. You will be trying to build a food pyramid for a shoreline ecotone, using as data the seven groups of organisms listed in the chart that represent the levels of the food chains possible for this shoreline ecotone. The program will ask you to enter your choices in the pyramid. If you are incorrect, it will let you know and ask you to try again. As you collect the data for the food pyramid, record it on a chart similar to the Pyramid data chart.

Organisms Chart

Group	Organisms
1	Algae, phytoplankton, protozoa
2	Bacteria, zooplankton
3	Bluefish, herring, sardines, squid
4	Cormorants, sea gulls
5	Crustaceans, insects
6	Mollusks, sea urchins, starfish, worms
7	Tuna, whales

Pyramid Data Chart

Pyramid #1 Organisms	Consumer Level	Pyramid #2 Organisms
	Upper	
	4th	
	3rd	
	2nd	
	Primary Producer	

PROGRAM

```
100  REM PYRAMID
110  DIM G(7),O$(7)
120  FOR X = 1 TO 7: READ G(X),O$(X): NEXT X
130  FOR CS = 1 TO 24: PRINT : NEXT CS
140  PRINT : PRINT "      USE THE GROUP
     LISTINGS OF SHORELINE": PRINT
     "ORGANISMS ON THE SCREEN AND IN THE
     TEXT": PRINT "TO COMPLETE THIS FOOD
     PYRAMID ACTIVITY."
150  PRINT : PRINT : PRINT "      THE
     ORGANISMS REPRESENT PART OF": PRINT "A
     SHORELINE ECOTONE. THE LISTS": PRINT
     "INCLUDE EXAMPLES OF CONSUMERS,":
     PRINT "DECOMPOSERS, AND PRODUCERS.
     THE": PRINT "CONSUMERS ALSO INCLUDE
     CARNIVORES,": PRINT "HERBIVORES, AND
     OMNIVORES."
160  FOR TD = 1 TO 7500: NEXT TD
165  FOR CS = 1 TO 24: PRINT : NEXT CS
170  PRINT : PRINT : PRINT "      INPUT TO THE
     COMPUTER, A NUMBER": PRINT
     "REPRESENTING THE GROUP OF
     ORGANISMS": PRINT "THAT YOU FEEL ARE
     'UPPER LEVEL' ": PRINT "CONSUMERS."
180  PRINT : PRINT : INPUT " ";UC
190  IF UC = 4 THEN GOTO 220
200  IF UC = 7 THEN GOTO 250
210  IF UC < 4 OR UC > 4 OR UC < 7 OR UC > 7
     THEN GOTO 165
220  UL$ = "CORMORANTS & SEAGULLS":CL$ =
     "UPPER":TC$ = "WATERBIRDS":U = 5:AR =
     36521
230  X$ = "CORMORANTS SEAGULLS,BLUEFISH
     HERRING SARDINES SQUID,MOLLUSKS
     URCHINS STARFISH WORMS,CRUSTACEANS
     INSECTS,BACTERIA ZOOPLANKTON,ALGAE
     PHYTOPLANKTON PROTOZOANS,"
240  GOTO 280
250  UL$ = "WHALES & TUNA":CL$ =
     "FOURTH":TC$ = "FISH & MAMMALS":U =
     4:AR = 3521
260  X$ = "WHALES TUNA,BLUEFISH HERRING
     SARDINES SQUID,CRUSTACEANS
     INSECTS,BACTERIA ZOOPLANKTON,ALGAE
     PHYTOPLANKTON PROTOZOANS,"
270  GOTO 280
280  FOR CS = 1 TO 24: PRINT : NEXT CS
290  PRINT : PRINT "      YOU SELECTED ";UL$:
     PRINT "AS ";CL$;" LEVEL CONSUMERS."
300  PRINT : PRINT "      ";U;" OTHER ORGANISM
     GROUPS LISTED ON": PRINT "ON THE SCREEN
     ARE NEEDED TO COMPLETE": PRINT "THE
     FOOD PYRAMID FOR THIS SHORELINE": PRINT
     "ECOTONE EXAMPLE.": PRINT
310  FOR X = 1 TO 7: PRINT G(X);"          ";O$(X):
     NEXT X
320  PRINT : PRINT "      INPUT ";U;" ORGANISM
     GROUP NUMBERS."
330  PRINT : INPUT "";SR: IF SR < > AR THEN
     PRINT "WRONG! TRY AGAIN!": FOR TD = 1
     TO 1500: NEXT TD: GOTO 280
340  IF SR = AR THEN FOR CS = 1 TO 24: PRINT
     NEXT CS
350  PRINT : PRINT "      GOOD WORK": PRINT :
     PRINT "      YOU HAVE CORRECTLY
     IDENTIFIED": PRINT "THE ORGANISM
     GROUPS AND ";TC$: PRINT "AS THE ";CL$;"
     LEVEL CONSUMERS OF THIS": PRINT
     "EXAMPLE SHORELINE FOOD PYRAMID."
355  FOR CS = 1 TO 5: PRINT : NEXT CS
360  S = 1
370  FOR X = 1 TO LEN (X$): IF MID$ (X$,X,1) = ","
     THEN PRINT MID$ (X$,S,X − S):S = X + 1
380  NEXT X
390  FOR CS = 1 TO 4: PRINT : NEXT CS
400  PRINT "      PRESS <RETURN/ENTER> TO
     CONTINUE": PRINT : INPUT "";ZZ$: GOTO 280
410  DATA 1,"ALGAE,PHYTOPLANKTON,
     PROTOZOANS"
420  DATA 2,"BACTERIA & ZOOPLANKTON"
430  DATA 3,"BLUEFISH, HERRING, SARDINES,
     SQUID"
440  DATA 4,"CORMORANTS & SEAGULLS"
450  DATA 5,"CRUSTACEANS & INSECTS"
460  DATA 6,"MOLLUSKS, URCHINS STARFISH,
     WORMS"
470  DATA 7,"TUNA & WHALES"
480  END
```

BITS OF INFORMATION

For those students interested in video games and other uses of computers, there never seems to be enough software. You can exchange and buy software through the Young People's Logo Association (YPLA). YPLA lists programs in a number of computer languages. For membership information, write YPLA, 1208 Hillside Drive, Richardson, Texas 75801.

21-4. The Deserts

At the end of this section you will be able to:

- Describe the physical conditions of the North American deserts.
- Explain how organisms have adapted to the desert.
- Give examples of how humans use and abuse desert lands.

What comes to your mind when you hear someone speak of a desert? An area that is very hot, dry, covered with sand, and without much life? You are partly right. There is more life than you might think, however. In fact, after the winter rains, the desert is full of color. See Fig. 21–12.

Fig. 21–12 The Sonoran desert of Organ Pipe National Monument in Arizona.

THE DESERT BIOME

In general, deserts exist where less than 38 cm of rain falls each year. Many deserts were formed because of high mountains. Look at Fig. 21–13. The general flow of weather across the United States is from west to east. In the western United States, there are several large mountain chains. Moist air moving eastward from the Pacific Ocean must rise up and over the mountains. As the air rises, it cools and loses the water it is carrying. The water falls as rain or snow on the westward slopes. By the time the air crosses the mountains, it has very little moisture left. As it goes down the eastern side, the air

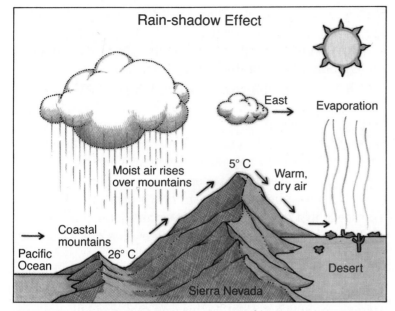

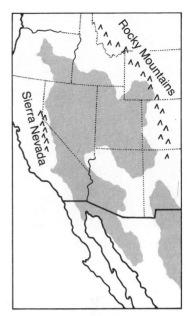

Fig. 21–13 The rainshadow effect (left) of the western mountains produces the deserts of the Southwest (right).

warms up. Water that is in the soil evaporates into the warm, dry air.

For this reason, the area between the Sierras and the Rockies is the driest area in the United States. It receives less than 40 cm of rain a year. Most of the rain falls during thunderstorms.

This area is called the Great Basin. It stretches from the Mexican border north to Oregon and Idaho. There is a great difference in temperature between the northern and southern areas. The northern desert has a yearly average temperature of about 10°C. The southern deserts have an annual average of over 20°C.

These low averages may surprise you. Remember that each of these deserts has a winter season. These cool periods lower the average annual temperature. It is not unusual to find winter snow in the Great Basin. Summer temperatures are very high. For example, Death Valley, California, has recorded temperatures of 57°C in the shade! Temperature changes in the desert vary widely from day to night. It may be 45°C during the day and drop below 0°C at night.

DESERT ADAPTATIONS

The most limiting factor for life in the desert is water. Many desert plants have adapted to collect and store water. For example, the leaves of cactuses have become small, sharp

Fig. 21–14 The spines of this Prickly Pear cactus are its leaves. The large, fleshy part is the stem.

Dormant In an inactive, resting stage.

Nocturnal Describing an animal that is usually active at night.

spines. See Fig. 21–14. Photosynthesis is carried on in the green stems. Water is stored in their tissues. The plants have a waxy covering on spines and stems. Roots go very deep underground. All these things help keep down water loss. The prickly spines also protect the cactus from animals.

Desert shrubs have small, thick leaves with waxy coverings. During dry spells, these shrubs drop their leaves to conserve water. Photosynthesis continues in the stem cells. Desert shrubs have large root systems. Sometimes the roots grow as deep as 30 m in search of water.

Shortly after a heavy rain, many of the desert plants flower. The seeds that develop may lie **dormant** during the long dry spells. When enough moisture is present, they quickly sprout, grow, blossom, and produce new seeds.

Animals have also adapted to the desert conditions. See Fig. 21–15. Many desert animals are **nocturnal.** *Nocturnal* animals are active at night. In this way, they avoid the hot sun. To save body water, reptiles excrete very little water. The kangaroo rat gets all the water it needs from the seeds it eats.

Fig. 21–15 Some members of the desert community. How does each animal keep cool?

The deserts contain many interesting food chains and webs. The producers are the plants already mentioned. The herbivores include mule deer, jack rabbits, wild pigs, bighorn sheep, insects, and birds. The carnivores include other birds, lizards, snakes, foxes, badgers, coyotes, and cougars.

HUMANS AND THE DESERTS

People have learned that desert soils are unusually fertile when water is available. Some desert areas have been irrigated. Pouring water into the desert has a negative side effect, however. The water dissolves the salts found in the soil. When the water evaporates, the salts coat the surface. Eventually, this makes it unfit for growing plants.

Deserts are also rich in valuable minerals. These include borax, gold, silver, and copper. In this country, thousands of people are discovering that the desert—with its clean, dry air—is a good place to live. This means that water must be piped in or drawn from deep, underground wells. Homes, cars, and businesses have to be air-conditioned. But the desert is a fragile biome, and people are often careless about how they affect it. Only careful planning and intelligent use will preserve this harsh but beautiful biome. See Fig. 21–16.

Fig. 21–16 The tracks of off-road vehicles erode the desert surface and create scars that last for many years.

SUMMARY

Deserts exist where less than 38 cm of rain falls. Living things that make the desert their home have adapted in many ways to conserve vital moisture. The desert resources of minerals and rich soil are making this biome more attractive to people. Careless use of the desert can cause many problems.

QUESTIONS

Use complete sentences to write your answers.

1. What causes the formation of deserts?
2. How are desert plants adapted to the lack of water?
3. State two ways animals adapt to desert conditions.
4. How have humans helped cause the expansion of desert areas?

SKILL-BUILDING ACTIVITY

COMPARING AND CONTRASTING

PURPOSE: To compare how desert and woodland plants are adapted to conserve water.

MATERIALS:

small flowering plant	3 large clear-plastic food bags
desert cactus	3 bag ties
	cobalt chloride paper

PROCEDURE:

A. Obtain a piece of cobalt chloride test paper.

   1. What color is the paper?

B. Place a drop of water on the paper.

   2. What color does the wet part of the paper turn?

C. Obtain a cactus plant and a small flowering plant. Place each in a separate clear-plastic bag. See Fig. 21–17.

D. Place a few strips of cobalt chloride paper into each bag. Seal the bags.

E. Place a strip of cobalt chloride paper in an empty bag. Seal this bag also.

F. Place all three bags in bright light for 15 minutes.

3. What is the purpose of the empty bag with the test paper?

4. In which bags did the test paper indicate the presence of the most water?

5. Is there any other evidence of water in the bags? If so, describe it.

6. Where do you suppose this water came from?

7. Which plant would need plenty of rainfall to survive? Explain why.

8. Which plant would survive best in a biome with low rainfall? Explain.

G. Remove the plants from the bags.

9. Which plant exposes the greater surface area to the sunlight?

10. How might this affect the amount of moisture this plant loses?

H. Feel the surface of both plants.

11. Describe any differences in structure between the plants.

CONCLUSION:

1. Summarize how these differences in structure help the cactus survive in the desert biome.

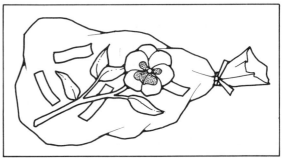

Fig. 21–17

21-5. The Tundra

At the end of this section you will be able to:

- Describe the physical conditions of the tundra biome.
- Name some plant and animal communities of the tundra.
- Discuss some of the environmental problems facing the tundra today.

It is early June. Great herds of caribous are moving northward into the tundra of northern Alaska and Canada. See Fig. 21–18. Here, the females will bear their young. The herds will spend the short summer months feeding on the low plants. In September, the herds will move south into the coniferous forests, where they will spend another winter.

Fig. 21–18 Migrating caribous cross a river in their spring journey to the northern tundra.

THE TUNDRA ENVIRONMENT

Winter temperatures in the tundra can drop as low as −40°C. Summer temperatures may reach 10°C, but on the whole, the tundra is a frigid world. Because of the short, cool, summer season, the tundra soil never completely thaws. Only the top 15 to 30 cm, called the **active layer**, thaws. This is the layer that supports tundra life. Beneath the *active layer* is the **permafrost.** The *permafrost* may be permanently frozen up to 60 m deep. This means that water has no way to drain

Active layer The thin, top layer of tundra that thaws during the short summer.

Permafrost The part of the tundra soil in which the ground water is permanently frozen.

off. For this reason, bogs are common. Rainfall is scarce. The average yearly rainfall on the tundra is less than 37.5 cm.

Further south, conditions similar to the arctic tundra may exist above the tree lines of high mountains. These areas are known as alpine tundras. Alpine tundra is found on the higher peaks of the Rocky Mountains.

LIFE ON THE TUNDRA

The arctic tundra has a very short growing season. It lasts about sixty days. Not too many kinds of plants can live where the temperatures are so low. See Fig. 21–19. In the spring, flowering plants bloom over large areas of the tundra. Mosses, lichens, grasses, and herbs thrive during the brief summer. Woody shrubs of willow, birch, and heath grow in drier areas. These shrubs are much smaller than those that grow in a warmer climate. Many tundra plants grow close to the ground. Here, they find warmer temperatures and are out of the cold, drying winds. The roots of tundra plants grow close to the surface because of the permafrost.

The best known of the large tundra animals are the caribous. These animals are relatives of the deer. Caribous are well-adapted to the tundra conditions. Their thick coats are made of hollow, insulating hairs. Their broad feet function as snow-shoes in winter. They also spread the animal's weight over the soggy summer ground. This makes it easier for them to walk on the tundra.

Small herbivores are common in the tundra. They spend the cold winters in tunnels under the snow. Lemmings, voles, hares, and ground squirrels find plenty of grass and roots to

Fig. 21–19 The tundra in summer.

eat in the tundra. Kodiak, grizzly, and black bears are among the largest of the tundra omnivores. These bears eat shoots, berries, insects, fish, and the kills of other animals.

Coyotes, wolves, and foxes are the main carnivores. Wolf packs often follow the caribou herds. They kill old and sick stragglers as well as calves.

In the alpine tundra, the largest herbivores are the Dall and bighorn sheep. These sure-footed mountain climbers often perch on narrow ledges. Carnivores, such as cougars and wolves, feed on the sheep and smaller herbivores.

Many tundra animals have adapted in special ways. For example, some animals, such as the arctic fox, have white fur in the winter. In the spring, their coat turns dark again. Most of the animals have a much heavier coat of fur in the winter.

Fig. 21–20 Large-scale oil drilling may disrupt the tundra biome.

PROBLEMS OF THE TUNDRA

The early Eskimos and Indians had only a small effect on the tundra. Recently, large deposits of oil, gas, and minerals have been discovered beneath the tundra. See Fig. 21–20. Dams are being built on lakes and rivers so that hydroelectric power can be made. Scientists are concerned about the effects of oil pipelines, drilling, and mining. They do not yet know if the large number of people working there will hurt the biome permanently.

SUMMARY

The tundra is a fragile biome. Its low temperatures and rainfall limit the plant and animal communities. The recent discoveries of oil, gas, and minerals may cause permanent changes in this biome.

QUESTIONS

Use complete sentences to write your answers.

1. What physical conditions make the tundra a harsh biome?
2. Describe some adaptations of animals and plants to the tundra biome.
3. How are conditions high on a mountain in Colorado like those of the tundra of northern Canada?
4. How do humans threaten the tundra biome?

SKILL-BUILDING ACTIVITY

PREDICTING BIOMES

PURPOSE: To use climatic data to predict the type of biome around selected cities.

MATERIALS:
climate data sheet (provided by your teacher)

PROCEDURE:

A. The climate data sheet contains data on temperature and rainfall for ten cities. For each city, the top line gives the average monthly temperature. The second line gives the monthly rainfall.

B. Determine the average yearly temperature for each city: Total the monthly readings that are above 0°C (+ values). Then total the monthly readings that are below 0°C (−values). Subtract the smaller total from the larger. Divide this answer by 12 to get the average yearly temperature.

C. Determine the total annual rainfall for each city by adding the monthly totals.

D. Using these totals and Data Table 21–2 on page 496, determine in which biome each city is located.

1. Which cities are in a desert biome?

2. What causes the large difference between the average yearly temperatures of these desert cities?

3. Which seasons are the wettest in the cold desert? In the hot desert?

4. Which cities are in a tundra biome? During which season does the tundra receive most of its rainfall?

5. Why is July the hottest month in all of these cities?

6. City I has a higher average temperature than city G. Why is city G in the desert while I is not?

E. Fig. 21–21 shows the locations and names of the ten cities. Match each city with the letter of its climate data (from the climate data sheet) and its biome.

7. What geographical feature created the desert west of Salt Lake City?

8. What might happen to Omaha if the annual rainfall increased by 5 to 10 cm?

Key to City Locations

1. Winnipeg, Manitoba, Canada
2. Phoenix, Arizona
3. Boston, Massachusetts
4. Omaha, Nebraska
5. Resolute, Northwest Territories, Canada
6. Anchorage, Alaska
7. Atlanta, Georgia
8. Churchill, Manitoba, Canada
9. Salt Lake City, Utah
10. Huron, South Dakota

Fig. 21–21

CONCLUSION:

1. What abiotic conditions determine the different biomes?

21-6. The Oceans

At the end of this section you will be able to:

- Describe the abiotic conditions found in oceans.
- Name the zones of the ocean and some living things found in each.

In December of 1982, the winds that usually blow westward across the Pacific Ocean at the equator did not arise. With no wind to hold it back, a current of very warm water flowed eastward toward South America. This current is called El Niño (The Child). The warm water it brought was very bad for the rich fishing grounds of the eastern Pacific. See Fig. 21–22.

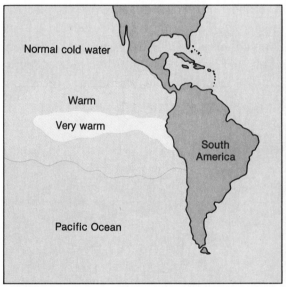

Fig. 21–22 *The El Niño current* (left) *creates very unusual conditions in the ocean of the eastern Pacific, such as flooding along the coast of Peru* (right).

THE OCEAN ENVIRONMENT

There are several physical factors that combine to make different life zones in the oceans. The amount of sunlight and the temperature are the most important. Photosynthesis can only take place in sunlight. Sunlight reaches only about 200 m down into the ocean water. This depth is affected by the *turbidity*, or cloudiness, of the water. If there is a lot of material in the water, not much sunlight will pass through.

Temperatures in arctic waters are much lower than in those at the equator. Ocean currents are very important because they carry water long distances. In this way, cold water is driven into warm water and warm water is brought to cold water. However, temperatures within a life zone seldom change more than a few degrees. Water temperatures over 10°C higher than normal caused the disaster of El Niño.

The ocean contains minerals that make it salty. In addition to minerals, sea water contains many other nutrients. These nutrients are most abundant close to shore. They are brought into the ocean by rivers. There are also areas where deep ocean currents bring the nutrients from the bottom up to the surface. The western coasts of most continents are such areas. These places are very rich in living things. See Fig. 21–23.

The western coast of South America is one such area. When the warm waters of El Niño came close to shore, cold water from the bottom could not reach the surface. Thus, no nutrients were brought up. The numbers of ocean producers were reduced. Food chains and webs were broken. Fishes left. Birds and mammals that usually eat the fishes did not reproduce. The humans that depend on sea life suffered too.

Fig. 21–23 The near shore zone of the ocean is a home for seabirds such as these cormorants.

THE LIFE ZONES OF OCEANS

The oceans can be divided into several major life zones. See Fig. 21–24. The *near shore* zone includes the edges of the land between high and low tidemarks and the shallow offshore water of the continental shelf. The *oceanic zone* includes the open ocean waters where there is enough light for photosyn-

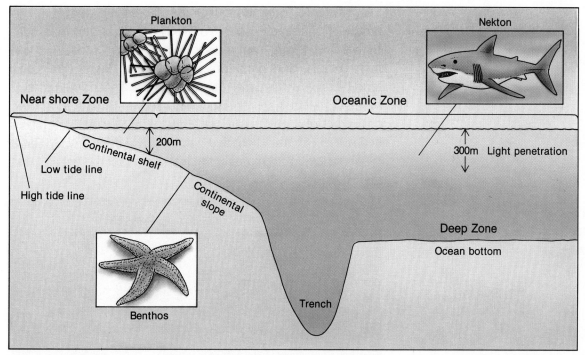

Fig. 21–24 Three major life zones of the ocean biome.

thesis. The depth of this zone varies with the turbidity of the water. The *deep zone* is under the oceanic zone. It extends all the way down to the ocean floor. Conditions in these zones vary greatly.

Life in the sea is more abundant and more varied than anywhere on land. Ocean life is usually divided into three groups. **Plankton** is those life forms that drift with the currents at or near the surface. *Plankton* includes the algae that are the ocean's main producers and the protozoa and larvae that feed on the algae. The **nekton** includes all the free-swimming fishes, reptiles, and mammals. The **benthos** is all those organisms that live on or attached to the sea floor, including seaweed, clams, starfish, crabs, and sea anemones.

The near shore zone has the most life. There is plenty of sunlight. Temperatures are fairly constant. Plenty of nutrients are washed in from the land or brought to the surface by rising currents. Such areas are the most productive for fishing.

The oceanic zone also has enough sunlight for photosynthesis. In areas where currents bring nutrients to the surface, it is very rich in life. Where these nutrients are missing, the oceanic zone can be as barren as a rocky mountaintop.

Plankton Tiny organisms that drift in surface currents or tides.

Nekton Free-swimming organisms, such as fishes, of the oceans.

Benthos Oceanic life forms that dwell on or attached to the ocean bottom.

Organic detritus The remains of dead organisms.

In the cold, dark waters of the deep zone, there are no producers. The major source of food is **organic detritus** (deh-**trite**-us) that falls from above. This is made up of dead animals, plants, and plankton. Some strange living things are found in the deep zone. See Fig. 21–25. These include lanternfish, hatchetfish, viperfish, and squid. The squid are the basic diet of a giant mammal, the sperm whale. Even in the deepest ocean trenches, living things have been found. There is still much that remains to be learned about the oceans and their inhabitants.

Fig. 21–25 A predator of the deep zone.

SUMMARY

Abiotic conditions in oceans are water temperature and the depth to which sunlight reaches. The major life zones in the oceans are the near shore, the oceanic, and the deep zones.

QUESTIONS

Use complete sentences to write your answers.

1. Why do the surface layers of the ocean, or the oceanic zone, contain the greatest numbers of fishes and other living things?

2. What are the three ocean life zones? What are the abiotic conditions in each zone?

3. What types of living organisms would be found in each zone?

TECHNOLOGY

EXOBIOLOGY: MAKING A HOME AWAY FROM HOME

Scientists are at work deciding just what humans need in order to live and work in space. Their field is exobiology—the study of extraterrestrial biology. Since space is cold and airless, an entire life-support system must be created. If orbiting space stations, sometimes called worldships, are to become a reality, they must be livable. They must also be able to sustain themselves without getting constant supplies from earth. To create a livable environment in space, scientists look to actual ecosystems on earth as a model.

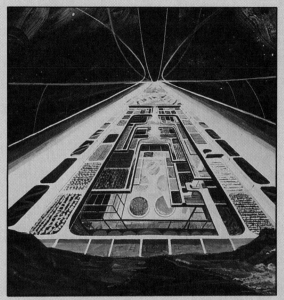

Think of an environment made of interlocking metal modules and platforms moving in orbit high above earth. One design for such a worldship calls for a gigantic rotating wheel in which 10,000 people would live and work. The heavy metal would shield the inside of the wheel from the harmful cosmic rays of space. What would the inside be like? There must be breathable air. An artificial atmosphere would be created that blends oxygen, carbon dioxide, nitrogen, and other gases. This mixture would be stored in tanks. After a while, however, the atmosphere would run out of oxygen and contain too much carbon dioxide. Therefore, plants would have to be included in the space station environment. Plants use carbon dioxide to form carbohydrates, and they split water molecules to form oxygen. Thus, plants would recycle carbon dioxide generated by the people on the space station. They would also be a food source.

Oxygen could also be supplied by tanks of algae. Algae provide a huge percentage of the oxygen in the earth's atmosphere.

In space, growth lights would probably be used. These artificial lights as well as all electricity would come from the same source as it does on spaceship *Earth*—the sun. Solar panels would convert the sun's energy to electricity. There are no cloudy days in space, so the flow of power would be steady. This power would be used to control the climate. Temperature and humidity (around 40 percent) would be comfortable. The water in the air would be part of the water recycled throughout the space station.

Systems that can purify waste water and dispose of harmful waste have already been developed for space shuttle missions. Applied on a larger scale, these same systems could serve the needs of the entire population of a space station.

Our extraterrestrial environment would have an acceptable atmosphere, water, a comfortable temperature, and light, as well as growing plants. Scientists are also exploring the possibilities of keeping animals and fish in space. What are some other factors to be considered in building an extraterrestrial environment? What kinds of problems could occur in a small environment where many people live and work?

CHAPTER REVIEW

VOCABULARY

On a separate piece of paper, match each term with the number of the statement that best explains it. Use each term only once.

permafrost   plankton   nekton   benthos
detritus   dormant   nocturnal   active layer
biome

1. An area with similar climate over a wide area with certain kinds of plants growing in it.
2. Describing an animal that is usually active at night.
3. The permanently frozen subsoil in the tundra biome.
4. The top several centimeters of tundra soil that thaws during the short summer season.
5. Describing a state of inactivity of living things.
6. Small organisms that drift with tides and currents.
7. Free-swimming ocean organisms.
8. Organisms that live on or attached to the ocean bottom.
9. The remains of organisms that fall to the ocean bottom.

QUESTIONS

Give brief but complete answers to each of the following questions. Unless otherwise indicated, use complete sentences to write your answers.

1. Which factors are most important in determining biomes?
2. What are the temperature and rainfall conditions of the biome in which you live?
3. What are the three types of organisms found in the ocean life zones? Give two examples from each group.
4. With which biome is each of the following animals associated? caribou, buffalo, antelope, raccoon, armadillo, squid, black bear
5. List the six land biomes of North America. State the climatic conditions found in each biome.

6. What are several of the uses people have found for the coniferous forest biome?
7. Forests develop where there is plenty of rainfall. Why then are there three different forest biomes?
8. Why are tundra plants mostly small and close to the ground?
9. How does the deciduous forest canopy affect the physical factors of the habitat beneath it?
10. Which of the biomes of North America has the greatest human population? Give reasons for this.

APPLYING SCIENCE

1. Humans are one of the few living things that can survive in any of the land biomes described here. How is that possible?
2. List, in order, the biomes you would expect to meet climbing from a rain forest at sea level to the top of a high mountain.
3. Describe the abiotic conditions facing organisms living on a rocky northern coastline between the high and low tidemarks.
4. Use the organisms discussed in the text to construct a food web for each biome.
5. Construct a desert or deciduous forest terrarium for the classroom. Collect the organisms from their habitats. Remember to maintain the abiotic conditions necessary for the organisms.

BIBLIOGRAPHY

Barrett, Ian. *Tundra and People;* Carson, James. *Deserts and People;* Horton, Catherine. *Grasslands and People.* Morristown, NJ: Silver Burdett, 1982.

Canby, Thomas. "El Niño: Global Disaster." *National Geographic,* February 1984.

Hargreaves, Pat, ed. *The Arctic.* Morristown, NJ: Silver Burdett, 1981.

Simon, Seymour. *From Shore to Ocean Floor: How Life Survives in the Sea.* New York: Watts, 1973.

White, Peter. "Tropical Rain Forests." *National Geographic,* January 1983.

Water rushes down from the White Mountains of New Hampshire. Here, it is clean and pure. What uses will people make of the water as it travels over land to the sea? How will it change?

HUMANS AND THE BIOSPHERE

CHAPTER GOALS

1. Distinguish between renewable resources and nonrenewable resources.
2. Describe the effect of an increasing human population on our natural resources.
3. Identify ways to conserve natural resources.
4. Discuss the causes for the extinction of a species.
5. Explain the causes and suggest some solutions for water and air pollution.

22-1. Conserving Our Natural Resources

At the end of this section you will be able to:

- ☐ Give examples of the use and the misuse of *natural resources.*
- ☐ Explain the need for conservation of *natural resources.*
- ☐ Give examples of *renewable resources* and *nonrenewable resources.*

Since you were born, the population of the world has increased by more than 1.25 *billion* people. This is more than five times as many people as live in the United States today. See Fig. 22–1 on page 522. This increasing number of humans has meant increased demands on their environments.

RENEWABLE RESOURCES

Human societies need a steady supply of materials and energy. Substances taken from the environment are called **natural resources.** They provide us with food, water, clothing, shelter, and energy. Animals, plants, soil, water, and coal are examples of *natural resources.*

Many natural resources are renewable. **Renewable resources** are those that the ecosystem can replace at about the same speed we use them. They include trees, wildlife, crops, livestock, and fishes. Plants and animals reproduce to

Natural resource Any natural substance that humans remove from the environment for their own use.

Renewable resource A resource that is replaced or recycled at the same rate we use it.

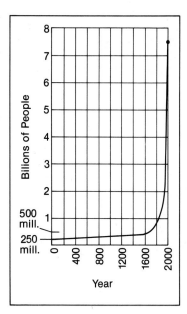

Fig. 22–1 Estimated world human population changes since A.D. 1.

Conservation The protection and wise use of natural resources.

Fig. 22–2 Forests are renewable resources. However, wise management is needed to keep them renewable.

replace their own kind. Nitrates, oxygen, and water are renewed every time they are *recycled* in the biosphere.

Forests are one example of a *renewable natural resource.* Humans harvest the trees. They use the lumber to build homes and to make paper. Many people in North America are now using wood for fuel in wood-burning stoves. This helps them reduce their oil and gas bills.

North America has vast areas covered in forests. The trees grow and reproduce to replace their species. Foresters plant even more seedlings in areas that have been cut over. See Fig. 22–2. We still grow more trees than we cut. But, as our needs increase, we will have to be careful. In other parts of the world, the forests are being cut down at a faster rate. The trees cannot renew themselves fast enough. These forest ecosystems are rapidly disappearing.

Water is one of our greatest resources. Unfortunately, humans have often treated water supplies as waste disposal systems. Beautiful, clean lakes and rivers have become open cesspools. Water stored underground is being used faster than it can be replaced by rain. Some groundwater is being poisoned by chemical wastes. Power plants and some factories use water as a coolant. Water is taken from rivers to cool generators. Hot water is then returned to normally cool waters. Fish and other organisms are killed by the warm water. Others survive but are unable to reproduce. We are becoming more and more aware of the need to take care of water and our other renewable resources. The wise use of natural resources is called **conservation.**

NONRENEWABLE RESOURCES

Nonrenewable resources are those that cannot be replaced at the speed with which we use them. They include metals, minerals, oil, and coal. Soil is also considered a *nonrenewable resource.* This is because it takes over 300 years for one cm of topsoil to develop. Each year about 6,200 kg of good soil per acre of cropland are carried away by wind and water erosion. In the past 300 years, about one third of the soil in the United States has been lost. Better soil *conservation* methods can help reduce such soil losses.

Minerals are also nonrenewable resources. The earth cannot replace the iron, copper, and aluminum we use. Some materials can be recycled. See Fig. 22–3. Many states encourage recycling by charging deposits on metal and glass containers. Does your state have such a law?

Nonrenewable resource A resource that is not replaced or recycled at the rate we use it.

The land itself is a natural resource. It is used for growing crops, building homes and roads, and for recreation. However, the amount of suitable land is limited. When one type of use increases, another type decreases. Urban areas increase by over 3,000 km^2 per year. Other projects, such as roadbuilding, gobble up over 1,000 km^2 of rural land each year.

Fig. 22–3 Bauxite is a nonrenewable resource. It is mined (left) *to get aluminum. Aluminum products can be recycled* (right) *instead of thrown away.*

Coastlines are also important land resources. Tidal marshes are "nurseries" for many species of water life and birds. Bays and sounds are important shellfish habitats. Offshore barrier islands protect these coastal areas from waves and the fury of violent winter storms. Each of these ecosystems is the source of one or more kind of renewable resource.

Fig. 22–4 The oceans have become a source of oil. This offshore oil rig drills into the rock at the ocean bottom.

We build homes near the water. We also use the barrier beaches for recreation. These activities interrupt the area's natural cycles. They also increase erosion. Natural beaches are also being eroded. Marshes are being filled for housing and shopping malls. Oil spills from offshore drilling and tankers threaten beaches and marine "nurseries." See Fig. 22–4. The resources we need from these areas will no longer be renewable.

ENERGY

Fuels such as oil, coal, and natural gas are nonrenewable resources. These substances formed very slowly over millions of years. We are using them at a faster rate than the earth can replace them. Known supplies of oil and gas will be used up in just a few hundred years. The coal supply will last a while longer.

People are also wasteful with energy. Cars with low gas mileage waste gasoline. Many of our throwaway containers are made of plastic. Plastic is made from oil. How much electricity do you waste in one day? Are all the lights, radios, and television sets wisely used in your home? Think of all the energy that factories must use to manufacture these appliances. The decreasing supply and the waste of fuels may be leading us toward an energy crisis.

SUMMARY

People demand more and more of the resources that our environment has to offer. Within limits, some resources are able to replace themselves as we use them. Other resources will never be replaced.

QUESTIONS

Use complete sentences to write your answers.

1. Make a list of the renewable and nonrenewable resources you used today.
2. What are several ways we misuse our resources?
3. What could you personally do to conserve energy? Suggest at least four things.
4. Why is soil considered a nonrenewable resource?

22-2. *Vanishing Wildlife*

At the end of this section you will be able to:

☐ Define the terms *extinct* and *endangered*.

☐ Explain why organisms become *extinct*.

☐ Identify ways to help endangered species survive.

In 1983, the blue pike was officially declared **extinct.** It was a fish long important to commercial and sport fishing in the Great Lakes. The blue pike was a victim of overfishing and pollution.

Until about the year 1800, about one species became *extinct* every 50 years or so. Nowadays, that rate is one species every year. Compare Fig. 22–5 with the graph of human population growth on page 522. What can you infer from this comparison?

WHY DO SPECIES BECOME EXTINCT?

In recent times, humans have been responsible for the extinction of many species. Since 1600, over 130 species of animals have become extinct worldwide. Since Colonial times, over 60 species of animals have become extinct in the United States alone. Those extinct animals include the passenger pigeon, the Labrador duck, the Carolina parakeet, and the heath hen. At present, the U.S. Department of the Interior lists over 760 species of animals and plants worldwide as **endangered** species. These organisms are recognized as being in danger of becoming extinct. About 240 of these endangered species are found only in the United States.

Why do humans cause the extinction of so many species? There are several reasons. The most important one is the destruction of habitats. As humans change the environment to meet their own needs, they change, even destroy, ecosystems. The changing of forests and grasslands to farmlands destroyed many plant species. In this way, the habitat of the heath hen was destroyed. Swamps and coastal marshlands have been filled to build shopping centers and houses. Birds and fishes living in these areas had no place else to go. They became either extinct or *endangered*.

Even now, over 400,000 acres of these wetlands are destroyed each year. As a result, the number of whooping cranes

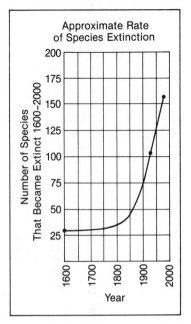

Fig. 22–5 *The increasing rate at which species are becoming extinct.*

Extinct Refers to an organism that no longer exists on earth.

Endangered In danger of becoming extinct.

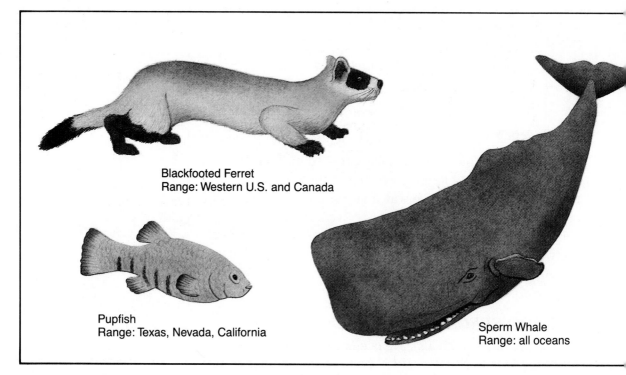

Blackfooted Ferret
Range: Western U.S. and Canada

Pupfish
Range: Texas, Nevada, California

Sperm Whale
Range: all oceans

dropped to only 17 birds by 1945. Owing to protection and careful attention, their numbers have slowly increased. By 1982, there were 114 whooping cranes in the United States.

Other reasons for extinction include overhunting and pollution. Millions of passenger pigeons and American bison were killed for food. The bison, beavers, and American alligators were hunted for their hides or furs. Each year, two to three million ducks, geese, and many bald eagles die of lead poisoning from shotgun pellets.

Still another reason for extinction is the introduction of a foreign species into an ecosystem. At times, this was deliberate. At other times, it was accidental. In some cases, the foreign species became a predator or parasite on the native species. For example, the American chestnut trees have almost been destroyed by a fungus from China. In other cases, the invader competes with a native species. The native species then becomes endangered. Starlings are birds imported from Europe. They have pushed out native bluebirds from their nesting spots. Bluebirds are now very hard to find.

Eventually, any of these pressures may reduce the size of the population to a critical level. The critical level differs for

California Condor
Range: Oregon, California, Baja California

Ocelot
Range: Texas, Arizona, Central America

Franklinia
Range: no longer
grows in the wild

Fig. 22–6 Some endangered species of North America.

each species. Once below this level, the population will probably not recover. There may be too few individuals to breed successfully. Predators can easily wipe out the few that are left. For these or other reasons, the species becomes extinct.

WHY SHOULD WE CARE?

Why should we be concerned if organisms become extinct? Species do become extinct naturally. They are usually replaced by a species better able to survive in the changing ecosystem. If we speed up the rate of extinction, however, we reduce the total number of species. This limits the variety of life on earth. Variations allow life to cope with the changes that are always occurring. Fig. 22–6 shows some endangered species.

Every organism occupies a niche in the food chains of its ecosystem. If we eliminate a species, will another occupy its niche? Or will that food chain collapse and threaten the whole food web? Remember, humans are at the top of many food chains. Are we then endangering ourselves? It has been said that each time we eliminate a species, we lose a little of ourselves. After all, we are all part of the same biosphere.

What is being done to conserve wildlife? Organizations, such as the Survival Service Commission, publish lists of endangered species. They also lobby for protective laws. Governments can set aside wildlife refuges and limit hunting. One of the best ways to protect wildlife is to preserve their habitats. Current U.S. laws require detailed reports when building projects threaten such habitats. Construction may be stopped.

Many such programs have had excellent success. Wild turkeys, once almost extinct, have been restored in 43 states. Atlantic salmon are spawning in New England rivers where they haven't been seen for a century. And the number of nesting pairs of bald eagles has increased from about 480 in 1963 to over 1,400 in 1981. See Fig. 22–7.

Fig. 22–7 Peregrine falcons are another endangered North American predator. These chicks have been incubated and hatched in a breeding program.

SUMMARY

In the last 200 years, species of plants and animals have disappeared from our land at a rapid rate. Many more are in danger of suffering the same fate. In most cases, human activities are responsible for these losses. But, there are things humans can do to slow the extinction of species.

QUESTIONS

Use complete sentences to write your answers.

1. What is the difference between an extinct species and an endangered species?
2. How do humans cause the extinction of species?
3. In what ways do humans destroy habitats?
4. How can humans slow down the rate of extinction?

INVESTIGATION

EXTINCTION OF A SPECIES

PURPOSE: To examine factors that may be involved in the extinction of a species.

MATERIALS:

cardboard box with cover (at least 30 cm by 45 cm)
100 paper organism pieces (50 identified as male, 50 as female) (*Note:*

Extra pieces should be available.)
1 "killer" piece, circular cardboard (about 14 cm in diameter)

PROCEDURE:

A. Copy Table 22–1 below on a separate piece of paper.

B. Place 50 male and 50 female organism pieces in the box. Close the lid and shake the box several times. Keep the box level so the pieces do not pile up in a corner.

C. Place the box on a flat surface. Open the cover carefully. Hold the "killer" piece level, about 60 cm above the center of the box. Drop the killer into the box. If necessary, repeat until it lies flat on the bottom of the box.

D. Any organism piece that is either fully or partly covered by the killer is "dead." It will be removed later. Of course, these "dead" organisms cannot reproduce.

E. Count the number of male-female pairs that overlap by any amount. Enter that number in the table column marked *Births.*

F. Now remove all the organisms that have been killed. Recall that this includes any organism that is either partly or fully covered by the killer. Enter that number in the table column marked *Killed.*

G. Complete *Year 1* in the table. Subtract the number killed from the 100 organisms you began with. This gives the number remaining. Add the number remaining and the number of births. This gives the total number of organisms present to begin *Year 2.*

H. Now place one organism piece for each birth back into the box. Use one-half male and the other half female.

I. Shake the box again. Repeat steps C through H until only one sex of organism remains or until the population becomes extinct.

J. Make a graph of the results using the *Year* and *Total* columns as data.

CONCLUSIONS:

1. What natural processes are represented by the "killer" disc?

2. What happened to the birthrate as the population decreased? Why?

3. If each mating pair produced two offspring, what might have happened?

4. What steps might be taken to prevent extinction of this species?

	Births	Killed	Total
Year 1			
Year 2			
Year 3			

Table 22–1

PROBLEM SOLVING

PURPOSE: To study some effects of heat pollution on fishes. Imagine that you are a biologist for your state wildlife commission. You have been assigned to investigate a fish kill in the Sippiwisset River.

Thousands of dead fish were found downstream from the Nanatuc Power Plant. Most of the fish were herring and perch. There were a few Atlantic salmon, bass, and shad also. The plant does not release any poisons or chemical wastes. However, it does use about 1,134 kL of water from the Sippiwisset River every minute to cool its generators. This water is then pumped back into the river.

As part of your research, you have taken temperature readings of the river near the power plant. See Fig. 22–8 for results.

You know that temperature is an important factor in water ecosystems. One reason is that the temperature determines how much oxygen can be dissolved in the water. As the temperature of the water increases, its ability to dissolve oxygen decreases. See Data Table 22–2. If the temperature gets too high, the fish do not get the oxygen they need and thus die. The temperature at which this happens differs from species to species. Data Table 22–3 lists this temperature for certain species.

MATERIALS:

drawing paper   colored pencils

PROCEDURE:

A. Trace the diagram in Fig. 22–8 on a separate piece of paper.

B. Shade in the parts of the river that have temperature ranges of 37°C, 36°C, and 35°C. Use a different color for each temperature zone. The temperature zones in the cooling canal have already been shaded. Using your sketch and the information in Data Table 22–3, answer the following questions.

   1. Examine Data Table 22–2 carefully. As the water temperature rises, what happens to the level of dissolved oxygen?

C. The river's temperature 1 km upstream from the plant is about 15°C.

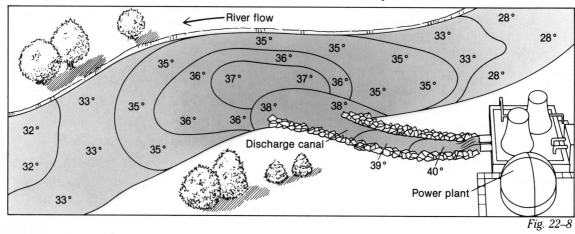

Fig. 22–8

AMOUNT OF OXYGEN THAT DISSOLVES IN WATER	
Temperature (°C)	Maximum Amount
0	14.6
5	12.7
10	11.3
15	10.1
20	9.1
25	8.3
30	7.5
35	6.9
40	6.5

Table 22–2

2. What is the usual level of dissolved oxygen (D.O. level) in the river?

3. Which of the species of fish listed in Data Table 22–3 could live in the part of the river pictured?

4. Can fish survive in waters with temperatures and D.O. levels that are not in their "preferred ranges"?

5. What two effects would the water from the discharge canal have on the river?

6. Why would any Atlantic salmon be in the river at all?

7. How might the hot water interrupt the life cycle of the salmon?

8. Which fish could probably survive in the part of the river just downstream from the discharge canal?

9. Which fish could survive in the discharge canal itself?

CONCLUSIONS:

1. What appears to be the probable cause of the fish kill?

2. Why were most of the dead fish herring and perch?

3. Would the warm water affect any other organisms in the river?

LIMITING FACTORS FOR CERTAIN FISHES			
Fish	Preferred Range	Deadly Temperature	Preferred Dissolved Oxygen Range
Herring	19–22°C	33°C	10^+ ppm*
Perch	20–22°C	33°C	10^+ ppm
Atlantic salmon	13–16°C	34°C	10^+ ppm
Bass	28–31°C	37°C	10^+ ppm
Shad	23–25°C	38°C	7^+ ppm
Carp	31–33°C	40°C	2^+ ppm
Shiners	19–21°C	34°C	10^+ ppm
Killifish	16–28°C	34°C	8^+ ppm

*ppm = parts oxygen in a million parts of water

Table 22–3

22-3. Polluting the Biosphere

At the end of this section you will be able to:

- ☐ Identify some causes of air and water pollution.
- ☐ Discuss the effects of pollution on living things.
- ☐ Describe ways to reduce pollution.

From a distance, the picture in Fig. 22–9 looks perfect. The dark green of the tall trees surrounds the deep blue of the crystal-clear lake. But everything is not perfect. There are few birds or frogs around. And the fish that once brought people by the dozens to catch them are almost gone. This lake, like thousands of others in eastern North America, is dying. It is being killed by acid rain.

Fig. 22–9 Many lakes in the Adirondack Mountains of New York are suffering from acid rain (left). Pollution from heavy industries to the west may be the cause (right).

ACID RAIN

Acid rain is now a worldwide problem. From South Africa to the Arctic; from California, through Europe, to China, scientists are reporting examples of its effects. Forests are dying. Lakes no longer support life. Ground water in some areas is acid enough to dissolve water pipes. This releases poisonous lead and copper into the drinking water. In other areas, acid rain dissolves soil nutrients. This reduces crop production. Acid rain also damages wood, paint, plastics, and many building stones and metals. It may also be related to some human health problems.

Acid rain is a form of both air and water pollution. What causes it? When various gases, such as carbon dioxide (CO_2), sulfur dioxide (SO_2), and nitrous oxide (N_2O), are dissolved in rainwater, weak acids develop. Many natural events produce these acids. Lightning, forest fires, and volcanoes all add some acids to the atmosphere. Therefore, rain is always slightly acid. However, many human activities—in particular the burning of coal and oil—produce large amounts of these gases. Automobiles, power plants, and other industries contribute large amounts. In industrial countries, as much as 95 percent of these **pollutants** may be from human activities. One study indicated that rainfall in the eastern United States is about 40 times more acid than "normal."

Because the weather knows no political boundaries, this problem is international. Gases produced in the central U.S. may form the acid rain that falls on eastern Canada. Canadian industry may be contributing to the problem in the eastern United States. Forests in Scandinavia show effects of pollution from Germany, and so on.

Because the problem is international, the solution must be also. Add to this the fact that not everyone agrees about the seriousness of the acid rain problem. Measuring these *pollutants* and their effects is difficult. It is hard to determine their sources. That is why it is so difficult to pass laws to control the problem.

Pollutant Any material, such as gases, particles, and chemicals, released into the air or water.

HAZARDOUS AND TOXIC WASTES

Americans produce a staggering 4 trillion kg of waste products each year. The garbage you throw away each day is part of this waste. So are the wastes of agriculture, mining, industries, schools, and businesses.

Disposing of these wastes has presented many problems. Many communities are unable to handle the enormous volume. Some of this waste is burned in incinerators. Some is recycled. But most is dumped. Some goes into landfills, some into the oceans. Some of it is simply discarded anywhere. Waste products are the main sources of the pollutants of our air and water.

More than 50 billion kg of that total waste are considered **hazardous wastes.** Wastes are called hazardous if they can

Hazardous waste Any waste that is capable of causing physical, chemical, or biological damage.

Toxic waste Any substance that is poisonous to living organisms.

be set on fire or react with other substances to produce other dangerous materials. *Hazardous wastes* are also those that can damage pipes and containers, or are poisonous to living things. This last group is often called **toxic wastes.**

Some of the most serious problems are created by the disposal of hazardous and *toxic wastes.* Radioactive wastes are extremely toxic. They are created by nuclear power plants, hospitals, and scientific research. A safe way to dispose of them has not yet been found. In 1983, radioactive material from old hospital equipment was accidentally recycled into steel rods and table legs. These products were then shipped to 40 states before they were discovered.

Fig. 22–10 At this site, low-level wastes from nuclear power plants are disposed of by burial under soil.

Many toxic chemicals are stored in secure dumps. See Fig. 22–10. But many other dumps either leak or are broken open accidentally. The poisons they contain can then be flushed out by rainwater. This happened in Love Canal, N.Y., in 1977. Sometimes toxic chemicals are illegally dumped into streams or sprayed along roadsides. Soil in North Carolina, Missouri, and other states has been contaminated in this way.

Cleanup efforts of the past few years have helped improve the quality of the air we breathe. Rivers and streams are also getting cleaner. There are relatively few of the "open sewer" type of rivers left in the United States. But the problem has not gone away. Much of it has gone underground. Forbidden to be dumped into rivers, many chemicals were instead pumped deep underground. It was assumed the earth could disperse these materials safely.

But that did not happen. The U.S. Environmental Protection Agency has determined that several thousand of the 22,000 waste disposal sites in the United States are probably dangerous. Water in Vermont shows evidence of a substance called chloroform. Wells in Florida and New Jersey are threatened by poisonous chemicals. Ground water in California and other states has been contaminated with pesticides.

WHAT CAN WE DO?

One of the first steps is already being taken. Laws are being passed at local, state, and national levels to control all types of pollutants. Automobiles and smokestack emissions, transport and disposal of hazardous wastes, plus clean air and water standards are all part of this legal approach. The important part is enforcement. See Fig. 22–11.

New ways to use and dispose of wastes are being developed. In some cases, the wastes of one process can be used in a different process. High-temperature furnaces—some on ships at sea—and new chemical processes to neutralize toxic wastes may help get rid of some wastes. New manufacturing processes that create less waste are also parts of the answer to cleaning up our environment.

Fig. 22–11 Many states require the exhaust of motor vehicles to be tested for pollutant levels. Those that give off too many pollutants must be repaired.

SUMMARY

As population and industries grow, pollution becomes more of a problem. Poor methods of disposal of toxic and hazardous wastes have created health problems. Although the quality of the air and water has improved somewhat recently, much remains to be done.

QUESTIONS

Use complete sentences to write your answers.

1. What are some causes of acid rain?
2. What are some effects of acid rain?
3. What is a hazardous waste?
4. What are some of the solutions to the pollution problems being investigated now?

INVESTIGATION

STUDYING ACID RAIN

PURPOSE: To determine the acidity of local water sources and its effect on organisms.

MATERIALS:

small jars with covers	uncooked rice and wheat
large collecting jar	microscope
wide-mouth funnel	glass slides
litmus paper (blue and red)	medicine dropper
pH test papers (wide-range and short-range)	mixed protozoan culture

PROCEDURE:

A. Collecting jars and covers should be thoroughly washed and rinsed. A final rinse with distilled water is recommended.

B. Collect water samples from home, school, a nearby stream or pond, and rainfall. To collect rainfall, place the funnel in the large jar. Place in a safe but open place. Transfer water to smaller jars for testing after each rain or snowfall.

C. Label each sample jar with the name of its source. Copy Table 22–4 below. Provide a line in your table for each source.

D. Cut separate 3 cm strips of red and blue litmus paper for each sample. Dip one end of each into the sample. If red litmus turns blue, sample is basic. If blue litmus turns red, sample is acidic. If there are no changes, the sample is neutral.

E. Test each water sample with red and blue litmus. Enter results in the table.

   1. Which water sources were acidic? Which were neutral?

F. Test each sample with a 3-cm strip of wide-range pH paper. Then use the short-range paper to establish the pH more accurately. Enter the pH in the table.

   2. Which source was the *most* acidic? (Which had the lowest pH reading?)

   3. Which source was the *most* basic? (Which had the highest pH reading?)

G. After testing is complete, add 1 grain each of uncooked rice and wheat to each sample with a pH of 7.0 or less. Add one dropper full of a mixed protozoan culture to each of these samples.

H. After 24–48 hours, make wet mounts from each sample. Examine them for microorganisms and record the results in the last column of the table.

   4. In which pH ranges did the microorganisms die?

   5. What might happen to food supplies and fishes in an acidified lake?

Water Source	Litmus Results	pH Results	Organisms Survived?

Table 22–4

CONCLUSIONS:

1. Is there an acid rain problem in your area?

2. Are surface waters in your area acidic?

TECHNOLOGY

A DIFFERENT KIND OF TRANSPLANT

For some animals born in zoos, the family resemblance is surely hard to see. That was the case of a baby zebra born to a mother horse. This was not a case of mistaken identity. It was really a case of switching "homes" for a week-old zebra embryo. The tiny embryo was taken from its mother's uterus and surgically inserted into a surrogate mother's uterus. (A *surrogate* is a substitute.) The horse and zebra are two different species, but they are closely related. Once transplanted, the zebra embryo grew larger, using nutrients coming through the surrogate mother's placenta. The infant zebra was then delivered normally. Why was this transplant attempted? One reason was to develop a technique to save rare animals that are near extinction. If embryos of vanishing animals can be successfully transferred to more common but related animals in a zoo, chances of survival are quite good.

Take, for example, the case of the bongo, a rare African antelope. Several bongo embryos were taken from their mothers in one zoo, then flown cross-country to another zoo. There, after only about ten hours between mothers, the embryos were implanted into foster mothers who were a more common variety of African antelopes. The result? A light-brown mother gives her darker, striped infant a gentle nudge. The two animals are both antelopes, but they are classified in two different genera.

Are animals born to a mother of a different species affected by this procedure? Pepi, who belongs to a species of relatively small monkeys, was the result of an embryo transfer from one species to another. His mother was a rhesus monkey, a larger species. Pepi weighed more than any baby of his species ever recorded. Scientists are studying him to see if there are any other factors, besides size, that his surrogate mother might be contributing to his growth and development.

Some implanted animals seem to get along just fine. This is the case with the Gowers from India, the world's largest cattle. At the Bronx Zoo in New York, a tiny Gower embryo was implanted in a surrogate mother—a Holstein cow. The Gower calf was born normally. Now it is the father of two baby Gowers.

Embryo implants may have applications other than saving vanishing species. For example, a farmer might want to add to or improve a herd of livestock, but might not want to transport large animals over a long distance. Instead, a cow embryo could be purchased and implanted into a smaller animal, like a rabbit, This rabbit is more easily transported, and would, upon arrival, give up the embryos that have hitched a ride with her. The embryos could then be transferred into foster mother cows. Can you think of other ways that embryo transplants could be used in research or in farming?

CHAPTER REVIEW

VOCABULARY

On a separate piece of paper, match each term with the number of the statement that best explains it. Use each term only once.

renewable resource   nonrenewable   toxic wastes   hazardous wastes
extinct   resource   endangered   natural resource
pollutant   conservation

1. A resource that is not replaced at the speed with which we use it.
2. Any material released into the air or water that does not belong there.
3. No longer existing on earth.
4. The protection and wise use of natural resources.
5. Pollutants that are poisonous to living things.
6. Wastes that can cause physical, chemical, or biological damage.
7. Any natural substance that humans remove from the environment for their own use.
8. A resource that is replaced at about the same rate that humans use it.
9. In danger of becoming extinct.

QUESTIONS

Give brief but complete answers to each of the following questions. Unless otherwise indicated, use complete sentences to write your answers.

1. What is the difference between a renewable and a nonrenewable resource? Give three examples of each.
2. Soil is continually being formed by natural processes. Why then is it considered a nonrenewable resource?
3. In what ways does our increasing population threaten our land resources?
4. List three ways in which people might protect an endangered species from extinction.
5. How might clothing fashions result in the extinction or endangering of species?
6. Why are coastal areas an important natural resource?
7. How does the expansion of urban and suburban areas contribute to the extinction of wildlife?

8. Explain the term conservation. Why is it necessary to conserve our natural resources?
9. How could using the oceans as dumps for our garbage and waste endanger that environment?
10. How are the "hidden pollutants" of acid rain and groundwater causes of contamination?

APPLYING SCIENCE

1. Heating homes with wood shifts energy consumption from a nonrenewable to a renewable resource. But what problems could this switch create?
2. DDT is a very long-lasting, toxic pesticide that was once used extensively on crops in North America. Later it was found in the body tissues of penguins in Antarctica. Explain how it might have gotten there.
3. Make a class bulletin board display. As a background, sketch a large equal-arm balance. Label the display: The Environment Hanging in the Balance. On one side, place articles dealing with conserving natural resources. On the other side, put articles dealing with misuse of natural resources.
4. Find out if there are any wildlife refuges in your state. If so, describe the area and the species of organisms there. Determine if any are endangered and if your state protects them.

BIBLIOGRAPHY

"Environmental Quality Index—(Current Year)," by the editors of *National Wildlife Magazine*, February/March (current year).

Harris, John, and Aleta Pahl. *Endangered Predators.* New York: Doubleday, 1976.

Hornblower, Margot. "How Dangerous Is Acid Rain?" *National Wildlife Magazine*, June/July 1983.

Pringle, Laurence. *What Shall We Do with the Land? Choices for America*. New York: Crowell, 1981.

Sitwell, Nigel. "Our Trees are Dying." *Science Digest*, September 1984.

APPENDIX A: METRIC UNITS

METRIC SYSTEM PREFIXES

Greater than 1:
kilo (k) = 1,000
hecto (h) = 100
deca (dk) = 10
Less than 1:
deci (d) = 0.1
centi (c) = 0.01
milli (m) = 0.001
micro (μ) = 0.000 001

METRIC-ENGLISH EQUIVALENTS

1 m = 39.37 inches
2.54 cm = 1 inch
1 km = .621 mile
1.61 km = 1 mile (5,280 feet)
1 kg = 2.2046 pounds
453.6 g = 1 pound
28.35 g = 1 ounce
1 L = 1.06 quarts
1 mL = 0.00106 quart

COMMONLY USED METRIC UNITS

Length: The basic unit of length in the metric system is the meter (m), which is slightly more than the height of a doorknob from the floor.
Examples: 1 kilometer (km) = 1,000 m
1 centimeter (cm) = 0.01 m
1 meter (m) = 100 cm
1 millimeter (mm) = 0.001 m
1 meter (m) = 1,000 mm

Mass: The basic unit of mass in the metric system is the gram (g), which is equal to the mass of a paper clip.
Examples: 1 kilogram (kg) = 1,000 g
1 milligram (mg) = 0.001 g
1 gram (g) = 1,000 mg

Volume: The basic unit of volume in the metric system is the liter (L), which is slightly more than four times the volume of one cup (8 ounces).
Examples: 1 milliliter (mL) = 0.001 L
1 liter (L) = 1,000 mL

APPENDIX B: USING A METER STICK OR METRIC RULER

When using a meter stick or metric ruler, keep the following procedures in mind:

1. Place the ruler firmly against the object you are measuring.
2. Line up a numbered line on the ruler with one end of the object. This number is the **first reading.** See Fig. B–1. The first reading is 1 cm.

3. The **final reading** is taken where the other end of the object lines up with the ruler. This reading may not be exactly on a numbered line on the ruler. See Fig. B–1. The final reading is 5.4 cm.
4. Subtract the first reading that you obtained from the final reading.

5.4 cm − 1.0 cm = 4.4 cm

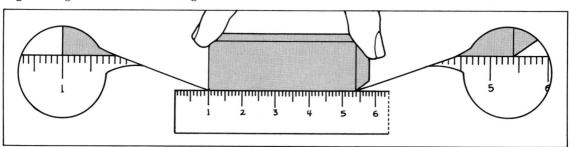

Fig. B–1.

APPENDIX C: USING A GRADUATED CYLINDER

The volume of a liquid is measured with a graduated cylinder. See Fig. C–2. When water and most other liquids are put into a glass graduated cylinder, the surface of the liquid is curved. In making volume measurements, it is important to read the mark closest to the bottom of the curve. See inset. The reading shown is 19.4 mL. To take the reading, make sure the cylinder is on a flat, level surface. Move your head so that your eye is level with the surface of the liquid.

Today, some graduated cylinders are made of plastic and do not show this curve. These should simply be read at eye level.

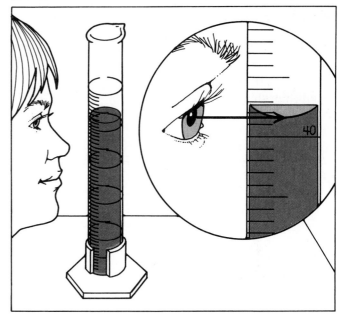

Fig. C–2.

APPENDIX D: HOW TO READ A LINE GRAPH

Fig. D–3 is a line graph. The following directions will help you interpret such graphs.
1. The horizontal line is the x-axis. It represents time in years.
2. The vertical line is the y-axis. It represents the numbers in thousands of hares and lynxes.
3. How many hares were there in 1885? To answer this question, you must first find the date 1885 along the x-axis. When you find it, then move straight up until you reach the solid line. From this point, look straight across to the y-axis. There you can read the number of hares in thousands.

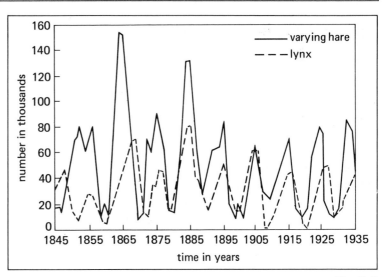

Fig. D–3.

APPENDIX E: USING A BALANCE

A balance is an instrument used in determining the mass of an object.

First, a balance must be checked with no mass on it. It should be placed on a level surface and adjusted until the indicator shows that both sides are level.

The balance should always be moving a little when it is used. When the pointer moves equally on both sides of the center of the scale, the instrument is "IN BALANCE." See Fig. E–4. The equal-arm balance in Fig. E–4 is based on the fact that the right and left sides of the instrument are the same size and shape. To use it:

1. Place the object of unknown mass on the left pan of the balance.

2. Place known masses on the right pan until the pointer again moves equally on both sides of the center of the scale.

3. If your balance has a numbered scale and a rider on that scale, you can use the rider to help balance the unknown mass.

4. The unknown mass is equivalent to the total of the known masses and the rider reading.

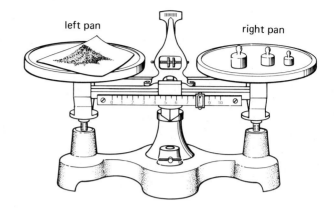
left pan   right pan

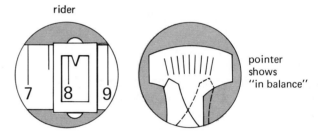
rider

pointer shows "in balance"

Fig. E–4.

Another kind of balance is the triple-beam balance shown in Fig. E–5. It has one pan and three movable riders. To use it:

1. Place the object of unknown mass on the pan.

2. Move the riders until the pointer moves equally on both sides of the center of the scale.

3. The unknown mass is equivalent to the total of the readings of the riders.

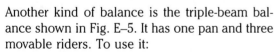

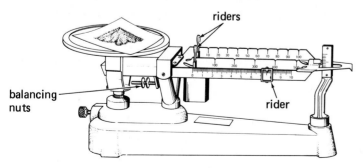
riders

balancing nuts

rider

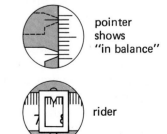
pointer shows "in balance"

rider

Fig. E–5.

GLOSSARY

A

Abiotic. Refers to the nonliving materials and energy in an ecosystem.

Absorption. The diffusion of water and dissolved materials into cells.

Acquired immunity. Resistance to reinfection with a disease after the body has once recovered from the disease.

Active layer. The thin top layer of tundra that thaws during the short summer.

Active transport. The process by which a cell must use energy to move materials into or out of the cell.

Adaptation. A trait or behavior that enables an organism to survive in its environment better than similar organisms.

Addiction. A physical and/or psychological dependence on a drug.

Algae. Plantlike protists that contain chlorophyll and make their own food.

Alveoli. Microscopic sacs in the lungs in which gas exchange takes place. (Singular: alveolus.)

Amino acids. About 20 different organic compounds that form proteins.

Angiosperms. Vascular plants with seeds enclosed in a protective organ.

Antibiotic. A chemical, produced by living things such as fungi, that is used to kill some disease-causing bacteria.

Antibodies. Chemicals made by the body to fight germs or other foreign bodies.

Anus. The opening of the digestive system through which solid wastes are passed out of the body.

Artery. A blood vessel that carries blood away from the heart. (Plural: arteries.)

Arthropod. An invertebrate animal that has jointed legs and an exoskeleton.

Asexual reproduction. Reproduction that requires only one parent.

Association neuron. A neuron that carries impulses from sensory neurons to motor neurons.

Atmosphere. A layer of gases surrounding a planet.

Atom. The smallest particle of an element that has the characteristics of that element.

Atria. The two upper chambers of the heart that receive blood from the veins.

B

Bacteria. Very small organisms with a simple cell structure and no nucleus.

Benthos. Oceanic life forms that dwell on or attached to the ocean bottom.

Biome. A large area with similar climate that supports a certain community of plants and animals.

Biosphere. That area at or near the earth's surface where life can exist.

Biotic. Refers to the living organisms in an ecosystem.

Birds. A group of vertebrate animals that have feathers.

Blood pressure. The amount of force caused by blood pushing on the artery walls.

Bone. A hard, living tissue made up of bone cells and deposits of calcium and phosphorus compounds.

Bronchi. Two breathing tubes that branch out of the trachea and lead to the lungs. (Singular: bronchus.)

Bronchial tubes. Large breathing tubes that branch out from the bronchi.

Budding. A type of asexual reproduction in which an outgrowth on the organism develops into an entirely new organism.

C

Cancer. A group of cells in the body in which there is loss of control of cell division and function.

Capillaries. Small blood vessels connecting arteries to veins.

Carbohydrates. Organic compounds, such as sugars and starches, that are sources of energy.

Cardiac muscle. An involuntary muscle found only in the heart and made up of tightly woven cells.

Carnivore. A consumer that eats only animal tissue.

Carrier. Someone who can spread a disease without showing symptoms of the disease.

Cartilage. A firm but flexible tissue that gives shape and support to the bodies of some animals.

Cell. The smallest organized unit of living protoplasm.

Cell division. The process by which a cell forms two new cells.

Cell membrane. The part of the cell that determines what enters and leaves the cell.

Cell wall. The rigid protective layer that surrounds the plant cell.

Central nervous system. Made up of the brain and spinal cord.

Cerebellum. Area of the human brain between the cerebrum and the medulla. It coordinates muscular activities and body balance.

Cerebrum. The largest region of the human brain. It is the center of human intelligence.

Chemical digestion. The process in which our foods are chemically broken down into simpler substances.

Chitin. The semihard compound that forms an exoskeleton.

Chlorophyll. Green material in green plants that uses sunlight to make food.

Chloroplasts. Oval-shaped structures containing chlorophyll.

Chordate. An animal having a nerve cord and gill slits at some time in its development.

Chromosomes. Rod-shaped structures found in the nucleus of the cell.

Classify. To arrange things into groups according to ways in which they are alike and not alike.

Climax community. The final stage of a community at the end of a succession.

Cochlea. The spiral canal of the inner ear that contains the receptors for hearing.

Coldblooded. A vertebrate whose body temperature changes according to the temperature of the environment.

Community. All the plants and animals living in a particular area.

Compounds. Substances composed of two or more different kinds of atoms.

Connective tissue. A type of tissue that supports, joins, and binds various parts of the body.

Conservation. The protection and wise use of natural resources.

Consumer. An organism that depends on other organisms for its food needs.

Control. The part of an experiment in which all factors remain the same.

Cotyledon. A seed leaf.

Cytoplasm. The protoplasm surrounding the nucleus of the cell. The cell's activities are carried out here.

D

Data. Information collected from observations.

Daughter cells. Cells that result from the division of a parent cell. They have the exact same chromosomes that the parent cell had.

Decomposer. An organism that obtains its food by breaking down dead and decaying organisms.

Density. The number of organisms found in a certain area at a given time.

Depressant. A drug that slows down the activities of the central nervous system.

Dermis. The thick, active layer of skin underneath the thinner epidermis.

Diaphragm. A strong sheet of muscle that forms the floor of the chest cavity.

Dicot. A plant that produces seeds with two halves.

Diffusion. The spreading out of molecules from a crowded area to a less crowded area.

Digestion. The process that breaks down food into a simpler form so that it can be used by cells.

Disease. A condition in which some part of a living thing is not working properly.

Disinfectant. A chemical capable of killing many of the germs that cause disease.

DNA. Deoxyribonucleic acid—a large molecule that contains the hereditary information of the cell and controls the cell's activities.

Dominant. Refers to the stronger of a pair of traits.

Dormant. In an inactive, resting stage.

Drug. Any substance, other than food, that changes the body or mind.

E

Echinoderm. Spiny-skinned invertebrate.

Ecological niche. The role an organism plays within its community.

Ecosystem. A group of organisms and their physical environment.

Egg. A reproductive cell from a female parent.

Elements. Substances composed of only one kind of atom.

Embryo. A developing organism in its earliest stages of development.

Endangered. In danger of becoming extinct.

Endocrine glands. Ductless glands that release hormones into the bloodstream.

Endoskeleton. An internal support structure in vertebrates, made of cartilage and/or bone.

Energy. Ability to produce motion or cause change.

Environment. Everything in the surroundings that affects the way an organism grows, lives, and behaves.

Enzyme. A substance that helps to cause or to control the speed of chemical changes in living things.

Epidermis. The outer layer of skin.

Epithelial tissue. A type of tissue that covers all the body surfaces.

Esophagus. The food tube that connects the throat and the stomach.

Evolution. The process by which organisms change and become different from generation to generation.

Excretion. The process by which an organism gets rid of metabolic wastes.

Exoskeleton. The hard outer covering that is typical of arthropods.

Experiment. An activity designed to test a hypothesis.

Extensor. A muscle that straightens, or extends, a joint.

External fertilization. The joining of egg and sperm outside the female's body.

Extinct. Refers to an organism that no longer exists on earth.

F

Fats. Organic compounds that are storage compounds. They provide twice as much energy as carbohydrates.

Feedback. An automatic "turn on" or "turn off" process that occurs when a certain level of a hormone or other substance is reached.

Fermentation. The release of useful energy from food without the use of oxygen.

Fertilization. The joining of a sperm cell and an egg cell.

Fission. When a single-celled organism divides to form two new cells.

Flexor. A muscle that bends a joint.

Food calorie. A measure of energy that foods provide.

Food chain. The passing of energy in the form of food from one organism to another.

Food web. All the feeding patterns in an ecosystem.

Formula. A combination of symbols and numbers that shows both the kind and amount of atoms in a molecule.

Fossil. The remains or imprint of an organism that once lived.

Fruit. A ripened ovary that contains the seeds.

Fungi. A kingdom of organisms that have no chlorophyll and absorb food molecules.

G

Gall bladder. A sac in which bile from the liver is stored.

Genes. Locations on chromosomes that determine hereditary traits.

Genetics. The science that studies the laws of heredity.

Genus. A group of related species differing only in a few ways.

Gill. Structure in some water-dwelling animals that absorbs oxygen from the water.

Gland. An organ that secretes substances having special functions in the body.

Glycogen. A starch that is made and stored in the liver.

Gram (g). The unit used to measure mass in the metric system.

Guard cells. Two bean-shaped cells in the epidermis that form a stoma.

Gymnosperms. Plants with seeds that are not protected by another covering in addition to the seed coat.

H

Habitat. The place in which an organism is usually found in a community.

Hazardous waste. Any waste that is capable of causing physical, chemical, or biological damage.

Hemoglobin. A protein that combines with oxygen, carrying it to all cells in the body.

Herb. A plant whose stem does not have a lot of supportive tissue.

Herbivore. A consumer that eats only plant tissue.

Heredity. The passing of traits from parent to offspring.

Hominidae. In the classification system, the family to which humans belong.

Hormone. A chemical messenger that regulates and balances body functions. Hormones are carried in the blood.

Hybrid. Refers to an organism or cell in which the genes for a trait are different.

Hypothesis. A statement that explains a group of related observations.

I

Impulse. Information from a stimulus that travels along neurons.

Incubate. To keep eggs warm so that they can mature.

Infection. An area where a microorganism is invading the tissue of a host.

Infectious disease. A disease caused by a microorganism. It can spread from person to person.

Inorganic. Refers to compounds that do not contain carbon.

Internal fertilization. The joining of egg and sperm inside the female's body.

Intestine. The hollow tube in which digested food is absorbed.

Invertebrates. Animals without backbones.

Involuntary muscles. Muscles that cannot be controlled at will.

J

Joint. A place where two bones meet.

K

Kidneys. Excretory organs that remove wastes from the blood.

Kingdom. The level of the classification system with the largest number of living things that are alike in some ways.

L

Large intestine. Part of the digestive tube in which wastes are stored until they can be eliminated from the body.

Life processes. Activities or processes carried on by all living organisms.

Ligaments. Tough strands of elastic tissue that connect bones at movable joints.

Limiting factor. A condition in the environment that may cause a change in the size of a population.

Liter (L). A unit of volume used in the metric system.

Liver. The largest gland in the body; it performs many important functions, one of which is to produce bile.

M

Mammal. An animal that has hair and feeds its young milk.

Mantle. The outer layer of tissue that produces the shell on a mollusk.

Marrow. Soft, fatty tissue in the center of some bones that produces special cells found in blood.

Mass. The measure of the amount of matter contained in an object.

Matter. Anything that takes up space and has mass.

Mechanical digestion. The process in which food is broken down into tiny pieces.

Medulla. An enlarged area of the brain stem. It controls the functions of the internal organs.

Meiosis. Cell division that reduces the number of chromosomes by half to produce sex cells.

Menstruation. The stage of the menstrual cycle in which the uterine lining and an unfertilized egg are released.

Metabolism. The sum of all the chemical processes that take place in an organism.

Metamorphosis. A series of major changes in the structure of an animal as it grows.

Meter (m). The basic unit of length in the metric system.

Migrate. Refers to seasonal movement of animals from one environment to another, usually in search of food.

Mitochondria. Structures in the cytoplasm that release energy from food.

Mitosis. The process in cell division in which the nuclear material divides.

Molecule. Smallest unit of a compound. Usually two or more atoms held together by energy.

Mollusk. Invertebrate animal, usually having a hard shell that surrounds and protects the body.

Monocot. A plant that produces seeds with one part.

Motor neuron. A neuron that carries impulses from the brain or spinal cord to a muscle or gland.

Mucus. A thick, sticky fluid covering many surfaces inside the body and in its natural openings.

Multicellular. Made up of more than one cell.

Muscular tissue. A type of tissue that provides for movement.

Mutation. A change in one or more genes that can result in a new trait.

N

Natural resource. Any natural substance that humans remove from the environment for their own use.

Natural selection. The process by which only the organisms that are the best suited to their environment survive. As a result, they pass their traits on to their offspring.

Nekton. Free-swimming organisms, such as fishes, of the oceans.

Nephrons. Special tubes in the kidneys that filter the blood and return essential substances to it.

Nervous tissue. A type of tissue that helps an organism respond to stimuli in the environment.

Neuron. A nerve cell that is the basic unit of the nervous system.

Nocturnal. Refers to an animal that is usually active at night.

Nonrenewable resource. A resource that is not replaced or recycled at the rate we use it.

Nonvascular. Lacking specialized tissue for transporting water, minerals, and food.

Nuclear membrane. Boundary separating the nucleus from the cytoplasm.

Nucleic acid. A chemical that controls activities in a cell and passes on traits to new cells.

Nucleus. The "control center" of the cell that directs all the cell's activities.

Nutrient. Substance that does one or more of the following: furnishes energy; builds, repairs, and maintains the body; and regulates body processes.

O

Observation. Any information that we gather by using our senses.

Omnivore. A consumer that eats both plant and animal tissue.

Organic. Refers to compounds that contain carbon.

Organic detritus. The remains of dead organisms.

Organism. A complete living thing.

Organs. Groups of different tissues working together to perform a specific function.

Osmosis. The diffusion of water through a selectively permeable membrane.

Ovaries. Female reproductive organs that produce eggs and a female sex hormone.

Ovulation. The process in which a mature egg is released from an ovary.

P

Pancreas. A gland whose secretions help digest food chemically.

Parasite. An organism that lives in or on another living organism and causes harm to its host.

Permafrost. The part of the soil in which the ground water is permanently frozen.

Phloem. Tubes that transport dissolved food materials.

Photosynthesis. Process by which green plants use light energy to make glucose and release oxygen.

Pistil. The female reproductive part of a flower.

Placenta. A structure that attaches the embryo to the wall of the uterus.

Plankton. Tiny organisms that drift in surface currents or tides.

Platelets. Tiny particles in the blood that help the blood to clot.

Pollen. Structure(s) containing the sperm cell(s) of a vascular, seed-bearing plant.

Pollination. The process in which pollen is transferred from the male organ to the female organ.

Pollutant. Any material, such as gases, particles, and chemicals, released into the air or water.

Population. A group of one kind of organism in a particular community.

Predator. An animal that hunts other animals for food.

Prey. Animals that are hunted by other animals as food.

Producer. A green plant able to make its own food by photosynthesis.

Proteins. Complex organic compounds that are the building blocks of organisms.

Protista. A kingdom of living organisms that are neither plants nor animals.

Protoplasm. Living material, found in a cell, capable of carrying on all the life processes. It is made up of organic and inorganic compounds.

Protozoans. Single-celled, animal-like protists.

Pulse. Regular stretching and relaxing of the artery walls caused by the beating of the heart.

Pure. Refers to an organism or cell in which the pair of genes required to produce a trait is identical.

R

Recessive. Refers to the weaker of a pair of traits.

Red blood cells. Cells in the blood that contain hemoglobin and carry oxygen.

Reflex. An automatic response to a stimulus in which the brain is not directly involved.

Regeneration. The ability of some living things to grow new parts.

Renewable resource. A resource that is replaced or recycled at the same rate we use it.

Respiration. The process in which oxygen combines with food to release energy for life activities.

Retina. The innnermost layer of the eye made up of receptors that are sensitive to light.

Ribosomes. Structures in the cytoplasm that make proteins.

Root cap. Cells at the tip of a root that are loosely packed and protect the root from being damaged as it grows.

Root hairs. Tiny, tubelike structures growing out of a root that allow the plant to take in large amounts of water and minerals.

S

Saliva. A liquid produced by the salivary glands in the mouth.

Saprophyte. An organism that gets its food from dead or decaying organisms.

Scientific law. A theory that has been proven correct over a long period of time.

Scientific method. The set of skills used to solve problems in an orderly way.

Seed. A complete, tiny young plant, surrounded by a stored food supply and protected by a seed coat.

Selectively permeable. Refers to a membrane that will allow some substances to pass through, but not others.

Sense organs. Structures of an organism, such as eyes and ears, that are sensitive to certain specific stimuli.

Sensory neuron. A neuron that carries impulses from sense organs to the spinal cord or brain.

Sex chromosomes. A pair of chromosomes that determines the sex of an individual.

Sex-linked gene. A gene that is carried only on the X sex chromosome.

Sexual reproduction. Reproduction that requires two parents.

Skeletal muscles. Voluntary muscles made up of cells that are shaped like long cylinders and have many nuclei.

Small intestine. Part of the digestive tube where food is digested and absorbed.

Smooth muscles. Involuntary muscles made up of cells that are long, thin, and have one nucleus.

Specialized. Refers to the fact that each cell has a particular function, and that different kinds of cells have different functions.

Species. A group of related organisms that are alike in many ways. They can mate and produce living young.

Sperm. A reproductive cell from a male parent.

Sphincter. A ring of muscle that closes an opening or a tube.

Spinal cord. A thick bundle of nerves that reaches from the medulla to near the base of the spine.

Sponge. A simple invertebrate animal that has one body opening.

Spore. A special reproductive cell of certain organisms that grows into a new organism when conditions are right.

Stamen. The male reproductive part of a flower.

Stimulant. A drug that speeds up body functions, especially the central nervous system.

Stimulus. Any change in the environment that causes a response in an organism. (Plural: stimuli.)

Stoma. Tiny opening in the epidermis. (Plural: stomata.)

Stomach. A muscular storage sac that helps break food into smaller pieces.

Succession. A series of changes occurring in the plant and animal community of an ecosystem.

System. A group of related organs performing a major function for an organism.

T

Taste buds. Special receptor cells on the tongue that are stimulated by flavors.

Tendons. Tough, nonelastic tissue that attaches some skeletal muscles to bones.

Territory. The area in which an animal or group of animals lives and breeds. The area is also defended by the animal or group of animals.

Testes. Male reproductive organs that produce sperm and male sex hormones.

Theory. A general statement based on hypotheses that have been tested many times.

Tissues. Groups of similar cells with similar functions.

Toxic waste. Any substance that is poisonous to living organisms.

Toxin. A poison produced by a microorganism. It causes harm to the host.

Trachea. A stiff, flexible breathing tube, also known as the windpipe.

Tropism. The growth of a plant in response to a stimulus.

Tube feet. Structures on the bottom surface of echinoderms, used for grasping.

U

Umbilical cord. A structure that connects the embryo to the placenta.

Urinary bladder. A muscular sac at the end of the ureters that stores urine.

V

Vaccination. The injection of a killed or weakened germ to give immunity to the disease caused by that germ.

Vaccine. A substance injected into the body to cause the body to produce antibodies.

Vacuoles. Storage areas located in the cytoplasm.

Variable. Any factor in an experiment that could affect the results.

Vascular. Having specialized tissue for transporting water, minerals, and food.

Vegetative propagation. When new plants are produced from a part of a plant other than a seed.

Vein. A blood vessel that carries blood toward the heart.

Ventricles. The two lower chambers of the heart that pump blood to the arteries.

Vertebrate. An animal that has a backbone.

Villi. Very small, finger like projections on the lining of the small intestine that absorb digested foods. (Singular: villus.)

Virus. A particle that is not a cell but can reproduce in a cell of a living organism.

Volume. The amount of space a solid or liquid takes up.

Voluntary muscles. Muscles that can be controlled at will.

W

Warmblooded. Refers to those vertebrates whose body temperature remains constant despite temperature changes in the environment.

Water cycle. The movement of water as a gas, liquid, or solid through all parts of the biosphere.

White blood cells. Cells in the blood that protect the body from infection.

Withdrawal. The addict's physical reaction to stopping the use of a drug.

Woody plant. A plant whose stem has a lot of supportive and transport tissue arranged in rings.

X

Xylem. Tubes that transport water and minerals upward in a plant.

INDEX

NOTE: Page numbers in **boldface** type refer to definitions, and those in *italic* refer to illustrations.

A

abdomen, 245, 298
abiotic ecosystems, **437,** *438,* 441
absorption, **332**–333
acid fermentation, *77*
acids; amino, **318,** 329, 376; fatty, 329; nucleic, 105
acid rain, 532–533, 536
acorn worms, 210
acquired characteristics, theory of, 471–472
acquired immunity, **141**
active layer (tundra), **509**
active transport, 64, **65,** 66
adaptation, environmental, 480–481, 482, 511; in deserts, 505–*506, 508*
addiction, **406;** to alcohol, 406–407; to nicotine, 408
addictive drugs, *405*–406
additives, food, 320–321
adolescence, 413
adrenal glands, *402,* 403
adult, 246, 247, 262, 418
aging process, *418*
agricultural extension service workers, 179 (careers)
agronomist, *5*
air pollution, 532, *535*
air bladders, 119
air sacs (lung), *275,* 362
albinism, *480*
albumin (egg), *276*
alcohol fermentation, 76–77
alcoholism, 406, *407*
algae, 112, **117**–121, 157, 201, 515; brown, 119; diatoms, 117; Euglena, *117;* green, 118–*119;* importance of, 119–120; photosynthesis in, 119; red, *119;* uses of, *120*
alligators, 268, 526
Alpine tundra, 510, 511
alveoli, **362**
amber, fossils in, 466, *467*
ameba, *44,* 113–*114,* 115; "false feet" of, *113*
amebalike cells, 211, 215
amino acids, **318,** 329, 376

amphibians, 255, 260–265; heart of, *363;* reproduction in, *262*
anaphase, mitosis, *83*
anemia, *343;* sickle-cell, *343*
angiogram, *338*
angiosperms, **174**–177; dicots, **176**–177, 178; monocots, **176**–177, 178; sexual reproduction in, 183–184
animal(s); burrowing, 498, 499; carbon dioxide–oxygen cycle and, *444*–445; characteristics of, 209–210; coldblooded, **255,** 266; competition among, 441; of coniferous forests, *494;* dependence on green plants, *201;* desert, *506*–507; endangered, **525,** *526, 528;* excretion in, 367; extinct, 468, 495, **525,** 527, 529; and food chains (of grasslands), 498–499; invertebrates, 208–**210,** 211–224, 228–243; and life processes, 31, **32**–34, 295; mammals. *See* mammals; nocturnal, *506;* population changes, **453,** *455,* 456, 526–527; trainer, 227 (careers); tundra, *510–511;* vertebrates, **210,** *252,* 254–255; warmblooded, **273**–274; water cycle and, *443*
animal cells, *48,* 49
animal community. *See* communities
animal succession, 459
Animalia, kingdom, *94*
Annelida, 222
antennae, 239, 245
anther, 184
antibiotics, **145**–146
antibodies, **140,** *141,* 340
anus, **221,** 223, 333; of bivalves, 231
aorta, 350
appendix, 332
arachnids, *238*
Aristotle, classification system of, 91
arteries, *317,* **348,** 350, 355; coronary, *354*
arteriosclerosis, 317

arthritis, 149
arthropods, *Arthropoda,* **237**–*239*
asexual reproduction, **115,** 167–*168,* 212, 215, 413
association neurons, **384,** 385, 388
astigmatism, 393
astronomer, *8*
Atlantic salmon, 528
atmosphere, *3*–4
atom(s), **36,** 39
atria, human, **353,** *354*
auditory nerve, 394
Aurelia, jellyfish, *216*
automobiles, air pollution and, *535*
average temperatures. *See* temperature
Aves, 274
axon, *383*

B

bacillus, *101*–102
backbone, 210, *254,* 301, *389*
bacteria, 2, **99**–103, 457; blue-green, *94,* 101; classification of, 101–102; decay, 102; diseases caused by, *99;* food for, 100–101; nitrogen-fixing, *446;* shapes of, *101;* structure of, 99–*100;* useful, *102*
balance, **5,** *21,* 395
bald eagles, 528
ball-and-socket joint, 302, *303*
bark, 158
barrier islands, 523–524
bats, 286
bauxite, *523*
Beaumont, Dr. William, and stomach function, *328*
bees, *244*
"bends," 361
benthos, **515**
bilateral symmetry, *255*
bile, *331,* 375
biochemist, 5, *36*
biogeography, 487
biologist, 1, 5; tropical, 461 (careers)
biome(s), *486,* **488**–516

echinoderms, **234;** characteristics of, *234–235, 236*
ecological niche(s), **440,** *441,* 527
ecologist, animal, 225 (careers)
ecosystem(s), 437, **438,** 441, 442, 450; coastal, 523–524; foreign species into, 526; forest, 522; ocean, 436; recycling in, *443, 444, 445–446*
egg, **161,** 167–168, 183, 184, 186, 246; amphibian, *262;* bird, *276;* fish, 257–258; reptile, *266,* 267
egg cells; in conifers, 172; in humans, *414–*415; egg-laying mammals, *283*
electron microscope, *104, 341*
element(s), **36;** common, *37;* symbols for, *37,* 38
elephants, 287
embryo; bird, 276; human, *416–*417; marsupial, *284;* plant, **187;** reptile, 267; transplant, 537 (technology)
emphysema, 365
endangered species, **525,** *526, 528*
endocrine glands, *400,* **401**
endocrine system, 400–*402,* 403, 413
endoskeleton, **254,** 299
energy, **2,** 36, 445; acid fermentation and, 77; active transport and, 65; carbohydrates and, 315; cells as releasers of, 74–*75;* and chemical bonds, 75; conservation of, 524; and ecosystems, 438–439; fats and, 316; food chains and, *448–*449, 451–*452;* mitochondria and, 75; and nonrenewable resources, *524;* proteins and, 317; pyramid of, 451–*452;* respiration and, 74–**75;** from sun, *444–*445, 448, 490
energy crisis, 524
entomologist, 249 (careers)
environment, 3; adaptation to, 480–481, 482, 505–*506,* 508, 511; influence on genes, 426; kinds of. *See* habitat(s); pollution of, *532–535*
environmental factors, 3
Environmental Protection Agency, 535
environmentalist, 461 (careers)
enzyme(s), **76,** 319; and digestion, 114, 215, 329
epidermal tissue (plant), 54

epidermis, **370;** of leaf, **198**
epiglottis, 324
epithelial tissue, **296**
erosion, 523, 524
esophagus, **324–**325
essential amino acids, *318*
Euglena, *45,* 117–118
evaporation, *443–*444
evergreens, 172
evolution, *464,* **471**
excretion, 32, **367–**372; in animals, 367; in bivalves, 231; blood vessels and, 370; of carbon dioxide, 367; lungs and, 367; of metabolic wastes, 367–368; in protozoans, *114;* skin as an organ of, 370–371; system of, 367–*369*
exobiology, 517 (technology)
exoskeleton, **238,** 299
experiment(s), **7,** 12–15; controlled, **13**
extensor, muscle, **307**
external fertilization, 236, **257–**258, 262, 413
extinct animals, 468, 495, **525,** 527, 529
eye(s); human, 292, *293;* of potatoes, 187–*188;* of univalves, 231
eyespots; of Euglena, 118; of planaria, 220; of starfish, 236

F

Fallopian tube, 415, 416
"false feet" of ameba, *113*
family classification, *96*
fat cells, 370
fats, **316–**317, 375, 376
fatty acids, 329
feathers, *272*
feces, 333, 375
feedback, **402,** *403*
fermentation, **76;** acid, 77; alcohol, 76
fern(s), 156, *167–168,* 169
fertilization, **161,** 183–184, 413, 416; chromosomes and, 414; in conifers, 172; external, 236; internal, **267**
fever, 141
fiber, 211, 329, 332, 333
field community, 440
filament, 184
Filidae, family, *96*
fish, 210, 255, 259; body plan of,

256–257; body temperature of, 255; bony, 255; cartilage, 255–*256;* circulatory system in, 257; jawless, 255–*256;* reproduction in, 257–258
fisher, mollusk, 249 (careers)
fission, *114–***115**
flagella, 100, 112, 118, 211
flagellates, 112, 113, 114, 115
flatworms, 210, *220–221*
Fleming, Alexander, penicillin, 145–*146*
flexor, muscle, 307
flower(s), 158, 175, 184, 189; dicot, 176; fertilization in, 174, 175; monocot, 176; parts of, *184;* reproductive organs of, 183–184
flowering plants, *174–*177
flukes, 220
flying, adaptations for, 273, *274*
food(s), 315–321, *444–*445; additives in, 320–321, 376; bacteria used in making, 102; chemical digestion of, 328, **329–**334; diffusion into a cell, 63–*64;* mechanical digestion of, **323–**327
food calorie, **321**
food chain(s), **448–**449, 450, 527; in coniferous forests, 494; on deserts, 507; and energy, *448–*449, 451–*452;* on grasslands, 498–499; in oceans, 514, 515, 516, 523; predator/prey relationship in, 454; in tundra, 510–511
food getting, 32, *113*
foodmaking in plants. *See* photosynthesis
foodmaking tissues (plant), 158
food pyramid(s), 450–452
food vacuole, 113, 114
food web(s), **450,** 527; on deserts, 507
forest biomes, 491–495
forest community, 439, 440
formula, **38**
fossils, *252,* **465,** *466, 467, 468–*470; age of, 467–*468;* time scale, 476 (computer program); types of, 465–*467*
freshwater ecosystems, 437–*438*
frogs, 210, *261;* lungs of, 260–261; metamorphosis of, *262*
frond, fern, 167
fruit(s), 174, **187**
Fucus, 119
fuels, as resources, 524

fungi, 112, **124,** 449; bracken, 126; club, 125, 126–*127;* as decomposers, **127;** and disease, 127; imperfect, 125, *127;* as kingdom, *94;* reproduction in, 124–125; sac, 125, *127;* threadlike, *125*

G

Galapagos finches, *479,* 481–482
Galapagos Islands, *464*
gall bladder, **331**
gas(es), exchange of; in arthropods, *239;* in earthworms, 223
gastric juice, 329, *331*
gene(s), **422**–423, 424, 425; mutated, 479–481; sex-linked, 427, **428**–*429*
genetic code, 423
genetic engineering, 203 (technology)
genetics, **420,** 423
genus, **93**–*94,* 95
geotropism, 192
germ theory of disease, 134
germination, *190*
gills, **229;** artificial, 269 (technology); of crustaceans, *238,* 239; of mollusks, 229, 230–231; in tadpoles, 262
gill slits, 253
ginkgos, *171*
gland(s), **329**–331; endocrine, 400, **401;** oil, 370; salivary, 323; sweat, 370
gliding joints, 302
glycerol, 329
glycogen, **375,** 376
gram (g), 18, *19,* **21**
grass(es), as pioneer, 458
grasshopper, 245, 247
grasslands, 488, *497–499, 500;* changes in, 499–500
Great Basin Desert, *505*
green algae, 118–*119*
green plants; animals' dependence on, *201;* carbon dioxide–oxygen cycle and, *444*–445; energy stored in, 451–452; foodmaking in. *See* photosynthesis; as producers, 449; *See also* plants
ground water, 444, 532
growth, 191; and repair, 32
growth responses, 191–192
growth tissue (plant), 158

guard cell(s), **199**
gymnosperms, **170**–173; sexual reproduction in, 183–184

H

habitat(s), **440,** *441;* deciduous forest, 493; grasslands, 198–199; human destruction of, 500, 511, 525; and succession, 457
hair cap moss, *160*
hair follicles, 370
hares, 286–287, *455*
Harvey, William, *347*
hatchet-footed mollusks, 230–231
hazardous wastes, **533,** *535*
head; in insects, 245; region, 297; in univalves, 231
heart; amphibian, 262, *263;* of birds, *275*–276; of earthworm, 223; of fish, 257; human, *338,* 347, 353, *354,* 355; insect, 246; reptile, 266
heart attack, 316, 351, *354*
heartbeat, 275, 309, 354–355
heat, and energy, 75–76
heath hen, extinction of, 525
hemoglobin, **341**
hemophilia, *343,* 428
herb(s), **191**
herb layer, 493
herbivore(s), **449,** 498, 510–511
heredity, **420,** 425–429; DNA and, **423;** environment and, 426; Mendel's experiments and, 420–423, 425; natural selection and, **473**–474
high blood pressure, 351
hinge joint, *302, 303*
hollow nerve cord, 253
hollow sac animals, 213
Hominidae, **288**
Homo sapiens, 96, 288, 295
hoofed mammals, 286, 287
Hooke, Robert, 43, 50
hormone(s), 401–403; functions of, *402;* sex, 415–416
horse tails, 166–167
human(s); and the biosphere, *520, 522, 524,* 525–527; and deserts, *507;* in ecosystem, 527–528; and forest use, 494–495; and grasslands, 499–500; and habitat destruction, 500, 511, 525; and pollution, 533–*535;* population

changes, *522;* and the tundra, *511*
human body; blood, 339–343; body regions, 297–298; body systems, *297;* brain, 387–389; circulatory system, 348, *349*–350, 401; digestive system, 323–334, 375; embryo, *416*–417; eye, 292, *293;* heart, *338,* 347, 353, *354,* 355; heredity, 420, 425–429; and life processes, 295; nervous system, *300,* 381–389; reproductive system, 414–*415,* 416; respiratory system, 362–*363,* 364–365; skeleton, 299–*300;* urinary system, 367–*369*
hybrid, **422,** 425
hydra, *214*–215, 216, 217
hydrogen, *38,* 39
hygiene, personal, 145
hypertension, 351
hyphae, 125
hypothesis, **7,** 475

I

ice, fossils in, 466–467
identical twins, 423, 426
immunity, 137–138, 340
imprint, fossil, 466, *467*
impulse(s), nerve, **383,** 384, *385*
incomplete metamorphosis, *246*
incomplete protein, *318*
incubate, **276**
incubation period, *276*
infection, **135**
infectious disease, 133–**134**
ink blot test, *387*
inner ear, *394*
inorganic compounds, **44**
insects, 210, *238,* 239, 244–246; characteristics of, 245–246; metamorphosis of, **246**–247; social organization of, *244,* 245; species of, 244; structure of, *245, 246;* success of, 244–245
insulin, 402
internal fertilization, **267,** 276, 413
interphase (cell division), *83*
intestinal juice, *331*
intestine(s), **221;** large, **333;** small, **329**–333
invertebrates; classification of, 230–232, 240; complex, 228–243; crossword puzzle, 242

algae, 119; carbon dioxide and, 200, *444;* chlorophyll and, 199–200; in desert plants, 506; equation for, 200; final products of, *200,* 202; growing season and, 492, 493; light and, 199–200; in oceans, 513, 514–515; water and, 200

phototropism, 192

phylum (classification), 94, 210

pine tree, *171, 184*

pioneer species, 457, 458

pistil, **184**

pituitary gland, 401

placenta, **417**

planaria, 220–*221,* 224

plankton, **515**

plant(s), *182;* adaptations of, 158; bryophytes, 159, 160; cells of, 50–51; characteristics of, 157–158; classification, *159;* desert, 505–506; flowering, 174–177; foodmaking by, 198–202; growing season of, 492, 493, *495,* 498, 510; growth responses in, 191–193; needs of, 157, 194; nitrogen cycle and, *445–446;* nonvascular, **159;** organs of, 158; phototropism in, 192; reproduction in, 183–184; seed, 170–177; structure of a, *158;* tissue. *See* plant tissue; tracheophytes, 159; transport system in, *54,* 194–196; tropisms in, **192;** tundra, 510; vascular, 158–159, 167; vascular system in, *158;* vegetative propagation of, **187**–188; water cycle and, *443;* woody, 191

plant cells, *50*–51; division of, 82–*83;* guard, 199; osmosis in, 69; structure of, 50–51; *See also* plant(s)

Plant (Plantae) kingdom, *94;* adaptations of, 158; characteristics of, 157

plant tissues; covering, 214; epidermal, 198; foodmaking, 199–200; growth, 191–192; vascular, 158, **159,** *177*

plasma, *340,* 341

plasma proteins, *340*

platelets, *340, 341*–**342**

platypus, *283*

Platyhelminthes, 220

polar bear, *439*

polar zone, 487, *488*

polio, 105

pollen, **172,** 183, 184

pollen tube, 186

pollination, 184, 185–186, 421

pollster, 107 (careers)

pollutant(s), 376, **533,** *535*

pollution, *532–535;* acid rain as, *532–533,* 536; air, 532, *535;* hazardous waste, **533**–534; toxic waste, 534; water, *532–533*

polyp, 214, *216*

population(s), **453,** *455,* 456; birth rate and, 453, 454; density of, 454; endangered, 526–527; human, changes in, 522; limiting factors in, **454,** 455, 505

population crash, 455

pores, 211, 212

Porifera, 210–212

pouched mammals, *284*

prairies, 498, *500*

predator(s), **454,** 499, 527, *528*

predator/prey cycle, 122 (computer program), 454

pregnancy, 408

prey, **454**

primary succession, 457

primates, 288–289

producer(s), **449,** 450–*451*

prophase (mitosis), *83*

protein(s), *317*–318; digestion of, 329–331

protein coat, 105

Protist Kingdom, *94,* **111,** 117, 157

protists, 111–116

protoplasm, **44;** compounds in, 44

protozoans, **112**–116, 413, 515; characteristics of, 113–*114,* 115; classification of, *112;* and disease, 136–137

pseudopods, *113*

puberty, 413

public health measures, *144*–145

pulse, 354–**355,** 356

Punnett Square, 424

pupa, 247

pupil, eye, 384, 393

pure (plant), **420**–421, 422

pyramids; of energy, 451–*452;* food, 450–452; of mass, *451;* of numbers, 450, *451;* shoreline, 502 (computer program)

Q

queen bee, *244*

R

rabbits, 286–287

radial symmetry, **234**

radioactive dating, 467–468

radioactive waste, *533*

rafflesia, 183

rain, *443*

rain forests, 488, 491, 494–*495*

rainfall, 497; in coniferous forests, 493; in deciduous forests, 492; on deserts, 504; on grasslands, 497; on rain forests, 491–492; on tundra, 510

rainshadow effect, *505*

rays (fish), *256*

receptor cells, *391*–392, 393, 394, 396, 397

recessive trait(s), **421**–422, 425, 428

recycling, *433–444, 445*–446, 522; of reusable materials, *523*

red blood cells, **340,** 342, *350,* 375

Redi, Francesco, experiments of, *42–43,* 46

reflex(es), **384**–*385,* 386; in newborn, *384*

regeneration, **212;** in planaria, 221;\in sponge, **212;** in starfish, 235

renewable resources, **521**

reproduction, 33; in ameba, *114;* in amphibians, *262;* asexual, **115;** in bacteria, 100; in birds, *276–277;* by budding **126,** 212, *214,* 215; in coelenterates, 215–*216;* in ferns, 167–168; in fish, 257–258; by fission, *114*–**115;** in fungi, 124–125; in mosses, 161–162; in paramecium, *114*–115; in planaria, 221; in plants, 158; in protozoans, *114*–115; in reptiles, *266, 267;* sexual. *See* sexual reproduction; in sponges, 212; by spore formation, **124**–125; in yeasts, 126

reproductive system, human, *297,* 414–*415,* 416

reptiles, 255, *266, 267, 268;* heart of, 266; reproduction in, *266,* **267;** types of, *267*–268

respiration, 32, 201, *360,* 361–366; and energy, 74–**75;** process of, equation showing, *75*

respiratory system, *297;* in arthropods, *239;* in birds, *275;* earthworm, 223; gills in, 229, 239;

succession, **457**–*458*, 459; in a field, *458*
sugar(s), *375*–376; double, 331; simple, 200, 329
sun, energy from, *444*–445, 448, 490
sunlight; importance of, 157, 199–*200;* and ocean life, 513, 514, 515
survival of the fittest, *469*, 473
Survival Service Commission, 527
swamps, 166
sweat glands, 370
sweat pores, 370
symbol(s), 38; for important elements, *37*
synapse, 383
systems, **55;** body, *297*

T

tadpole, *262*
tapeworms, 220
tar, cigarette, 407
tar pit, fossils, 466
taste buds, **392**
technician; laboratory, 57 (careers); chemical, food industry, 335 (careers); dialysis, 377 (careers)
teeth; human, and digestion, 323–324, 329; of mammals, *285*
telophase, mitosis, *83*
temperate zone, 487, *488*
temperature(s), 437, 439, *495;* body, 142, *367*, 371, 373; in coniferous forests, 493; in deciduous forests, 492; on deserts, 505; of Earth, 3; factors affecting, 488; measurement of, 23, 24; ocean, 513–514; in rain forest, 492; on tundra, 509
tendons, 304, **307**
tentacles; of coelenterates, 214–215; of mollusks, 231, 232
territory, **481**
testes, **416**
thalidomide, 408
theory, scientific, **10**
therapist, respiratory, 377 (careers)
thermometer, *23*, 138
thorax, 245
thyroid gland, *402*
tidal marshes, 523–524

tissue(s), **54**, 295, *296*–297; bone, **301;** connective, **296;** covering, 214; epithelial, 54; muscle, **296;** nerve, **296;** plant. *See* plant tissues
toads, 261
tobacco, 407–*408*
tongue, *392*
tortoise, *267*
toxic wastes, 533, **534,** 535
toxins, 135
trachea, **362**
tracheophytes, 159
traits, **421**–422, 425, 428–*429;* human, 425; Mendel's study of, 420–423, 425; and mutations, 479–481
transplant, embryo, 537 (technology)
transport theory, 196
transport tissue, 167
transport system in plants, 194–196, 197
tree line, 510
trees; bark of, 158; conifers, *172*–*173*, 493–494; deciduous, 492–493; growth rings on, 170, 191
trichina, *145*
trichinosis, 137, 145
tropical rain forests, *491*
tropical zone, 487, *488*
tropism(s), **192**
trunk region, body, 297, 298
tube feet of starfish, **235**
tube worms, *2*
tubes, breathing, *238*, 239
tuberculosis, 134, 135, *136*
tundra, 488, *489*, 509–*510, 511;* animal adaptation to, 510–511; environment, 509–510; life on, 510–511; problems of, 511
turbidity, 513, 515
turkeys, endangered species, 528
turtles, *266*, 267
twins, 423, 426

U

ultrasound, 357 (technology)
umbilical cord, **417**
underground stem, 187
univalves, *230, 231*
unsaturated fats, 316–*317*
urea, 368, 370, 376
ureter, 368

urethra, 369
urinary bladder, **368**
urinary system, human, 367–369
urine, 368, 369
uterus, 415, 416–417

V

vaccination, **144**
vaccine, **144**
vacuoles, **49,** 113–114
vagina, 415
valves, 348, 350, 353, *354*
variables, **13**
vascular plants, 158–159, 167
vascular system in plants, 175, *195*
vascular tissue, plant, 158, **159,** *177*
vegetative propagation, **187**–*188*
veins; human, 348, 350; leaf, 195, 198
vent(s), ocean, *2*
ventricles, heart, **353,** *354,* 355
vertebrae, bones, 301, *389*
vertebrates, **210,** *252,* 254–255; classes of, 255, 280 (computer program); coldblooded, 255–268; warmblooded, 272–288
veterinarian, 289 (careers)
villi **332**
virus(es), **104**–*105, 106;* diseases caused by, *104;* life processes of, 105–106
vitamins, 319–*320*
voice box, 362
volume, **22**
voluntary muscles, 388

W

Wallace, Alfred, and natural selection, 473
warmblooded animals, **273**–274
warts, 104
wastes; diffusion of, *64,* 236; excretion of, 318, 367–371; hazardous, **533;** metabolic, 367, 375; nitrogen, 376, 446; radioactive, *534;* toxic, **534**
water, 4, 437, 438, 439, 501; and climate, 488; diffusion of, 68–70; human-body content of, 4; importance to life, 157; in irrigation, 507; ocean. *See* oceans;

PHOTO CREDITS

Chapter 1: p. xii David Doubilet; pp. 2, 3 Al Giddings/Ocean Images; p. 5(l) Dan McCoy/Rainbow; (r) Arthur Sroot/Stock Shop; p. 6(t) E. R. Degginger/Animals Animals; (b) Michael Wells; p. 8 C. Lockwood/Animals Animals; p. 9(t) Tom McHugh/Photo Researchers; (br) Burt Glinn/Magnum; p. 12 Marc Bekoff; p. 18 Michael Melford/Wheeler Pictures; p. 19 Randy Taylor/Sygma; p. 20 NASA; p. 21 Courtesy of Mettler Instrument Corp.; p. 25 Robotics, Inc.

UNIT 1: pp. 28–29 David Doubilet **Chapter 2:** p. 30 Zig Leszczynski/Animals Animals; p. 32 Joe McDonald/Animals Animals; (r) David Madison/Bruce Coleman; p. 33 E. R. Degginger/Bruce Coleman; p. 36 Michael Heron/Woodfin Camp & Assoc.; p. 43 New York Public Library Picture Collection; p. 45(l) Walker/Photo Researchers; (c) J. Pickett-Heaps/Photo Researchers; (r) J. Pickett-Heaps/Photo Researchers; p. 47 Tom Bean/DRK Photo; p. 48 K. R. Porter/Photo Researchers; p. 50 Biophoto Assoc./Photo Researchers; p. 53(t) Culver Pictures; (bl) John Gerlach/DRK Photo; (br) Stephen J. Krasemann/DRK Photo; p. 54(l) M. I. Walker/Photo Researchers; (c) Eric V. Grave/Photo Researchers; (r) Runk/Schoenberger/Grant Heilman; p. 57(l) The Stock Market; (r) Chuck Keeler Jr./After Image Inc. **Chapter 3:** p. 60 Odyseaus/Peter Arnold; pp. 62, 63 HRW Photos/Richard Haynes; p. 65(r) Rocky Thies/Duomo; 65(l) Adam J. Stoltman/Duomo; p. 66 Bruno J. Zehnder; p. 68 HRW Photo/Carol Michel; p. 70 Steve Proehl/The Image Bank; p. 74(both) HRW Photo/Bill Kontzias; p. 76 National Food Processors Assoc.; p. 77 Richard Wood/Taurus Photos; p. 78(both) HRW Photo/Charles Biasiny-Rivera; p. 79 Fran Allan/Animals Animals; p. 80(t) M. Abbey/Bio-Tec Images; all others M. Abbey/Photo Researchers; p. 82 M. Abbey/Photo Researchers; p. 83(all) M. Abbey/Bio-Tec Images; p. 84(t) J. F. Gennaro, L. R. Grillome/Photo Researchers; (b) Manfred Kage/Peter Arnold, Inc.; p. 85 HRW Photo/Russell Dian and Robert J. Ellison/Photo Researchers.

UNIT 2: pp. 88–89 Wardene Weiser/Bruce Coleman **Chapter 4:** p. 90 Chip Clark; p. 93 D. Klesenski/Int'national Stock Photo; p. 95(l) Z. Leszczynski/Animals Animals; (r) Harold R. Hungerford; p. 96(l) Stephen J. Krasemann/DRK Photo; (r) K. Gunnar/Bruce Coleman Inc.; p. 97(l) Arthus-Bertrand/Peter Arnold Inc.; (r) Bruce Coleman; p. 99(l) M. Abbey/Photo Researchers Inc.; (r) Andrew H. Knoll; p. 101(l & c) M. Abbey/Photo Researchers Inc.; (r) Runk/Schoenberger/Grant Heilman; p. 102 Hank Morgan/Rainbow; p. 104(t) Lee D. Simon/Photo Researchers Inc.; (l) Heather Davis/Photo Researchers Inc.; (r) Dr. Gopal Murti/Photo Researchers; p. 107(r) Gerhard Gscheidle/Peter Arnold, Inc.; (r) HRW Photo/Yoav Levy/Phototake. **Chapter 5:** p. 110 Biophoto Assoc./Photo Researchers Inc.; p. 113(all) M. Abbey/Photo Researchers Inc.; p. 114(t) Biophoto Assoc./Photo Researchers; (b) Taurus Photos; p. 117(t) David Doubilet; (b) Runk/Schoenberger/Grant Heilman; p. 118 Eric Gravé/Photo Researchers Inc.; p. 119(l) Manfred P. Kage/Peter Arnold; (c) Anne Wertheim/Animals Animals; (r) E. R. Degginger/Animals Animals; p. 120(l) HRW Photo/Yoav Levy; (r) Roy Morsch/The Stock Market; p. 124 E. R. Degginger/Earth Scenes; p. 125 Manfred Kage/Peter Arnold; p. 126 Runk/Schoenberger/Grant Heilman; p. 127 A. Davies/Bruce Coleman Inc.; p. 129 Anne Wertheim/Animals Animals; p. 131 Pat Lynch/Photo Researchers Inc. **Chapter 6:** p. 132 Howard Sochurek; p. 134 Society of American Bacteriologists; p. 137 Peter Ward/Bruce Coleman; p. 139 Diane Koos Gentry/Black Star; p. 140 D. Fawcett/E. Shelton/Photo Researchers; p. 143 Granger Collection; p. 144(l) Peter Angelo Simon/Phototake; (r) Lionel Atwell/The Stock Shop; p. 145 Russ Kinne/Photo Researchers; p. 146(l) Martin Rotker/Taurus Photos; (r) Culver Pictures; p. 148 Grace Moore/Taurus Photos; p. 149 William Rivelli/Image Bank; p. 150(both) Martin M. Rotker/Taurus Photos; p. 151 Roger Ressmeyer/Wheeler Pictures.

UNIT 3: pp. 154–155 Lisl Dennis/The Image Bank **Chapter 7:** p. 156 W. H. Hodge/Peter Arnold; p. 158 Biomedia Assoc.; p. 160(l) W. H. Hodge/Peter Arnold; (r) Michael P. Gadomski/Bruce Coleman; p. 162 Paul Moylett/Taurus Photos; p. 167(both) W. H. Hodge/Peter Arnold; p. 169 Adrian Davies/Bruce Coleman; p. 170 Sandra Grant/Photo Researchers; p. 171(l) Zig Leszczynski/Earth Scenes; (r) E. R. Degginger/Bruce Coleman; p. 173 Tom Tracy/The Stock Shop; p. 175(l) George D. Lepp/Bio-Tec Images; (r) Barry L. Runk/Grant Heilman; p. 179(l) Roger Tully/After-Image; (r) Read D. Brugger. **Chapter 8:** p. 182 G. Rogers/The Image Bank; p. 184 Walter H. Hodge/Peter Arnold; p. 185(l) Rosalie LaRue Faubian/Bruce Coleman; (r) James Bell/Photo Researchers; p. 188(l & r) Runk/Schoenberger/Grant Heilman; (lc) HRW Photo/Ken Karp; (rc) Farrell Grehan/Photo Researchers; p. 190 George F. Godfrey/Earth Scenes; p. 192(l) HRW Photo/Daniel C. Wasp; (r) HRW Photo/Ken Karp; p. 194 Stephen J. Krasemann/D.R.K. Photo; p. 196 Manfred Kage/Peter Arnold Inc.; p. 199 Carolina Biological Supply; p. 201 John Dominis/Wheeler Pictures; p. 203 Phototake.

UNIT 4: pp. 206–207 Silvester/Rapho/Photo Researchers **Chapter 9:** p. 208 Anne L. Doubilet; p. 211 Carl Roessler/Animals Animals; p. 212 Van Bucher/Photo Researchers; p. 213 James M. Cribb/Bruce Coleman Inc.; p. 214 Kim Taylor/Bruce Coleman Inc.; p. 215 OSF/Animals Animals; p. 220 Carolina Biological Supply; p. 222 Photo Researchers; p. 223 Runk/Schoenberger/Grant Heilman; p. 225(l) HRW Photo/Yoav Levy/Phototake; (r & c) Susan Kukan/Photo Researchers.

Chapter 10: p. 228 Fred Bavendam/Peter Arnold; p. 230(l) Zig Leszczynski/Animals Animals; (c) J. Foott/ Bruce Coleman; (r) Fred Bavendam/Peter Arnold; p. 232 Jeff Rotman; p. 234(all) E. R. Degginger; p. 235(l) Tom McHugh/Photo Researchers; (r) Carl Roessler/ Bruce Coleman; p. 237(t) The Armeuries, H. M. Tower of London; (b) Rod Borland/Bruce Coleman; p. 244 Grant Heilman; p. 249(l) HRW Photo/Yoav Levy; (r) E. R. Degginger/Bruce Coleman. **Chapter 11:** p. 252 Gert Wagner/Discovery Magazine 1982; p. 254(t) James M. Cribb/Bruce Coleman; (b) OSR/Animals Animals; p. 255(l) James Hanken/Photo Researchers; p. 256(t) Hans Reinhard/Bruce Coleman; (l) Tom McHugh/ Photo Researchers; p. 256(r) Carl Roessler/Animals Animals; p. 257 Jeff Rotman; p. 261(l & c) Z. Leszczynski/Animals Animals; (r) Andrew Odum/Peter Arnold; p. 265 Runk/Schoenberger/Grant Heilman; p. 266 Z. Leszczynski/Animals Animals; p. 267(l) William E. Ferguson; (r) Tom McHugh/Photo Researchers; p. 268(l) Tom McHugh/Photo Researchers; (r) Z. Leszczynski/Animals Animals; p. 269 Scott D. Taylor/Duke U. Marine Lab. **Chapter 12:** p. 272 Dan Guravitch, 1981; p. 274 Andreas Feininger/Life Mag., Time, Inc.; p. 277(l) Clyde H. Smith/Peter Arnold; (r) Stephen Delton/Photo Researchers; p. 283(t) Peter Arnold Inc.; (b) Tom McHugh/Photo Researchers; p. 284(l) Carolina Biological Supply; (r) Leonard Lee Rue III/Photo Researchers; p. 286 Marty Stouffer Prod./Animals Animals; p. 288 Charlie Cole/Black Star; p. 289(l) Magnum; (r) HRW Photo/Richard Haynes.

UNIT 5: pp. 292–293 "Spectrum Ballet," 1981 Alvin Ailey Dance Co. / Photo by Johann Elbers **Chapter 13:** p. 294 Alexander Hubrich/Image Bank; p. 298 HRW Photo/Yoav Levy/Phototake; p. 301 Manfred Kage/Peter Arnold; p. 302 Pat Miller/Medichrome; p. 304 David York/Medichrome; p. 306 Biofoto Assoc./Photo Researchers; p. 307(t) Manfred Kage/Peter Arnold; (b) HRW Photo/Yoav Levy/Phototake; p. 308 Ida Wyman; p. 309 Warren Morgan/Focus on Sports; p. 311 Chuck O'Rear/WestLight. **Chapter 14:** p. 314 Matthew Klein; p. 316(both) HRW Photos/Ken Karp; p. 317(t) HRW Photos/Ken Karp; (b, both) Norman Rotker/Phototake; p. 318(t) HRW Photo/Russell Dian; (b) Janeart Ltd./ Chuck Fishmar/The Image Bank; p. 321 Richard T. Nowitz/Phototake; p. 322 HRW Photo/Richard Haynes; p. 324 HRW Photo/Yoav Levy; p. 329 HRW Photo/Richard Haynes; p. 330 HRW Photo/Yoav Levy/Phototake; p. 332 Manfred Kage/Peter Arnold; p. 335(l) John Dominis/Wheeler Pictures; (r) Joan Menschenfreund/International Stock. **Chapter 15:** p. 338 © Howard Sochurek; p. 340 © Bill Longcore/Photo Researchers; p. 341 (l) Manfred Kage/Peter Arnold; (r) Grant Heilman; p. 342 Martin Rotker/Phototake; p. 347 The Granger Collection; p. 349 HRW Photo/Richard Haynes; p. 350 Lennart Nilsson, from *Behold Man;* p. 351 Richard Laird/Medichrome; p. 353 Brad Nelson/U. of Utah Medical Center; Black Star; p. 355 Yoav Levy/Phototake; p.

357 © Tobey Sanford/Wheeler Pictures. **Chapter 16:** p. 360 Carl Roessler/Animals Animals; p. 363 HRW Photo/Yoav Levy/Phototake; p. 367 Robert Barclay/ Grant Heilman; p. 369 HRW Photo/Richard Haynes; p. 371 David Brownell/The Image Bank; p. 373 HRW Photo/Russell Dian; p. 376 Mark Bolster/International Stock; p. 377(l) Roger Allan Lee/After-Image; (r) Sarah Feinsmith/Photo Associates News Service. **Chapter 17:** p. 380 R. Phillips/The Image Bank; p. 382 HRW Photo/Russell Dian; p. 383 Ward's Natural Science Establishment; p. 384 Jeffrey Reed/Medichrome; p. 387 Stan Goldblatt/Photo Researchers; p. 391 St. Elizabeth Hospital, Appleton, WI; p. 395 Johnson Space Center, Houston; p. 404 HRW Photo/Russell Dian; p. 406 HRW Photo/Richard Haynes; p. 407 National Council on Alcoholism/HRW Photo/Richard Haynes; p. 408 American Cancer Society/HRW Photo/Richard Haynes; p. 409 Dan McCoy/Rainbow. **Chapter 18:** p. 412 ERIKA; p. 418 Jim Balog/Black Star; p. 423 Science Source/Photo Researchers; p. 425 Peter Tomann/The Image Bank; p. 427(both) Martin M. Rotker/Taurus Photos; p. 428(both) HRW Photo/Russell Dian; p. 431(l) Erich Hartmann/Magnum Photos; (r) Paul Fusco/Magnum Photos.

UNIT 6: pp. 434–435 Steve Krongard/The Image Bank **Chapter 19:** p. 436 Stephen J. Krasemann/DRK Photo; p. 439(l) Wayne Lenkiner/DRK Photo; (r) Stephen J. Krasemann/DRK Photo; p. 441 Z. Leszczynski/Animals Animals; p. 442 Henry Ausloos/Animals Animals; p. 446 Breck Kent/Earth Scenes; p. 453 Jim Brandenburg/ Woodfin Camp; p. 455 L. David Mech; p. 457 VESTAL/ Earth Images; p. 459 Ted Levin/Animals Animals; p. 461(l) Raymond Mendez/Animals Animals; (r) Kevin Galvin/Picture Group. **Chapter 20:** p. 464 C. C. Lockwood; p. 467(tl) Breck Kent; (tr) Richard Kolar/Earth Scenes; (cl) Breck Kent; (cr) Raymond A. Mendez/ Animals Animals; p. 469(both) American Museum of Natural History; p. 471 The Bettmann Archive; p. 473 © Richard Kolar/Animals Animals; p. 474 Dwight Kuhn; p. 480 M. Austerman/Animals Animals; p. 483 Alfred Pasieka/Taurus Photos. **Chapter 21:** pp. 486, 493, 494 C. C. Lockwood; p. 495 Doug Wechsler; p. 497 Jim Brandenburg/Woodfin Camp; p. 500 Fermilab; p. 504 Stephen J. Krasemann/DRK; p. 506 Doug Wechsler; p. 507 A. Blank/Bruce Coleman; p. 509 Dan Budnik/ Woodfin Camp; p. 510 Stephen J. Krasemann/Peter Arnold, Inc.; p. 511 Stephen J. Krasemann/DRK; p. 513 Thomas Nebbia/Woodfin Camp; p. 514 Townsend Dickinson/Photo Researchers; p. 516 O S F/Animals Animals; p. 517 NASA. **Chapter 22:** p. 520 Manuel Rodriguez; p. 522 Erich Hartmann/Magnum; p. 523(l) David Moore/Black Star; (r) Yoram Lehmann/Peter Arnold; p. 524 Jim Pickerell/Black Star; p. 528 Peregrine Fund; p. 532(l) Peter Arnold; (r) © Yoram Lehmann/ Peter Arnold; p. 534 James Mason/Black Star; p. 535 HRW Photo/Yoav Levy/Phototake; p. 537 Paul Schuhmann/The Courier Journal and the Louisville Times.